Malassezia and the Skin

Teun Boekhout • Eveline Guého-Kellermann
Peter Mayser • Aristea Velegraki (Eds)

Malassezia and the Skin

Science and Clinical Practice

Springer

Teun Boekhout, PhD
Centraalbureau voor
Schimmelcultures
Uppsalalaan 8
3584 CT Utrecht
The Netherlands
t.boekhout@cbs.knaw.nl

Eveline Guého-Kellermann, PhD
INSERM
5 rue de la Huchette
61400 Mauves sur Huisne
France
e.gueho@orange.fr

Peter Mayser, MD
Department of Dermatology and Andrology,
Justus Liebig University Giessen,
Gaffykstrasse 14,
35385 Giessen, Germany
peter.mayser@derma.med.uni-giessen.de

Aristea Velegraki, PhD
Professor
Mycology Laboratory, Medical School,
National and Kapodistrian University
of Athens,
Athens 11527, Greece
aveleg@med.uoa.gr

ISBN 978-3-642-03615-6 ISBN 978-3-642-03616-3 (eBook)

DOI 10.1007/978-3-642-03616-3

Springer Heidelberg Dordrecht London New York

Library of Congress Control Number: 2009933985

Cover design: eStudioCalamar, Figueres/Berlin

Printed on acid-free paper

Springer is part of Springer Science+Business Media (www.springer.com)

Foreword

It has been known for many years that the *Malassezia* yeasts are associated with a number of different human diseases ranging from pityriasis versicolor to seborrhoeic dermatitis. However the evolving history of their taxonomy and pathogenicity, and the management of the diseases that they cause has been a long, and often difficult, journey. Their fastidious growth requirements defied the initial attempts to culture these organisms on laboratory media and their true identification and the relationship between different species only became apparent with the application of modern molecular techniques. Likewise although recognised in the 19th century as potential causes of human infection, piecing together the complex and, in certain cases, still uncertain relationships to different human diseases has taken many years. Recognised initially as causes of infection of the skin, they are now known to be superficial commensals as well as potential causes of infections in domestic animals and more serious human conditions such as fungemia. They have also been implicated in the pathogenesis of allergic and other inflammatory diseases.

Given this complex, yet fascinating, history it seems appropriate to bring together current thought on these yeasts, their structure and function and their association with both human and animal disease states. This book provides such a view of the genus *Malassezia* and the diseases caused by its members. In accomplishing this task the book takes the reader through the history of the genus *Malassezia* and critically reviews its taxonomy, physiology and the diseases caused, both directly or indirectly. It is intended to provide an up-to-date account of these organisms that will appeal to the specialist scientist and student as well as the practicising microbiologist or physician.

London, UK Roderick J. Hay

Contents

Contributors

H. Ruth Ashbee, PhD
Mycology Reference Center,
Department of Microbiology,
Leeds General Infirmary,
Leeds, LS1 3EX, UK
h.r.ashbee@leeds.ac.uk

Dominik Begerow, PhD
Ruhr-Universität Bochum,
Geobottanik, ND 03/174,
Universitätsstr. 150,
44801 Bochum, Germany
dominik.begerow@rub.d

Teun Boekhout, PhD
Centraalbureau voor Schimmelcultures,
Uppsalalaan 8, 3584 CT Utrecht,
The Netherlands
t.boekhout@cbs.knaw.nl

Ross Bond, DVM, PhD
Senior Lecturer in Veterinary Dermatology,
Department of Veterinary Clinical Sciences,
Royal Veterinary College, Hawkshead Lane,
North Mymms, Hatfield,
Herts AL9 7TA, UK
rbond@rvc.ac.uk

F. Javier Cabañes, DVM
Veterinary Mycology Group,
Department of Animal Health and Anatomy,
Faculty of Veterinary,
Universitat Autònoma de Barcelona,
08193 Bellaterra, Barcelona, Spain
javier.cabanes@uab.es

Reto Crameri, PhD
Swiss Institute of Allergy and Asthma
Research (SIAF),
Obere Strasse 22, 7270 Davos,
Switzerland
crameri@siaf.uzh.ch

Vicente Crespo Erchiga, MD
Department of Dermatology,
Regional University Hospital Carlos Haya,
Plaza del Hospital Civil s/n,
29009 Málaga, Spain
vicente.crespo.sspa@juntadeandalucia.es

Tom Dawson, PhD
The Procter & Gamble Co,
Miami Valley Innovation Center,
Cincinnati, OH 45253-8707, USA
saunders.cw@pg.com

Yvonne DeAngelis, BSc
The Procter & Gamble Co
Miami Valley Innovation Center
Cincinnati, OH 45253-8707, USA
saunders.cw@pg.com

George Gaitanis, DM, PhD
Department of Skin and Venereal Diseases,
University Hospital of Ioannina,
University of Ioannina Medical School,
S. Niarhos Avenue,
45500 Ioannina, Greece
ggaitanis@med.uoa.gr

Andreas Groll, DM, PhD
Infectious Disease Research Program,
Center for Bone Marrow Transplantation and Department of Pediatric Hematology/Oncology
Children's University Hospital Muenster,
Albert-Schweitzer-Strasse 33,
48149 Muenster, Germany
grollan@ukmuenster.de

Eveline Guého-Kellermann, PhD
INSERM
5 rue de la Huchette
61400 Mauves sur Huisne, France
e.gueho@orange.fr

Jacques Guillot, DVM, PhD
Department of Parasitology-Mycology,
Joint Research Unit BIPAR (Biologie Moléculaire et Immunologie Parasitaires et Fongiques) AFSSA, ENVA, UPVM,
Ecole Nationale Vétérinaire d'Alfort,
94704 Maisons-Alfort, France
jguillot@vet-alfort.fr

Suzana Hadina, DVM, PhD
Department of Microbiology and Infectious Diseases with Clinic,
Faculty of Veterinary Medicine,
Heinzelova 55, 10 000 Zagreb, Croatia
suzana.hadina@vef.hr

Roderick J. Hay, MD
Professor of cutaneous infection,
Skin Infection Clinic,
Dermatology Department,
Kings College Hospital,
London SE5 9RS, UK
roderick.hay@ifd.org

Peter Mayser, MD
Department of Dermatology and Andrology,
Justus Liebig University Giessen,
Gaffykstrasse 14, 35385 Giessen,
Germany
peter.mayser@derma.med.uni-giessen.de

Gillian Midgley, PhD
35 Lebanon Park,
Twickenham TW1 3DH, UK
mail@bryn-jones.net

Vincent A. Robert, PhD
Bioinformatics group leader,
Centraalbureau voor Schimmelcultures,
Uppsalalaan 8, 3584 CT Utrecht,
The Netherlands
v.robert@cbs.knaw.nl
www.cbs.knaw.nl

Charles D. Saunders, PhD
The Procter & Gamble Co,
Miami Valley Innovation Center,
Cincinnati, OH 45253-8707, USA
saunders.cw@pg.com

Annika Scheynius, MD, PhD
Department of Medicine Solna,
Clinical Allergy Research Unit,
L2:04, Karolinska Institutet and University Hospital,
171 76, Stockholm, Sweden
annika.scheynius@ki.se

Takashi Sugita, PhD
Department of Microbiology,
Meiji Pharmaceutical University,
2-522-1 Noshio, Kiyose,
Tokyo 204-8588, Japan
sugita@my-pharm.ac.jp

Athanasios Tragiannidis, MD, PhD
Second Department of Pediatrics,
Hematology and Oncology Unit,
Aristotle University Thessaloniki,
AHEPA Hospital, S. Kiriakidi 1 str,
54636 Thessaloniki, Greece
atriagia@auth.gr

Aristea Velegraki, PhD
Mycology Laboratory, Medical School,
National and Kapodistrian
University of Athens,
Athens 11527, Greece
aveleg@med.uoa.gr

Jun Xu, PhD
The Procter & Gamble Co,
Miami Valley Innovation Center,
Cincinnati, OH 45253-8707, USA
saunders.cw@pg.com

Introduction: *Malassezia* Yeasts from a Historical Perspective

1

Roderick J. Hay and Gillian Midgley

Core Messages

› The history of the recognition of the significance of the *Malassezia* yeasts and their potential role in causing disease in humans and animals is one of the more intriguing medical stories of the last 150 years. It contains all the elements of a classic mystery, from the early proposed association between an unusual looking organism seen in the skin diseases, seborrhoeic dermatitis and pityriasis versicolor, to its reclassification into a number of different species on the basis of molecular taxonomy. In the intervening period, other important discoveries were made. Its presence on normal skin indicated that it was a skin commensal, yet the argument over its true identity and nomenclature form a key part of this story. It was accepted at first as a cause of seborrhoeic dermatitis, but then rejected before the argument swung back in favour of a causal link in the late 1980s. While it has always been accepted as a cause of pityriasis versicolor, the new molecular work has allowed for the initiation of investigations to relate different clinical forms to individual species. The association of *Malassezia* species with animal disease has been an important development in veterinary medicine as skin disease caused by these organisms is very common in certain breeds. More recently, it has been reported that *Malassezia* species can be a cause of systemic infection, particularly in new born infants, and that it may also contribute to allergic disorders such as atopic dermatitis. As is often the case, the molecular work has outpaced the definition of clinical states, and it will be necessary to re-examine the status of individual diseases and reconsider whether they should be subdivided further. Early difficulty in the isolation of these lipophilic yeasts in the laboratory held back research, but the introduction of molecular methods has expedited work on the genus in recent years. This has contributed to our understanding of its pathogenesis, epidemiology and even its management. For instance, it has provided a more robust method of designing intervention strategies, such as the use of appropriate antifungals, as different species exhibit different drug sensitivity patterns.

R. J. Hay (✉)
Professor of cutaneous infection, Skin Infection Clinic, Dermatology Department,
Kings College Hospital, London SE5 9RS, UK
e-mail: roderick.hay@ifd.org

T. Boekhout et al. (eds.), *Malassezia and the Skin*
DOI: 10.1007/978-3-642-03616-3_1, © Springer Verlag Berlin Heidelberg 2010

1.1 Definition

Malassezia yeasts are lipophilic organisms, which have been recognised for over a century as members of the normal human cutaneous flora, and also as agents of certain skin diseases [1, 2]. In addition, since the 1980s, they have been reported as causing opportunistic systemic infections [3]. These yeasts have frequently been found on a diverse range of other warm-blooded animals (Chap. 3.3), and their wider distribution in nature is now being explored using molecular techniques [4–9]. Characteristic features of the genus *Malassezia* include a distinctive morphology and an affinity for lipids in culture. Electron microscopic observations have shown that these yeasts have a thick, multi-layered cell wall with buds formed successively from a single locus on the parent cell [10]. This leads to the formation of a prominent bud scar that gives the typical bottle shape to the parental cell and the bud, thus enabling the easy recognition of the yeasts in skin material (Chap. 2.3). They have the physiological property of being able to utilise lipids and, with *M. pachydermatis* as the exception, the species show an absolute requirement for lipids in culture media (Chap. 2.1). The genus is currently classified in the order Malasseziales among the Exobasidiomycetes (Basidiomycota) (Chap. 2.2).

1.2 Taxonomy

The earliest description of *Malassezia* is that by Eichstedt in 1846 [1] when he recognised the fungal nature of the disease, pityriasis versicolor (PV). He observed the presence of yeasts and filaments in material from patients, but the organism was not named until Robin identified it as *Microsporon furfur* [11] in 1853. This name was replaced by Baillon, when he created a new genus *Malassezia* in 1889, and the binomial *Malassezia furfur* was established [12].

Contemporary reports at this time revealed an interest in scaly conditions of the scalp including seborrhoeic dermatitis (SD), and dandruff (pityriasis capitis) where the presence of yeasts without filaments was seen not only in these patients but also in apparently healthy skin. Rivolta noted double contoured organisms in scales from psoriasis lesions in 1873 [2] placing them within the genus *Cryptococcus*, and in the following year, Malassez described both spherical and oval "spores" in scaly lesions of the scalp [13]. These spores were named as two *Saccharomyces* species, *S. ovalis* and *S. sphaericus*, by Bizzozero in 1884 [14] but other workers [15, 16] considered them to represent a single species and it was Sabouraud in 1904 who assigned them to a new genus *Pityrosporum.* Although he recognised the similarities between PV and pityriasis capitis, both in clinical signs and in the superficial site of colonisation by the yeasts, the presence of filaments in PV scales and the lack of them in material from other conditions led to his reluctance to place *M. furfur* and *Pityrosporum* in a single genus [17]. The yeasts observed in scaly scalps were therefore named by him as *Pityrosporum malassezii* but this was altered later to *P. ovale* by

Castellani and Chalmers in 1913 [18] who acknowledged the earlier specific name given by Bizzozero [14].

Subsequent studies by Panja however, recognised similarities between *Malassezia* and *Pityrosporum* and he suggested a single genus for these yeasts, naming them *M. furfur* and *M. ovalis*, with a third species, *M. tropica*, as a cause of a tropical variant of PV, pityriasis flava [19]. This concept was not accepted by workers at that time, and it was found convenient to reserve the name *M. furfur* to describe the diagnostic microscopic feature of yeasts together with filaments in PV scales, and to continue to use *Pityrosporum* to identify the yeasts found in other conditions and in culture studies. Although various combinations for these yeasts appeared in the literature up to the 1930s, as reviewed by Ingham and Cunningham [20], the majority of authors maintained the binomial *Pityrosporum ovale*. This species, under its various names continued to include yeasts with a diverse morphology, including spherical cells [15, 16, 21].

During the years 1925–1955, more species of *Pityrosporum* were described. In 1925, Weidman [22] isolated yeasts from the skin of a rhinoceros, which resembled *P. ovale* in morphology but could be distinguished by the lack of dependence on lipids for growth. He named it *P. pachydermatis*. A proposal by Dodge, in 1935 [23], that the designation, *Malassezia pachydermatis,* should be adopted for this yeast, was not accepted at the time. A similar yeast was described by Gustafson in 1955 from otitis externa in dogs, which he named *P. canis* [24] but this was later made a synonym with *P. pachydermatis* [17, 25]. This species is not a normal member of the human mycobiota but it has been found in a variety of animal hosts (Chap. 3.3). In the series of publications titled "The Yeasts, a taxonomic study", the first edition of 1952 listed two species of *Pityrosporum*, *P. ovale* and *P. pachydermatis* [26]. A third species *P. orbiculare* [27] was added to the genus *Pityrosporum* [17] in the second edition of 1970. This species was lipid-dependent and isolated from human skin and was distinguished by the spherical shape of the yeasts. In this 1970 publication, Slooff gave a very comprehensive historical review [17]. She acknowledged the similarity between *Pityrosporum* and *Malassezia*, but since the parasitic phase *Malassezia furfur* could not be obtained in culture, the genus *Pityrosporum* was maintained. The lipophilic yeasts were therefore included in two genera, *Malassezia* and *Pityrosporum*, which was not a satisfactory situation, but one that continued for more than a decade. Further evidence was forthcoming revealing similarities between *M. furfur*, the organism seen in PV, and the yeast *P. orbiculare* both in their morphology, such as ultra-structure of the cell wall [28–30], and in serological relationships [31, 32]. An additional factor was the demonstration of filament production by *P. orbiculare in vitro* [33–36]. Therefore in 1984, in the third edition of "The Yeasts" [37], and in 1986, at The International Commission on the Taxonomy of Fungi [38], a single generic name was proposed for these yeasts and *Malassezia* was accepted as correct. This name had precedence since it was created in 1889 [12], 15 years before Sabouraud's *Pityrosporum* of 1904 [15]. Although this decision was recognised by taxonomists, *"Pityrosporum"* continued to be used, particularly in the dermatological literature [39–42].

Yarrow and Ahearn [37] placed both of the former lipid-dependent *Pityrosporum* yeasts in a single species, *M. furfur*. This was consistent with the view held by a number of workers that morphological differences in the *Pityrosporum* species were due to the instability of the cell shape [17, 43–45] and also to common antigenic properties [32, 46, 47]. This opinion was not universally accepted, and it presented a problem for authors wishing to distinguish yeasts according to differences in their morphology, revealed by variation in the size and shape of the cell, the ability to produce filaments and colonial features. Publications, therefore either referred to *Pityrosporum* spp., or divided *M. furfur* into several varieties [48, 49]. These authors were able to demonstrate additional factors such as physiological and serological properties of these organisms, which supported their proposed divisions within the genus.

Once molecular characteristics were established, it became apparent that the varieties reported in the literature did in fact, represent a diversity of species within the *Malassezia* genus. A new species, *M. sympodialis*, was described in 1990 [50], being differentiated by DNA/DNA reassociation experiments and values of the G+C content of nuclear DNA. This species was incorporated within the genus in the fourth edition of "The Yeasts" [51]. Later, a number of karyotypes, determined by pulsed field gel electrophoresis (PFGE), were reported which could be related to phenotypic variants [52–54]. The analysis of large subunit (LSU) rRNA sequencing published in 1995 by Guillot and Guého confirmed the divisions already suspected within the genus and indicated further possible taxa [55]. Subsequently, four new species were established in 1996 where the molecular data could be correlated with physiological properties, such as the ability to utilise different lipid supplements, differences in the catalase reaction, and variation in temperature tolerance [56].

Schemes were developed which enabled the identification of species using phenotypic characters [57, 58] and this has led to publications on ecological studies showing the distribution of species according to pathological conditions and geographical areas [59–61]. Additional molecular methods were subsequently introduced which not only allowed identification of species *in situ* [62–64], but also led to the revelation of further *Malassezia* species. These have been from human skin in Japan, *M. dermatis* [65], *M. japonica* [66], *M. yamatoensis* [67], and from animal sources, *M. nana* [68], *M. equina* and *M. caprae* [69]. Currently, thirteen species of *Malassezia*, are recognised and have been included in Chap. 2.1 and in the fifth edition of "The Yeasts: a taxonomic study" [70].

1.3 Culture Studies

Lipids have been incorporated in culture media for the isolation of *Malassezia/Pityrosporum*, since the earliest studies on their recovery, as formulae in use at that time included natural products such as milk, eggs, lanolin, meat infusion and butter, [19, 21, 71–73] all of which would have supplied some lipid content. The validity of cultures isolated by Oudemans and Pekelharing in 1885 [72], van Hoorn in 1896 [73] and Kraus in 1913 [71] is now a matter of opinion but those of Castellani in 1925 [74] and Panja in 1927 [19] were studied further by other laboratories and are therefore credited with being confirmed isolates of *Malassezia/Pityrosporum* [17, 21, 27]. The work of Ota and Huang [21] and Benham [75] stressed that lipid was, in fact, essential for the growth of *P. ovale*. Benham preferred the inclusion of butter but it was olive oil, or its component oleic acid that was used by the majority of subsequent authors [26, 27, 76, 77]. Even when cultures were successful, the optimum conditions for the maintenance of isolates were not established and many of these were often short lived so that exchange of cultures and comparison of isolates between laboratories were slow to develop. Investigations were performed by several authors into the precise supplements needed for the culture of lipophilic yeasts. Shifrine and Marr [78], Meinhof and Braun-Falco [79] and Wilde and Stewart [77] demonstrated the effect of individual fatty acids, the critical factor being the number of carbon atoms in the molecule. Further to this, Martin Scott in 1952 [80] had demonstrated the positive effect of the inclusion of a bile salt in the medium and these studies led to the development of various formulae for the isolation and maintenance of these lipid-dependent yeasts, which remain in use to the present day [49, 81–83].

Further physiological conditions found to be important for success in the isolation and maintenance of *Malassezia* cultures, were the use of freshly prepared media, humid conditions and a temperature between 30 and 35°C [49, 83]. Although some of the currently recognised species can survive even at temperatures above 37°C, many isolates from human skin, at least in temperate zones, belong to species such as *M. globosa* and *M. restricta,* which are more limited in their temperature tolerance (Chap. 2.1). It was also shown that frequent transfers onto fresh media and the avoidance of low temperatures for storage (e.g. refrigeration at 0–4°C) are both necessary to maintain the viability of cultures *in vitro* an exception being the use of liquid nitrogen, where even the most susceptible species were found to survive [49].

Descriptions of the colony characteristics of *Malassezia* cultures were very limited until media were developed to provide substantial growth. Formulae with an emulsion of lipid in the agar rather than an oily overlay allowed the formation of discrete colonies with reproducible features. Illustrations by Panja in 1927 [19] and Martin-Scott in 1952 [80] showed cultures with a rough surface and the various colony forms reported later by van Abbe in 1964 [83] on Dixon agar can be readily identified with those in later publications using a similar medium [44, 49, 56]. These features are now included in the standard species descriptions (Chap. 2.1).

From the earliest records of the lipophilic yeasts in tissue, and in cultures, a diversity in the shape of the cells was recognised, with descriptions of oval, globose, bottle shaped and elongated yeasts [1, 2, 13–16, 73]. Although in the illustrations in Benham's publication [75] and in the first edition of "The Yeasts, a taxonomic study" in 1952 [26], *P. ovale* was shown with fairly uniform oval cells, with the acceptance of Gordon's *P. orbiculare* in 1970 [27], spherical cells were again included in the genus description [17]. The characteristic feature of repetitive budding from a thick walled cell, which gives the typical bottle shape with a pronounced bud scar was seen in historical publications but was described more clearly in later years with electron microscopy (EM) studies. The cell wall of *Malassezia* was seen to be composed of several layers and a remarkable feature is the appearance of spiralling ridges on the inner surface. This can be visualised by the light microscope and was first noted, as long ago as 1899 by Matakieff [84] but later shown with precision by illustrations of the ultra-structure (Chap. 2.3). The production of filaments in *Malassezia* cultures was reported infrequently in the past [33, 35, 36, 71, 85, 86], and is found to be a typical feature of only two species, namely *M. furfur* and *M. globosa.* Saadatzadeh in 2001 obtained atypical tortuous filaments in *M. sympodialis* [87], but they have never been observed *in vivo* associated with this species.

The property of differences in fatty acid requirements of variants of *Malassezia* reported by Shifrine and Marr [78] and Meinhof and Braun-Falco [79] has been employed more recently in species identification. Using the assimilation patterns of Tweens (which have different fatty acids as major components), along with catalase activity and the tolerance of temperatures above 37°C, schemes have been devised to confirm the identity of the species by cultural methods [56, 58]. Two additional reactions, the growth with Cremophor EL (castor oil) and the presence of ß-glucosidase activity, were found useful to aid in species identification [57, 70, 88].

1.4 Ecology

Numerous surveys to determine the extent of *Malassezia* colonisation on human skin have been undertaken using both direct microscopy of tissue and by culture [89, 90]. They revealed that these species are universally present in the population, with colonisation occurring in infancy, and reaching highest concentrations after puberty and in early adulthood [81, 91–94]. Many have focused on the scalp in relation to SD and dandruff [80, 83, 95] and investigations covering multiple sites have shown the highest concentrations of these yeasts to be in areas rich in sebaceous glands [76, 94]. In these reports, the individual species of *Malassezia* were often identified as far as the current knowledge allowed. However, after the newer species of *Malassezia* were established in 1996 [56], and molecular methods expanded to identify the species in tissue as well as in culture, the numerous studies reviewed by Ashbee in 2007 [89] have indicated the distribution of each species from healthy skin and their relationship with dermatological conditions.

Surveys on animals have focused on the recovery of *M. pachydermatis,* which shows a wide range of animal hosts (Chap. 3.3). Several of the remaining *Malassezia* species have been reported from animals, but the extent of the spectrum of the thirteen species in both human and animal skin, as well as the non-living environment, is still being determined (Chaps. 2.1 and 3.3).

1.5 Associated Diseases

Malassezia species have been associated with a number of different diseases in both humans and animals [96]. These range from "true" infections, where there is invasion of tissue, such as PV, or the much rarer cases of systemic *Malassezia* infection, to conditions where *Malassezia* appear to play a pathogenic role, but through indirect or immunological mechanisms such as in SD. In addition, there have been diseases that have been ascribed to *Malassezia* but with insufficient scientific evidence, such as confluent and reticulate papillomatosis [97] and onychomycosis. In the latter instance, *Malassezia* are almost certainly, colonists of subungual space.

The history of the descriptions of these diseases is difficult to disentangle, partly because of the use of similar terms to describe what we now know to be distinct disease states. Braun-Falco et al. [98], for instance, used the term pityriasis to describe a number of different conditions including scaling in the scalp. Generally, for centuries, the term pityriasis had been used to describe scaling conditions distinct from psoriasis or ichthyosis; the alternative word porrigo was also often used for similar scaling conditions

1.5.1 Pityriasis Versicolor (PV)

PV is a superficial infection of the skin in which the organisms, usually *M. globosa*, develop filaments and cause minor pathological damage to the stratum corneum of the epidermis through tissue penetration and resultant inflammation. PV is one of the commonest skin diseases in a wide range of geographic areas, but is most frequent in the tropics, with the highest incidence in older children and young adults. It is also seen regularly in northern climates, often in patients who have taken their holidays in a warmer environment. While this infection may occur in otherwise healthy individuals, it is also seen in some immunocompromised patients, such as those receiving long-term therapy with systemic corticosteroids and other immunosuppressive drugs; however, it is not a marker of HIV/AIDS infection. PV presents with scaly hypo- and hyper-pigmented patches, which, with time, become confluent over the trunk, head and proximal limbs. It may be accompanied by mild symptoms of irritation. Atypical forms spreading to the proximal limbs or producing atrophic skin changes have been described. The infection is best diagnosed by direct microscopy of skin scales; culture of scales has no place in the laboratory diagnosis. Treatment with a range of antifungals such as imidazoles is usually effective, although adequate application of topical medications to wide surface areas of the skin is technically challenging. In patients who are not residents of tropical regions, relapse is uncommon; however, if predisposing factors are not removed, re-infection may occur.

PV was a term used by the late eighteenth and early nineteenth century physicians interested in skin disease, such as Willan [99]. But the first recognition that PV, as a distinct form of scaly dermatosis, was caused by a yeast is attributed to Eichstedt in 1846 [1, 3, 17]; he was the first to observe that patients with PV had both yeasts and filaments in skin samples. At this time, culture was not possible and the causal connection was based on identification of the characteristic morphology *in vivo*. The history of the subsequent nomenclature has been described previously. However, the relationship between specific yeasts and PV may appear confusing with Baillon's renaming of the organism in 1889 as *Malassezia furfur* and the subsequent use of the genus *Pityrosporum*. Nowadays, with the more recent classification, it is becoming clear that *Malassezia globosa* is the principle cause of PV [100], although other *Malassezia* species, including the newly genetically defined *M. furfur,* may also be involved. Many of the earlier clinical studies of this condition, after its recognition as an infection, were documented by Castellani [101] who maintained that there were a number of distinct forms of PV distinguishable by colouration or distribution. Pityriasis versicolor tropicalis, for instance, was the name given to the disease where there was extensive depigmentation associated with lesions. Similar clinical variations were outlined by other authors [102, 103], although today, the main distinct entities, recognised as clinically relevant, are the atrophic form and a variety of PV associated with oval yeasts and an atypical distribution (Chap. 6.1).

1.5.2 Seborrhoeic Dermatitis (SD)

SD is a scaly skin disease affecting the scalp, face and chest where there are scaly erythematous patches. These are best found in the scalp, in the nasolabial folds, eyebrows, behind the ears and on the chest over the sternum. It is a major, if not the only, cause of scalp scaling or dandruff. The pathogenetic role of *Malassezia* in SD remains something of a mystery, as there is no evidence that the organisms invade the skin. However, our understanding of the explicit link between SD and *Malassezia* is based on observational evidence that removal of the yeasts from the skin, using an antifungal agent, usually leads to remission, and relapse is associated with re-emergence of organisms on the skin surface. There is no experimental model for SD and not all patients respond to antifungals. Despite intensive investigations, the pathogenetic mechanisms involved remain obscure. Attempts to demonstrate the involvement of specific immunological or biochemical pathways have not been successful. However, besides affecting healthy patients, SD is common in the early phase of AIDS [104] and also in patients with chronic neurological illness, such as Parkinson's disease [105]. No defects in T-cell function in SD patients have been demonstrated that would explain the link with HIV. There is some evidence that there are higher antibody levels to *Malassezia* in SD patients. However, the true pathological basis for SD remains unclear. Intuitively, the mechanism would appear to be the failure to suppress an inflammatory response to a surface commensal yeast, which, as a consequence, is activated in SD patients. However, at present, such ideas are speculative. Diagnosis, therefore remains a clinical one based on the characteristic clinical morphology and distribution of lesions. Treatment with antifungal shampoos, creams or oral antifungals is usually successful, although there are exceptions.

As mentioned previously, scaling of the scalp associated with varying degrees of inflammation and erythema had been known for centuries. But the connection between fungi and SD was first proposed by Malassez in 1874 [13]. He had observed the morphology of the yeasts, which were described as spores, that are present in the scaly scalp in patients with dandruff. This view was supported by Sabouraud who recognised that these were indeed yeasts and believed that they were the cause of pityriasis capitis or dandruff. He also pointed out the similarity between these organisms and those that caused PV and named them *Pityrosporum malassezii.* While dermatologists in the early part of the twentieth century supported an infective aetiology for SD, this was called into question by a number of later scientists who were interested in the development of inflammation in the skin. They proposed that the yeasts were incidental to a primary inflammatory dermatosis, which resulted in increased cell turnover in the epidermis, such as psoriasis, causing scaling and inflammation [106, 107]. For many years, the connection of yeast with SD was forgotten, until the development of the oral antifungal, ketoconazole, led to the observation that patients with dandruff receiving this drug began to improve. This stimulated more work and in 1984,

Shuster [108] proposed that the yeast theory was indeed the correct one and that there was sufficient evidence in existence to make an explicit connection between the presence of *Malassezia* yeasts and SD. In developing his argument, Shuster pointed out that a critical piece of evidence used in favour of the hyperproliferative theory was the failure of patients with dandruff to respond to topical amphotericin B [107]. However, he also pointed out the importance of a study that showed that patients with dandruff, who had responded to nystatin, relapsed when nystatin-resistant cells of *Malassezia* were reintroduced [109].

The association between SD and a number of other conditions remains perplexing. For instance, in the nineteenth century, it was associated with syphilis, [110] although it is not always clear whether the lesions of these two distinct diseases were always separated. Likewise, the pathogenetic explanation for the association between SD and neurodegenerative disorders, such as Parkinson's disease and latterly HIV/AIDS, is unknown.

1.5.3 *Malassezia* Folliculitis

Malassezia folliculitis is a scattered pustular inflammatory rash that is mainly seen on the upper trunk, back and front [111]. The pustules are not associated with comedones (blackheads) and are itchy. The condition is often triggered by sun exposure, but may also occur in severely ill patients admitted to hospital [112]. There is also a pustular form of SD, again usually on the chest or back where the pustules are small and surrounded by erythema. Biopsy of lesion reveals large numbers of yeast within hair follicles and surrounding inflammation. Generally, oral therapy with either itraconazole or ketoconazole is effective.

1.5.4 Atopic Dermatitis (AD)

Malassezia species have also been implicated in the development of a form of AD affecting the face and scalp [113, 114]. This phenomenon is mainly recorded in young adults rather than in children and affects merely the face and not the flexures. The evidence that there is a connection with *Malassezia* is based on the observation that a) patients respond to antifungal therapy, and b) there is a high prevalence of specific antibody (IgE) to *Malassezia* antigens in affected patients. The positive effect of treatment is difficult to understand on the basis of any explanation, other than that of removal of yeasts from the skin surface, as azole antifungals without activity against *Staphylococcus aureus,* such as ketoconazole, are as effective as those with antibacterial properties. There are precedents for the role of IgE mediated immunity in eliciting flares in AD, as a similar mechanism has been proposed as contributory to the much commoner, *Staphylococcus*–associated, flares of eczema. In the latter example, there is evidence that other immunologically mediated bacterial triggers may be involved. *S. aureus*, for instance, is a source of superantigens that may directly cause T cell activation. So, the mechanism whereby *Malassezia* plays a role in aggravating AD is not known [115].

1.5.5 Confluent and Reticulate Papillomatosis

Confluent and reticulate papillomatosis [97] is a rare condition affecting the upper trunk where patients develop a network of linear, noninflammatory scaly lesions [116, 117]. Biopsies of lesions may reveal the presence of *Malassezia* yeasts, but the evidence that they play a role in the causation of this condition is weak. Firstly, there is neither satisfactory pathogenetic explanation nor firm evidence that lesions respond to antifungals. The condition is very uncommon and, therefore, controlled comparative therapeutic studies would be hard to perform. Yet, generally patients respond to minocycline, again difficult to explain on the basis of a central role for *Malassezia* spp.

1.5.6 Other Conditions

Malassezia have been implicated in other conditions such as blepharitis and otitis externa. Again, the problem comes in relating pathogenesis to the organisms, as lipophilic yeasts can be found in both the eye-lashes and the external auditory meatus. In addition, blepharitis and scaling of the external ear are seen in patients with SD and so the link may be similar to that seen with SD patients. There is a form of psoriasis, sometimes referred to as sebopsoriasis, where the clinical distribution of lesions can resemble those seen in SD. Patients with this form of psoriasis have been reported to improve on antifungals, thus suggesting that *Malassezia* (or another fungus) may play a role [118]. It is not clear, though, whether this is a distinct entity or that it represents an overlap between SD and psoriasis – or even psoriasis co-localising (Koebner's effect) with SD [119].

1.5.7 Systemic Infection

Rarely, *Malassezia* species have been described as causes of systemic infection, usually in neonates. The first case in 1981 reported an association with the use of intravenous lipid feeds and care in a neonatal ITU [120]. Transmission from staff or other patients or their relatives has been postulated [121]. The usual cause is *M. pachydermatis*. Case clustering has been reported suggesting the potential for multiple transmission. This condition is fatal if untreated, and patients respond, if treated in a timely manner with systemic antifungals. However, given that this is an unusual pathogen, late diagnosis is often the problem with this infection.

1.6 Perspectives

The recent elucidation of the genomes of *M. globosa* and *M. restricta* [122] will have a significant impact on future studies of *Malassezia* yeasts. First of all, the data generated will be a major source of reference for many future comparative studies of the genus, e.g., those that aim to understand fundamental biological processes, such as the intriguing possibility of a functional mating system, to understand the diversity of allergens and the regulation of their coding genes, and, more generally, comparative pathogenomics, such as the regulation of the expression of genes involved in pathogenicity, inflammation, lipid utilisation *in vivo* and *in vitro*, etc. Secondly, the genome may harbour important new drug targets that may contribute to the control of *Malassezia*-related diseases. Thirdly, the genome data may contribute to the selection of genomic markers that can be used to develop a multilocus sequence typing (MLST) scheme.

Moreover, the availability of the genome sequence will also allow the development of gene replacement techniques, as well as the generation of well defined genetic knock outs. Such constructs will be essential to boost the progress in the field of *Malassezia* pathophysiology. Eventually, this may result in the emergence of a comprehensive systems approach to understand *Malassezia*-related diseases in relation to human genetics.

1.7 Currently Recognised Species

1. *Malassezia furfur* (Robin) Baillon (1889)
2. *Malassezia pachydermatis* (Weidman) Dodge (1925)
3. *Malassezia sympodialis* Simmons & Guého (1990)
4. *Malassezia globosa* Midgley, Guého & Guillot (1996)
5. *Malassezia obtusa* Midgley, Guillot & Guého (1996)
6. *Malassezia slooffiae* Guillot, Midgley & Guého (1996)
7. *Malassezia restricta* Guého, Guillot & Midgley (1996)
8. *Malassezia dermatis* Sugita, Takashima, Nishikawa & Shinoda (2002)
9. *Malassezia japonica* Sugita, Takashima, Kodama, Tsuboi & Nishikawa (2003)
10. *Malassezia nana* Hirai, Kano, Makimura, Yamaguchi & Hasegawa (2004)
11. *Malassezia yamatoensis* Sugita, Tajima, Takashima, Amaya, Saito, Tsuboi & Nishikawa (2004)
12. *Malassezia caprae* Cabañes & Boekhout (2007)
13. *Malassezia equina* Cabañes & Boekhout (2007)

References

1. Eichstedt E (1846) Pilzbildung in der Pityriasis versicolor. Froriep Neue Notiz Natur Heilk 39:270
2. Rivolta S (1873) Dei parassiti vegetali. Giulio Speirani e figli, Torino, pp 469–471
3. Marcon MJ, Powell DA (1992) Human infections due to *Malassezia* spp. Clin Microbiol Rev 5:101–119
4. Bjelland T, Ekman S (2005) Fungal diversity in rock beneath a crustose lichen as revealed by molecular markers. Microbiol Ecol 49:598–603
5. Fell JW, Scorzetti G, Connell L et al (2006) Biodiversity of micro-eukaryotes in Antarctic Dry Valley soils with <5% soil moisture. Soil Biol Biochem 38:3107–3119

6. Gandra RF, Gambale W, de Cássia Garcia Simão R et al (2008) *Malassezia* spp. in acoustic meatus of bats (*Molossus molossus*) of the Amazon Region, Brazil. Mycopathol 165:21–26
7. Lai X, Cao L, Tan H et al (2007) Fungal communities from methane hydrate-bearing deep-sea marine sediments in South China Sea. ISME J 1:756–762
8. Renker C, Alphei J, Buscot F (2003) Soil nematodes associated with the mammal pathogenic fungal genus *Malassezia* (Basidiomycota: Ustilaginomycetes) in Central European forests. Biol Fertil Soils 37:70–72
9. Zhang N, Suh SO, Blackwell M (2003) Microorganisms in the gut of beetles: evidence from molecular cloning. J Invertebr Pathol 84:226–233
10. Swift JA, Dunbar SF (1965) Ultrastructure of *Pityrosporum ovale* and *Pityrosporum canis*. Nature 206:1174–1175
11. Robin C (1853) Histoire naturelle des végétaux parasites qui croissent sur l'homme et les animaux vivants. Baillière, Paris
12. Baillon H (1889) Traité de botanique médicale cryptogamique. Octave Doin, Paris, pp 234–235
13. Malassez L (1874) Note sur le champignon du pityriasis simple. Arch Physiol 1:451–459
14. Bizzozero J (1884) Über die Mykrophyten der normalen Oberhaut des Menschen. Virchows Arch Pathol Anat Physiol 98:441–459
15. Sabouraud R (1904) Maladies du cuir chevelu. II. Les maladies desquamatives. Masson & Cie, Paris
16. Unna PG (1891) Die Färbung der Mikroorganismen im Horngewebe. Monatschr F Prakt Dermatol 13:225–237
17. Slooff WC (1970) Genus 6 *Pityrosporum* Sabouraud. In: Lodder J (ed) The Yeasts, a taxonomic study, 2nd edn. North-Holland, Amsterdam, pp 1167–1186
18. Castellani A, Chalmers AJ (1913) Manual of tropical medicine, 2nd edn. Baillière, Tindal and Cox, London, p 1747
19. Panja G (1927) The *Malassezia* of the skin, their cultivation, morphology and species. Trans 7th Cong Far Eastern Ass Trop Med 2:442–455
20. Ingham E, Cunningham AC (1993) *Malassezia furfur*. J Med Vet Mycol 31:265–288
21. Ota M, Huang PT (1933) Sur les champignons du genre *Pityrosporum* Sabouraud. Ann Parasitol Hum Comp 11:49–69
22. Weidman FD (1925) Exfoliative dermatitis in the Indian rhinoceros (*Rhinoceros unicornis*), with description of a new species: *Pityrosporum pachydermatis*. In: Fox H (ed) Rep Lab Museum Comp Pathol Zoo Soc Philadelphia, Philadelphia, pp 36–43
23. Dodge CW (1935) Medical mycology. Fungous diseases of men and other mammals. Mosby, St Louis
24. Gustafson BA (1955) Otitis externa in the dog. A bacteriological and experimental study. Thesis. Royal Veterinary College Sweden, Stockholm
25. Fraser G (1961) *Pityrosporum pachydermatis* Weidman of canine origin. Trans Br Mycol Soc 44:441–448
26. Lodder J, Kerger-van-Rij NJW (1952) Genus 3 *Pityrosporum* Sabouraud. In: The Yeasts, a taxonomic study, 1st edn. North-Holland, Amsterdam, pp 440–445
27. Gordon MA (1951) The lipophilic mycoflora of the skin. I. *In vitro* culture of *Pityrosporum orbiculare* n. sp. Mycologia 43:524–535
28. Barfatani M, Munn RJ, Schjeide OA (1964) An ultrastructure study of *Pityrosporum orbiculare*. J Invest Dermatol 43:231–233
29. Keddie FM, Shadomy S (1963) Etiological significance of *Pityrosporum orbiculare* in tinea versicolor. Sabouraudia 3:21–25
30. Pierard J, Dockx P (1972) The ultrastructure of tinea versicolor and *Malassezia furfur*. Int J Dermatol 11:116–124
31. Sternberg TH, Keddie MD (1961) Immunofluorescence studies in tinea versicolor. Arch Derm 84:999–1003

32. Tanaka M, Imamura S (1979) Immunological studies on *Pityrosporum* genus and *Malassezia furfur*. J Invest Dermatol 73:321–324
33. Burke RC (1961) Tinea versicolor: susceptibility factors and experimental infection in human beings. J Invest Dermatol 36:389–401
34. Caprilli F, Mercantini R, Nazzaro Porro M et al (1971) Contributo allo studio del genere *Pityrosporum*. Boll 1st Dermatol S Gallicano 7:91–116
35. Dorn M, Roehnert K (1977) Dimorphism of *Pityrosporum orbiculare* in a defined culture medium. J Invest Derm 69:244–248
36. Nazzaro Porro M, Passi S, Caprilli F et al (1977) Induction of hyphae in cultures of *Pityrosporum* by cholesterol and cholesterol esters. J Invest Dermatol 69:355–360
37. Yarrow D, Ahearn D (1984) Genus 7 *Malassezia* Baillon. In: Kreger-van Rij NJW (ed) The Yeasts, a taxonomic study, 3rd edn. Elsevier Science, Amsterdam, pp 882–885
38. Cannon PF (1986) International Commission on the taxonomy of fungi (ICTF): name changes in fungi of microbiological, industrial and medical importance. Microbiol Sci 3:285–287
39. Clift DC, Dodd HJ, Kirby JDT et al (1988) Seborrheic dermatitis and malignancy. Acta Dermatol Venereol (Stockh) 68:48–52
40. Faergemann J (1992) *Pityrosporum* infections. In: Elewski BE (ed) Cutaneous fungal infections. Igaku-Shoin, New York, pp 69–83
41. Faergemann J (1999) *Pityrosporum* species as a cause of allergy and infection. Allergy 54:413–419
42. Wessels MW, Doekes G, van Ieperen-van Kijk AG et al (1991) IgE antibodies to *Pityrosporum ovale* in atopic dermatitis. Br J Dermatol 125:227–232
43. Faergemann J (1985) Lipophilic yeasts in skin disease. Sem Derm 4:173–184
44. Randjandiche M (1976) Polymorphisme de *Pityrosporum ovale* (Bizzozero) Castellani & Chalmers *in vivo* and *in vitro*. Bull Soc Fr Mycol Méd 5:79–84
45. Salkin IF, Gordon MA (1977) Polymorphism of *Malassezia furfur*. Can J Microbiol 23:471–475
46. Bruneau SM, Guinet RMF (1984) Quantitative immunoelectrophoretic study of genus *Pityrosporum* Sabouraud. Mykosen 27:123–136
47. Faergemann J, Tjernlund U, Scheynius A et al (1982) Antigenic similarities and differences in genus *Pityrosporum*. J Invest Dermatol 78:28–31
48. Cunningham AC, Leeming JP, Ingham E et al (1990) Differentiation of three serovars of *Malassezia furfur*. J Appl Bacteriol 68:439–446
49. Midgley G (1989) The diversity of *Pityrosporum (Malassezia)* yeasts *in vivo* and *in vitro*. Mycopathologia 106:143–153
50. Simmons RB, Guého E (1990) A new species of *Malassezia*. Mycol Res 94:1146–1149
51. Ahearn DG, Simons RB (1998) *Malassezia* Baillon. In: Kurtzman CP, Fell JW (eds) The Yeasts, a taxonomic study, 4th edn. Elsevier, Amsterdam, pp 782–784
52. Boekhout T, Renting M, Scheffers WA et al (1993) The use of karyotyping in the systematics of yeasts. Antonie van Leeuwenhoek 63:157–163
53. Howell SA, Quin C, Midgley G (1993) Karyotypes of oval cell forms of *Malassezia furfur*. Mycoses 36:263–266
54. Senczek D, Siesenop U, Böhm KH (1999) Characterization of *Malassezia* species by means of phenotypic characteristics and detection of electrophoretic karyotypes by pulsed-field gel electrophoresis (PFGE). Mycoses 42:409–414
55. Guillot J, Guého E (1995) The diversity of *Malassezia* yeasts confirmed by rRNA sequence and nuclear DNA comparisons. Antonie van Leeuwenhoek 67:297–314
56. Guého E, Midgley G, Guillot J (1996) The genus *Malassezia* with description of four new species. Antonie van Leeuwenhoek 69:337–355
57. Crespo Erchiga V, Guého E (2005) Superficial diseases caused by *Malassezia* species. In: Merz WG, Hay RJ(eds) Topley and Wilson's microbiology and microbial infections, medical mycology, 10th edn, vol 5. Hodder Arnold, London, pp 202–219

58. Guillot J, Guého E, Lesourd M et al (1996) Identification of *Malassezia* species. A practical approach. J Mycol Méd 6:103–110
59. Crespo Erchiga V, Ojeda Martos A, Vera Casano A et al (1999) Isolation and identification of *Malassezia* spp. in pityriasis versicolor, seborrheic dermatitis and healthy skin. Rev Iberoam Micol 16:S16–S21
60. Deesi Z (2006) The ecology and pathogenicity of the lipophilic yeasts (*Malassezia* species) – a study in Dubaï, United Arab Emirates. PhD Thesis, University of London, London
61. Gupta AK, Kohli Y, Faergemann J et al (2001) Epidemiology of *Malassezia* yeasts associated with pityriasis versicolor in Ontario, Canada. Med Mycol 39:199–206
62. Gaitanis G, Velegraki A, Frangoulis E et al (2002) Identification of *Malassezia* species from patient skin scales by PCR-RFLP. Clin Microbiol Infect 8:162–173
63. Gemmer CM, DeAngelis YM, Theelen B et al (2002) Fast, non-invasive method for molecular detection and differentiation of *Malassezia* yeast species on human skin and application of the method to dandruff microbiology. J Clin Microbiol 40:3350–3357
64. Sugita T, Suto H, Unno T et al (2001) Molecular analysis of *Malassezia* microflora on the skin of atopic dermatitis patients and healthy subjects. J Clin Microbiol 39:3486–3490
65. Sugita T, Takashima M, Shinoda T et al (2002) New yeast species, *Malassezia dermatis,* isolated from patients with atopic dermatitis. J Clin Microbiol 40:1363–1367
66. Sugita T, Takashima M, Kodama M et al (2003) Description of a new species, *Malassezia japonica*, and its detection in patients with atopic dermatitis and healthy subjects. J Clin Microbiol 41:4695–4699
67. Sugita T, Tajima M, Takashima M et al (2004) A new yeast, *Malassezia yamatoensis*, isolated from a patient with seborrheic dermatitis, and its distribution in patients and healthy subjects. Microbiol Immunol 48:579–583
68. Hirai A, Kano R, Makimura K et al (2004) *Malassezia nana* sp. nov., a novel lipid-dependent yeast species isolated from animals. Int J Syst Evol Microbiol 54:623–627
69. Cabañes FJ, Theelen B, Castellá G et al (2007) Two new lipid-dependent *Malassezia* species from domestic animals. FEMS Yeast Res 7:1064–1076
70. Guého E, Batra R, Boekhout T (2010) The genus *Malassezia* Baillon. In: Kurtzman CP, Fell JW, Boekhout T (eds) The Yeasts, a taxonomic study, 5th edn. Elsevier, Amsterdam (in press)
71. Kraus A (1913) Über das Wesen des sogenannten Unnaschen Flaschen-bazillus. Arch Derm Syph (Wien) 116:723–736
72. Oudemans CAJA, Pekelharing CA (1885) *Saccharomyces capillitii* Oudemans en Pekelharing: een spruitzwam van de behaarde hoofdhuid. Nederlandsch. Tijdsch Geneeskunde 21: 997–1005
73. van Hoorn W (1896) Mikroorganismen bei der Seborrhoe. Monatschr F Prakt Dermatol 23:525–530
74. Castellani A (1925) Notes on the three new yeast-like organisms and a new bacillus with remarks on the clinical conditions from which they have been isolated: furunculosis blastomycetica, macroglossia blastomycetica, stomatitis cryptococcobacillaris. J Trop Med Hyg 28:217–233
75. Benham RW (1939) The cultural characteristics of *Pityrosporum ovale* – a lipohilic fungus. J Invest Dermatol 2:187–203
76. Roberts SOB (1969) *Pityrosporum orbiculare*: Incidence and distribution on clinically normal skin. Br J Dermatol 81:264–269
77. Wilde PF, Stewart PS (1968) A study of the fatty acid metabolism of the yeasts *Pityrosporum ovale*. Biochem J 108:225–231
78. Shifrine M, Marr AG (1963) The requirement of fatty acids by *Pityrosporum ovale*. J Gen Microbiol 32:263–270
79. Meinhof W, Braun-Falco O (1967) Über lipotrophe Hefen auf der menschlichen Haut und deren Bedarf an bestimmten Fettsäuren. Fette-Seifen Anstrichmittel 69:861–864
80. Martin-Scott I (1952) The *Pityrosporum ovale*. Brit J Derm 64:257–273

81. Faergemann J, Fredriksson T (1980) Age incidence of *Pityrosporum orbiculare* on human skin. Acta Derm Venereol (Stockh) 60:531–533
82. Leeming JP, Notman FH (1987) Improved methods for isolation and enumeration of *Malassezia furfur* from human skin. J Clin Microbiol 25:2017–2019
83. van Abbe NJ (1964) The investigation of dandruff. J Soc Cosmetic Chem 15:609–630
84. Matakieff E (1899) Le pityriasis versicolor et son parasite. Thèse Fac Méd Univ, Nancy
85. Midgley G (1993) Morphological variation in *Malassezia* and its significance in pityriasis versicolor. In: van den Bossche H, Odds FC, Kerridge D (eds) Dimorphic fungi in biology and medicine. Plenum, New York, pp 267–277
86. Randjandiche M (1976) Obtention de la forme mycélienne de *Pityrosporum ovale*. Bull Soc Fr Mycol Méd 5:85–90
87. Saadatzadeh MR, Ashbee HR, Holland KT et al (2001) Production of the mycelial phase of *Malassezia in vitro*. Med Mycol 39:487–493
88. Mayser P, Haze P, Papavassilis C et al (1997) Differentiation of *Malassezia* species: selectivity of cremophor EL, castor oil and ricinoleic acid for *M. furfur*. Br J Dermatol 137:208–213
89. Ashbee HR (2007) Update on the genus *Malassezia*. Med Mycol 45:287–303
90. Ashbee HR, Evans EG (2002) Immunology of diseases associated with *Malassezia* species. Clin Microbiol Rev 15:21–57
91. Leeming JP, Notman FH, Holland KT (1989) The distribution and ecology of *Malassezia furfur* and cutaneous bacteria on human skin. J Appl Bacteriol 67:47–52
92. Noble WC, Midgley G (1978) Scalp carriage of *Pityrosporum* species: the effect of physiological maturity, sex and race. Sabouraudia 16:229–232
93. Randjandiche M (1981) Presence of *Pityrosporum ovale* in the ear of newborn infants. Sabouraudia 19:143–145
94. Vermander F, De Vroey Ch (1985) Dénombrement de *Pityrosporum ovale* par directe empreinte. Bull Soc Fr Mycol Méd 14:23–27
95. McGinley KJ, Leyden JJ, Marples RR et al (1975) Quantitative microbiology of the scalp in non-dandruff, dandruff and seborrheic dermatitis. J Invest Dermatol 64:401–405
96. Crespo Erchiga V, Delgado Florencio V (2002) *Malassezia* species in skin disease. Curr Op Infect Dis 15:133–142
97. Gougerot H, Carteaud A (1927) Papillomatose pigmentée innominée. Cas pour diagnostic. Bull Soc Fr Dermatol Syphil 34:712–719
98. Braun-Falco O, Plewig G, Wolff HH (2000) Dermatitis. In: Braun-Falco O, Plewig G, Wolff HH (eds) Dermatology, 2nd edn. Springer, Berlin, pp 465–466
99. Willan R (1801) Description and treatment of the cutaneous diseases. Order II. Scaly diseases of the skin, London, pp 192–196
100. Crespo-Erchiga V, Ojeda Martos A, Vera Casaño A et al (1999) *Malassezia globosa* as the causative agent of pityriasis versicolor. Brit J Dermatol 143:799–803
101. Castellani A (1935) Pityriasis versicolor tropicalis. Proc R Soc Med 28:1558
102. Bumgarner FE, Burke RC (1949) Pityriasis versicolor; atypical clinical and mycologic ariations. Arch Derm Syphilol 59:192–195
103. Roberts SOB (1969) Pityriasis versicolor: a clinical and mycological investigation. Br J Dermatol 81:315–326
104. Soeprono FF, Schinella RA, Cockerell CJ et al (1986) Seborrheic-like dermatitis of acquired immunodeficiency syndrome. A clinicopathologic study. J Am Acad Dermatol 14:242–248
105. Mastrolonardo M, Diaferio A, Logroscino G (2003) Seborrheic dermatitis, increased sebum excretion, and Parkinson's disease: a survey of (im)possible links. Med Hypotheses 60: 907–911
106. Ackerman AB, Kligman AM (1969) Some observations on dandruff. J Soc Cosmet Chem 21:81–85

107. Leyden JJ, McGinley KJ, Kligman AM (1976) The role of micro-organisms in dandruff. Arch Dermatol 112:333
108. Shuster S (1984) The aetiology of dandruff and the mode of action of therapeutic agents. Br J Dermatol 111:235–242
109. Gosse RM, Vanderwyk RW (1969) The relationship of a nystatin-resistant strain of *Pityrosporum ovale* to dandruff. J Soc Cosmet Chem 20:603–606
110. Unna PG (1888) Syphilis and eczema seborrhoicum. Br J Dermatol 1:37–45
111. Bäck O, Faergemann J, Hörnqvist R (1985) *Pityrosporum* folliculitis: a common disease of the young and middle-aged. J Am Acad Dermatol 12:56–61
112. Archer-Dubon C, Icaza-Chivez ME, Orozco-Topete R et al (1999) An epidemic outbreak of *Malassezia* folliculitis in three adult patients in an intensive care unit: a previously unrecognized nosocomial infection. Int J Dermatol 38:453–456
113. Hjorth N, Clemmensen OJ (1983) Treatment of dermatitis of the head and neck with ketoconazole in patients with type 1 hypersensitivity for *Pityrosporum orbiculare*. Semin Dermatol 2:26–29
114. Waersted A, Hjorth N (1985) *Pityrosporum orbiculare*-a pathogenic factor in atopic dermatitis of the face, scalp and neck? Acta Derm Venereol Suppl (Stockh) 114:146–148
115. Johanssen C, Eshagi H, Linder MT et al (2002) Positive atopy patch test reaction to *Malassezia furfur* in atopic dermatitis correlates with a T helper 2-like peripheral blood mononuclear cell response. J Invest Dermatol 18:1044–1051
116. Faergemann J, Fredriksson T, Nathorst-Windahl G (1980) One case of confluent and reticulate papillomatosis (Gougerot–Carteaud). Acta Derm Venereol (Stockh) 60:269–271
117. Roberts SOB, Lachapelle JM (1969) Confluent and reticulate papillomatosis (Gougerot–Carteaud) and *Pityrosporum orbiculare*. Br J Dermatol 81:841–343
118. Rosenberg EW, Belew PW (1982) Improvement of psoriasis of the scalp with ketoconazole. Arch Dermatol 118:370–371
119. Narang T, Dogra S, Kaur I et al (2007) *Malassezia* and psoriasis: Koebner's phenomenon or direct causation? J Eur Ac Dermatol Venereol 21:1111–1112
120. Redline RW, Dahms BB (1981) *Malassezia* pulmonary vasculitis in an infant on long-term intralipid therapy. N Engl J Med 305:1395–1399
121. Sizun J, Karangwa A, Giroux JD et al (1994) *Malassezia furfur*-related colonization and infection of central venous catheters. A prospective study in a pediatric intensive care unit. Int Care Med 20:496–499
122. Xu J, Saunders CW, Hu P et al (2007) Comparative genomics of dandruff-associated *Malassezia* species reveals convergent and divergent virulence traits with plant and human fungal pathogens. Proc Nat Acad Sc USA 104:18730–18735

Biodiversity, Phylogeny and Ultrastructure 2

Eveline Guého-Kellermann, Teun Boekhout and Dominik Begerow

Core Messages

- This chapter presents and discusses all techniques and media used to isolate, maintain, preserve, and identify the 13 species that are presently included in the genus. Each species is described morphologically, including features of the colonies and microscopic characteristics of the yeast cells, either with or without filaments; physiologically, including the growth at 37 and 40°C, three enzymatic activities, namely catalase, β-glucosidase and urease, and growth with 5 individual lipid supplements, namely Tween 20, 40, 60 and 80, and Cremophor EL. Their ecological preferences and role in human and veterinary pathology are also discussed.
- For quite a long time, the genus was known to be related to the Basidiomycota, despite the absence of a sexual state. The phylogeny, based on sequencing of the D1/D2 variable domains of the ribosomal DNA and the ITS regions, as presented in the chapter, confirmed the basidiomycetous nature of these yeasts, which occupy an isolated position among the Ustilaginomycetes. The relationship to the Basidiomycetes is also supported by monopolar and percurrent budding and the multilamellar cell wall ultrastructure. Some characteristics of this cell wall, which is unparalleled in the world of fungi, together with the lipophily demonstrate the uniqueness of this genus in the fungal kingdom.

E. Guého-Kellermann (✉)
5, rue de la Huchette, 61400, Mauves sur Huisne, France
e-mail: e.gueho@orange.fr

T. Boekhout et al. (eds.), *Malassezia and the Skin*
DOI: 10.1007/978-3-642-03616-3_2, © Springer Verlag Berlin Heidelberg 2010

2.1 Isolation, Identification and Biodiversity of *Malassezia* Yeasts

Eveline Guého-Kellermann and Teun Boekhout

The genus *Malassezia* was created for a fungus, *M. furfur*, which was seen in lesions of pityriasis versicolor (PV). Unfortunately, it took a long time to understand the lipid dependence of this fungus and, consequently, to obtain and maintain its culture *in vitro*. Due to the lipid requirements, conventional laboratory techniques used for the identification of yeasts could not be applied to this fungus. Despite the description of numerous species, their accurate identification was not feasible, and the taxonomy of the genus remained a controversial subject for decades. The development of molecular techniques allowed the unequivocal separation of species, and then new laboratory methods were developed to characterize these taxa.

2.1.1 Isolation of *Malassezia* Yeasts from Humans and Animals and their Maintenance

The genus *Malassezia,* created by Baillon in 1889 [1] and also known under the generic name *Pityrosporum* created by Sabouraud in 1904 [2], comprises lipophilic and lipid dependent yeasts that require long chain fatty acid (C12 up to C24) supplementation to grow and survive. Slooff [3] in her overview of the history of the genus considered that Panja had been the first to obtain a culture of *Malassezia* on Petroff's egg medium with 0.004–0.005% gentian violet [4, 5]. Shifrine and Marr [6] obtained cultures by adding several fatty acids, in particular oleic acid, to Sabouraud agar. These media, however, were disappointing, because growth was inconstant and resulted in rapid loss of cultures (see Chap. 1). Van Abbe [7] was more successful when he recommended the complex medium created by Dixon. This Dixon's agar (DA) is still in use, according to its original formula, or in a modified version (modified Dixon agar, mDA) as proposed by Midgley [8]. Next to DA, Leeming and Notman [9] proposed a medium, Leeming and Notham agar (LNA) that allowed growth of these nutrient-demanding microorganisms. This medium, elaborated after testing the different compounds separately, allows for isolation and maintenance of all *Malassezia* yeasts. Therefore, it is now largely used by most researchers working with *Malassezia* yeasts. All these complex media contain Ox bile, but the LNA replaces Tween 40, used in the Dixon formula, by Tween 60. According to the assimilation pattern of the 13 species presently described (Plates 2.1 and 2.2), Tween 60 seems to be more efficiently utilized, thus favoring growth of most species, whereas Ox bile, as demonstrated by Japanese authors [10], is an essential, if not sufficient compound, for good growth of *Malassezia* yeasts. Even *M. pachydermatis,* the less demanding species, requires growth media that are enriched with peptone (i.e., Sabouraud medium), which contains short chain fatty acids. On such media, however, viability is lost rapidly, except if the culture is transferred regularly (about every month). Lorenzini and de Bernadis [11] showed that the addition of Tween 80 enhanced the isolation of *M. pachydermatis* from clinical materials significantly.

Malassezia yeasts belong to the normal cutaneous mycobiota of humans and animals, and the skin lipids most likely contain the nutrients required. The optimal growth

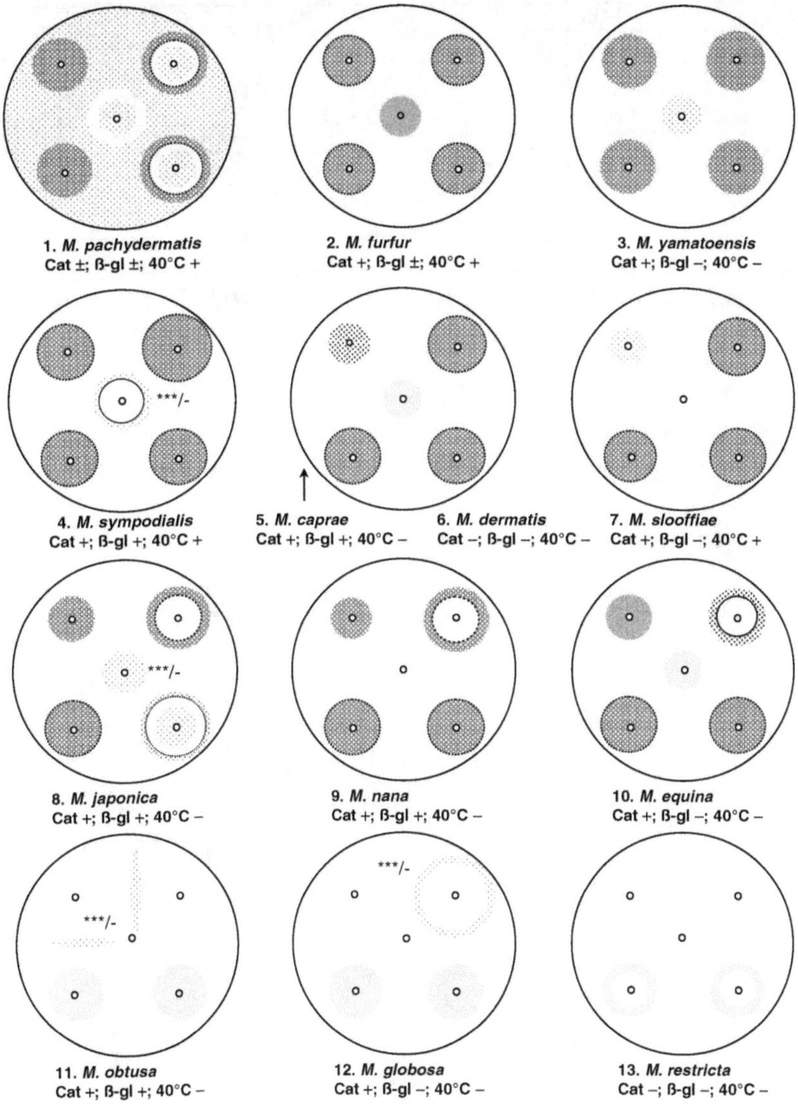

Plate 2.1 Key characteristics of *Malassezia* species, 1 *M. pachydermatis* T; 2. *M. furfur* T; 3. *M. yamatoensis* T; 4. *M. sympodialis* T; 5. *M. caprae* T; (6). *M. dermatis* (not drawn) T; 7. *M. slooffiae* T; 8. *M. japonica* T; 9. *M. nana* T; 10. *M. equina* T; 11. *M. obtusa* T; 12. *M. globosa* T; 13. *M. restricta* T. The 13 species are arranged from the left to the right according to their decreasing physiological and biochemical capacities. **o**: well, 2 mm diameter; top right Tween 20, clockwise Tween 40, 60, 80, Cremophor EL in the centre; : growth very weak or delayed secondary growth within the inhibition area after diffusion of the supplement (*M. pachydermatis* and *M. japonica*); : mild growth and growth of *M. pachydermatis* on GPA, apart of lipid supplements; : very good growth; : precipitate within the agar around the lipid supplements, -/***: absence of growth or colonies present only with recent isolates; *Cat*: catalase activity; *β-gl*: ß-glucosidase activity; 40°C: growth at 40°C, T: type strain of the species

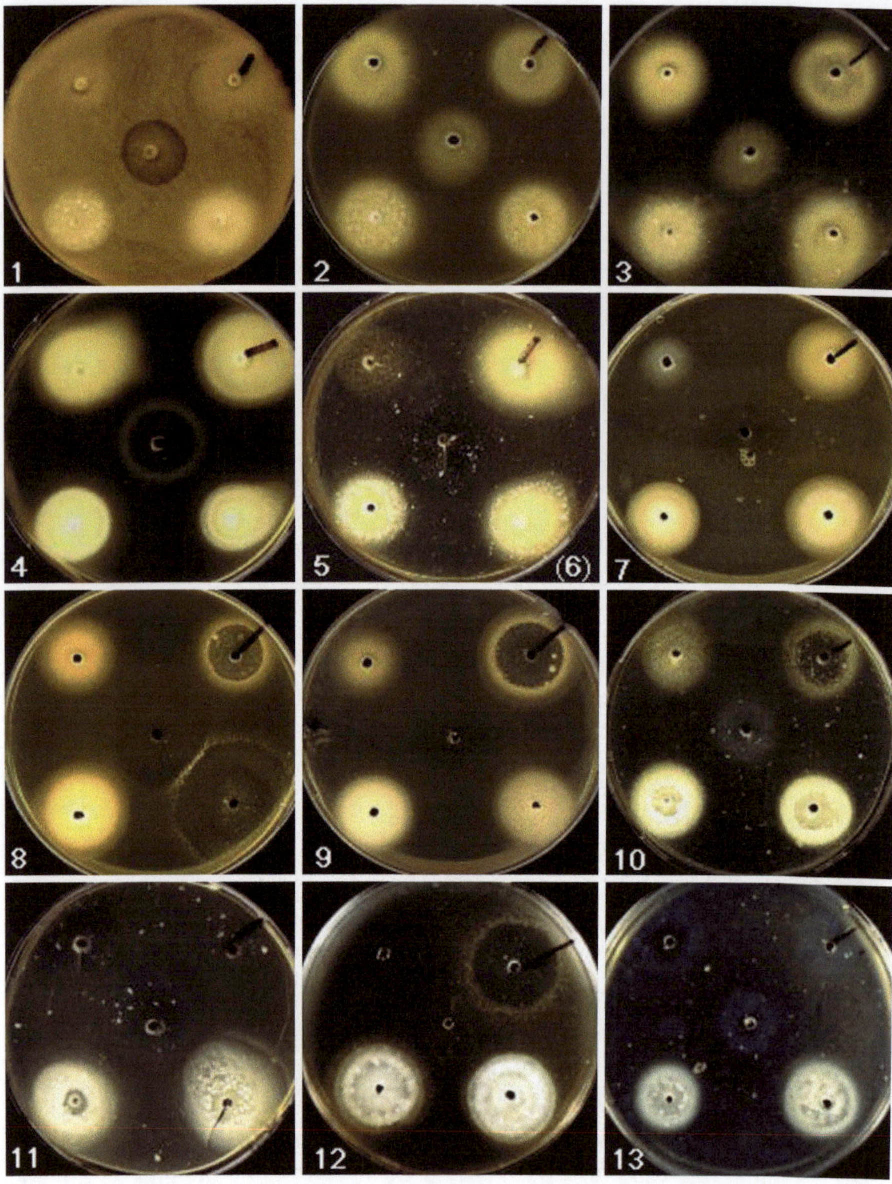

Plate 2.2 Assimilation pattern of *Malassezia* species Tween 20 (with a mark), 40, 60, 80 and Cremophor EL supplementation. *1 M. pachydermatis* (wild isolate from dog); *2 M. furfur*; *3 M. yamatoensis*; *4 M. sympodialis*; *5 M. caprae*; *6 M. dermatis* (not shown); *7 M. slooffiae*; *8 M. japonica*; *9 M. nana*; *10 M. equina*; *11 M. obtusa*; *12 M. globosa*; *13 M. restricta*. The 13 species are arranged from the left to the right according to their increasing lipid requirements

temperature is around 32–34°C; thus, both characteristics seem sufficient to preclude their presence in the environment. Surprisingly, the two fastidious species *M. globosa* and *M. restricta* have been identified by PCR; unfortunately, however, they were not obtained in culture from substrates, such as nematodes, in forest soils in Germany [12], sand stone beneath a crustose lichen in Norway [13], soils of Antarctica Dry Valleys [14], and even from methane hydrate-bearing deep-sea marine sediments in he South China sea [15].

2.1.1.1
Isolation

Below we describe the methods and media used to isolate *Malassezia* yeasts.

2.1.1.1.1
Methods

The samples, collected from skin, scalp, nails, hair, blood, catheter, or any other human or animal source, are transferred as soon as possible onto one or the other selective media to avoid dehydration of the yeasts. During transportation, moisture must be maintained as high as possible, using for instance a plastic bag or box. The samples, distributed onto the selective media in 9 cm Petri dishes, are incubated in a moist environment at 32–34°C, for at least 2 weeks.

2.1.1.1.2
Selective Media

a) *Sabouraud agar plus olive oil:* Mix 20 g glucose, 10 g bacteriological peptone and 10 mL virgin olive oil, 0.5 g chloramphenicol, 0.5 g cycloheximide in 1 L of demineralised water, adjust pH to 6.0, and add 12–15 g agar. Heat to dissolve the agar. Sterilize by autoclaving at 120°C for 15 min and aliquot as required. Addition of other oils or fatty acids, such as oleic acid, can be tested using the same recipe.
b) *Dixon agar (DA):* Mix 60 g malt extract, 20 g dessicated ox bile (Oxgall, BD Difco), 10 mL Tween 40, 2.5 g glycerol monooleate, 0.5 g chloramphenicol, 0.5 g cycloheximide in 1 L of demineralised water, adjust the pH to 6.0, and add 12–15 g agar. Sterilize by autoclaving at 115°C for 15 min, and aliquot as required.
c) *Modified Dixon agar (mDA):* Mix 36 g malt extract, 10 g bacteriological peptone, 20 g dessicated ox bile, 10 mL Tween 40, 2 mL glycerol, 2 g oleic acid, 0.5 g chloramphenicol, 0.5 g cycloheximide in 1 L of demineralised water, adjust the pH to 6.0, and add 12–15 g agar. Dissolve the agar by heating and sterilize by autoclaving at 115°C for 15 min, and aliquot as convenient.
d) *Leeming and Notman agar (LNA):* Mix 10 g bacteriological peptone (Oxoid), 0.1 g yeast extract, 5 g glucose, 8 g dessicated ox bile, 1 mL glycerol, 0.5 g glycerol monostearate, 0.5 g Tween 60, 10 mL whole fat cow milk, 0.5 g chloramphenicol, 0.5 g cycloheximide in 1 L of demineralised water, adjust the pH to 6.0, and add 12–15 g of agar. Sterilize by autoclaving at 110°C for 15 min, and aliquot as convenient.

2.1.1.1.3 Remarks

1) For an exhaustive survey, the samples, either from humans or animals, must be inoculated only onto a selective complex medium. Indeed, *M. globosa, M. obtusa,* and *M. restricta* are highly lipid-dependent, and a few isolates of primary cultures of *M. pachydermatis* do not grow on Sabouraud agar [16]. In clinical practice, the Sabouraud agar supplemented with olive oil, which can be prepared easily and rapidly, is not recommended because only *M. furfur*, *M. pachydermatis* and *M. yamatoensis* grow well on this medium [17].
2) Clinicians are also used to incubating *Malassezia* yeasts at 37°C, as this temperature is considered selective for pathogenic microorganisms. These yeasts, however, belong to the cutaneous mycobiota, and thus are ecologically adapted to a lower temperature. Because *M. globosa, M. obtusa,* and *M. restricta,* and also *M. caprae* and *M. equina,* which originated from animals, have a maximum growth temperature at 37°C [17, 18], the incubation temperature should never exceed 35°C, with an optimum between 32 and 34°C. *Malassezia* yeasts do not survive temperatures below 28°C very long, so, materials obtained from collects must not be maintained in a refrigerator before culturing. Use of a high incubation temperature and the utilization of olive oil, which does not allow the growth of most species, may explain why the knowledge of the genus remained limited to a few species for so long.
3) For epidemiological surveys, cultures must be made onto Petri dishes rather than tubes, because the latter do not allow a good separation of colonies. In the same way, the dark Dixon agars facilitate visualization of any mixed growth of *Malassezia* species, or any skin sample contaminated by other micro-organisms, such as bacteria or *Candida* spp.
4) For surveys of *Malassezia* spp. on animals, it is recommended to double the concentration of antibiotics and to use only selective media, because animal fur or/and skin are covered by a large quantity of micro-organisms. Besides, some isolates of *M. pachydermatis* have been shown to be lipid-dependent [16], and it is now well recognized that the veterinary *Malassezia* mycobiota are no longer limited to this unique species.

2.1.1.2 Maintenance of Cultures

Purified *Malassezia* isolates can be maintained on slant cultures in an incubator with a moist environment between 30 and 32°C. Cultures do not survive at room temperature very long. In routine work, they must be transferred on fresh medium every two months, but this may be one month for *M. obtusa* and *M. restricta*.

With the exception of the fastidious species *M. globosa*, *M. obtusa* and *M. restricta*, the other species can be preserved by lyophilisation. Probably, all species may survive freezing at −80°C ([19], Guého unpublished data).

a) *Lyophilization*: Cells of 4–5-day-old cultures of *Malassezia* spp. are suspended in liquid Dixon medium supplemented with 15% glycerol, and lyophilizates are stored in a refrigerator at 4°C.

b) *Freezing −80°C:* Cells of 4–5-day-old cultures of all *Malassezia* spp. are suspended in liquid mDixon medium supplemented with 15% of glycerol and aliquoted by 1 mL into 2 mL freezing Eppendorf tubes. Where possible, tubes are cooled down by −1°C per min in a progressive freezer up to −80°C, and are then stored at −80°C. To revive the yeasts, tubes are melted in a 37°C water bath, centrifuged to eliminate the medium, and subcultured by spreading the cells sparsely onto DA, mDA or mLNA medium.
c) *Liquid nitrogen:* Maintenance in liquid nitrogen was found to be the most satisfactory method of preservation for *M. pachydermatis* [20]. Therefore, it may be interesting to investigate this method of preservation for the lipid-dependent species as well.

2.1.2
Identification of *Malassezia* Yeasts Using Routine Laboratory Methods

After the species had been recognized by means of rRNA sequencing [21], it became possible to recognize their morphological, biochemical and physiological characteristics [22–26]. The subsequent description of an additional 6 species (*M. caprae, M. dermatis, M. equina, M. japonica, M. nana,* and *M. yamatoensis*) gave the opportunity to update this protocol [17], which now includes the characterization of urease -, catalase -, and β-glucosidase activities, growth at 37°C and 40°C, and the capability to grow with five water soluble lipid supplements, namely Tweens 20, 40, 60, 80 and Cremophor EL (castor oil).

2.1.2.1
Characterization of Urease Activity Using Bacto Urea Broth

Malassezia yeasts belong to Basidiomycota (see Chap. 2.2) which, in contrast to Ascomycota, are capable of hydrolysing urea. With *Malassezia* yeasts, this test is not used to separate species but rather to eliminate cultures that are contaminated by bacteria or ascomycetous yeasts, such as *Candida* spp. which are quite common on the skin. All *Malassezia* yeasts give a positive staining diazonium blue B reaction (DBB) [17, 27], but the urease test is easier to perform and provides more reliable information. The DBB staining reaction is described, with all other laboratory techniques, in the 5th edition of "The Yeasts, a taxonomic study" [28].

2.1.2.1.1
Method

A loopful of cells from 4- to 5-day-old cultures are suspended in urea broth and incubated at 37°C, irrespective of whether the yeast can grow at this temperature. The urease expression gives a bright pink to violet coloration. The reaction can be read after 1–4 h of incubation, or after 24 h in case of a doubtful reading. Any isolate giving a yellow color can be eliminated, or must be further purified if possible.

2.1.2.1.2
Medium

Difco Bacto Urea R Broth is dissolved in sterile demineralised water and aliquoted aseptically into 0.5 mL volumes in tubes. The tubes can be stored in a freezer for up to 6 months. The ready-made urea-indole medium (bioMérieux) can be used as well.

2.1.2.1.3
Remark

The reaction may give a doubtful reading on solid medium, such as Christensen's urea agar (numerous strains of *M. pachydermatis* missing an urease activity on this medium were observed by A. Velegraki, unpublished data).

2.1.2.2
Characterization of the Catalase Activity Using Hydrogen Peroxide

This test is commonly used in bacteriology as a first step in identification. It appeared to be useful within the genus *Malassezia* as well.

2.1.2.2.1
Method

Catalase activity of *Malassezia* yeasts is determined by adding a drop of hydrogen peroxide onto a culture smear on a glass slide or in a small seeded glass tube. The enzyme catalyzes the decomposition of peroxides formed during oxidation reactions. A positive result is indicated by effervescence, caused by the liberation of free oxygen. It is recommended to use only glass for this test in order to avoid doubtful results.

2.1.2.2.2
Reagent

The test is performed using 10–20% (vol. instead P% in France) hydrogen peroxide or the commercial reagent ID color Catalase (bioMérieux), which makes the reaction easier to read and more stable, owing to the presence of a thickener.

2.1.2.3
Characterization of β-Glucosidase Activity Using the Esculin Medium

Certain *Malassezia* species possess a β-glucosidase that is able to hydrolyse the glucosidic bond of esculin, thus liberating glucose and esculetin. The phenol moiety reacts with the iron to give a black color.

2.1.2.3.1
Method

The esculin medium in tubes is inoculated by stabbing the yeast culture centrally into the medium using a platinum wire, and is incubated at 37°C, irrespective of whether the yeast can grow at this temperature. There is no need to screw the cap down tightly. Examine daily for 5 days. A positive reaction is indicated by blackening of the medium, whereas absence of blackening indicates lack of β-glucosidase activity.

2.1.2.3.2
Media

Esculin agar (EA) Mix 10 g bacteriological peptone, 1 g ferric ammonium citrate, 1 g esculin per 1 L demineralised water, adjust the pH to 7.4, and add 15 g agar. Dissolve by heating and distribute in 6 mL volumes in tubes. Sterilize by autoclaving at 115°C for 15 min. Store at 2–6°C for 2 years.

2.1.2.3.3
Remarks

1) The esculin medium may be used directly in Petri dishes, but then the reaction can be slower, thus resulting in doubtful answers.
2) The ready-made Esculin Iron Agar (esculin 0.1 g, ferric ammonium citrate 0.5 g, agar 15 g, demineralised water 1 l) (Fluka) can been used as well, but with addition of 1% peptone. Furthermore, the esculin concentration is lower in this latter medium, and such differences in the composition of esculin media can explain discrepancies in results obtained.
3) Japanese authors [29, 30] have combined the esculin test to a growth medium (Tween 60-esculin agar), but this ready-made medium does not allow good reading of the β-glucosidase activity and may also increase the cost of identification.
4) Commercial bacteriological esculin tubes (bio-Rad) can be used as well, but may increase the cost.

2.1.2.4
Growth at 37°C and 40°C

Subcultures are incubated using one or the other selective medium at 37 and 40°C to get supplementary key characteristics.

2.1.2.4.1
Remark

Growth at 37°C may give doubtful responses, since several species are limited to this temperature, as indicated above.

2.1.2.5
Utilization of Tweens 20, 40, 60, 80, and Cremophor EL (see Scheme 2.1 and Plates 2.1 and 2.2)

Lipophilic and lipid-dependent *Malassezia* yeasts require complex media that are enriched with lipids. These particular requirements were found to be useful to separate species [25, 26]. Strains are tested for their capacity to grow on Sabouraud agar (GPA), separately supplemented with Tweens 20, 40, 60, 80, and Cremophor EL (CrEL or castor oil) as an unique lipid source. These five water-soluble compounds can be tested together, using their capacity to diffuse into a solid basic medium. Many insoluble lipids have been tested, but so far without improving the identification.

2.1.2.5.1
Methods

Two loops of a 4–5-day-old *Malassezia* culture are suspended in 3.0 mL of sterile demineralised water. This inoculum is added to 18 mL of a molten Sabouraud agar maintained at 50°C, and the mixture is poured immediately in a 9-cm Petri dish (Scheme 2.1). After complete solidification, wells are made with a 2-mm diameter punch, 4 devoted to test the growth using Tweens 20, 40, 60 and 80 clockwise around, with a mark to indicate the position of Tween 20, and a fifth hole in the center to test growth with Cremophor EL. The wells are filled with approximately 15 µL of each product (Sigma), which are not sterilized but aliquoted in 2 mL Eppendorf tubes. The dishes are incubated for 7–10 days (for good pictures) at 32–34°C in a moist environment, and turned upside down on the second day to delay their dehydration.

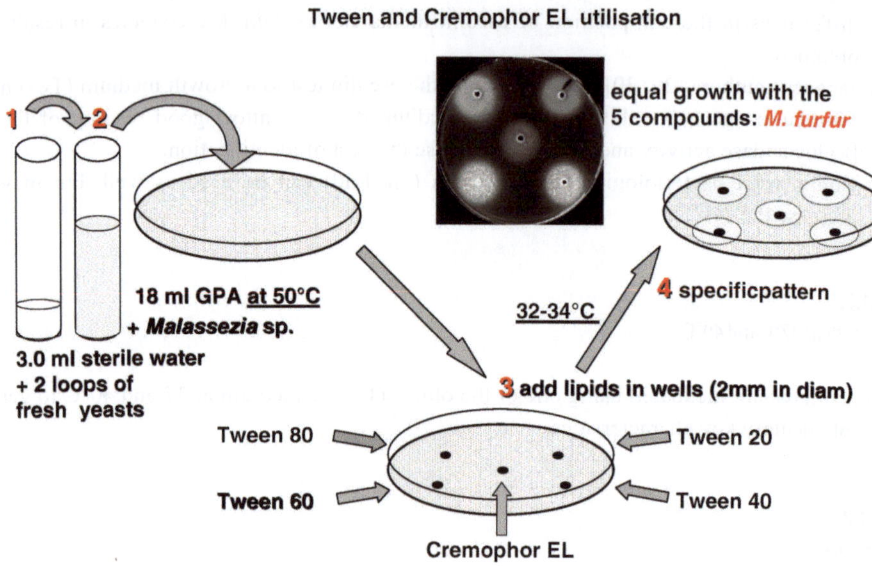

Scheme 2.1 Method to evaluate growth of *Malassezia* spp. with the five individual lipid supplements

2.1.2.5.2 Medium

a) *Glucose-peptone agar (GPA) (also named Sabouraud agar (SA)).* Mix 20 g glucose and 10 g bacteriological peptone in 1 L of demineralised water, adjust the pH to 6.0, and add 18 g agar. Heat to dissolve the agar, distribute in 18 mL-volumes in tubes and sterilize by autoclaving at 120°C for 15 min.

2.1.2.5.3 Remarks

1) To avoid any contamination, 0.5 g cycloheximide per L can be maintained in SA. The assimilation patterns can be scanned, but then the wells are filled with Tween supplements counter-clockwise. For pictures, always add a black sheet as bottom or cover, when camera or scanner are used, respectively.
2) Using the diffusion method, each soluble lipid supplement gives a concentration gradient, which, consequently, may provide supplementary information. Indeed, growth can start close to the well with a high concentration of the supplement, thus giving a full disk of colonies (i.e., *M. furfur, M. slooffiae*) (Scheme 2.1, Plates 2.1-2, 7 and 2.2-2, 7). In contrast, growth may appear mainly at some distance from the well after dilution of the supplement, resulting in a ring of colonies (i.e., Tween 20 of *M. japonica,* the type strain of *M. nana,* and *M. equina*) (Plates 2.1 and 2.2. 8, 9 and 10, respectively). Growth can also start at some distance from the well after diffusion of the supplement, but with secondary growth progression back towards the well, resulting in complete or incomplete centripetal growth (i.e., Tween 20 of *M. caprae, M. dermatis, M. sympodialis,* or *M. yamatoensis*; see also Tween 20 of *M. sympodialis* in Fig. 2.4f). Then supplements can generate a neat inhibition area, and later, a secondary growth zone within this area (i.e., Tween 20, 40, and Cr EL of *M. pachydermatis* or Tween 40 of *M. japonica*) (Fig. 2.1f, Plates 2.1-1, 8 and 2.2-1, 8).
3) Since the first description of a practical method to identify *Malassezia* yeasts [25], several improvements have been suggested, including assimilation of Cremophor EL and characterization of β-glucosidase activity [17, 22, 24, 26, 31]. However, the addition of six new species (viz. *M. caprae, M. dermatis, M. equina, M. nana, M. japonica* and *M. yamatoensis*) resulted in similar physiological patterns of several *Malassezia* species, and thus in a doubtful identification, e.g., of isolates belonging to *M. caprae* and *M. nana* [17]. In these cases, the identification should be confirmed by sequence analysis of the D1/D2 domains of the LSU rRNA gene and/or the ITS1+2 regions, in order to reduce the risk of misidentifications (see Chap. 3.1). Furthermore, attention should be given to the source of such isolates, whether human or animal, as well as their precise location on them.
4) The Japanese system of identification [29, 30] requires, after the catalase test with 3% hydrogen peroxide has been performed, subculturing of the isolate on several expensive media (CHROMagar-*Malassezia*, SDA, Cremophor EL agar, Tween 60-esculin Agar). Moreover, this system needs to be evaluated with all currently known species, and additional isolates from various sources for each of them.

2.1.3
Biodiversity of *Malassezia* Yeasts

Characteristics of all species have been re-evaluated for the 5th edition of "The yeasts: a taxonomic study" [17]. In this book, species are systematically arranged alphabetically [32]. In this chapter, we list them according to their decreasing biochemical and physiological capabilities as shown in Plates 2.1 and 2.2. This option somewhat parallels the chronological order of discovery, since the oldest described species are not too demanding in their grow requirements (i.e., *M. furfur, M. pachydermatis* and *M. sympodialis*). Urease and DBB staining reactions are positive for all species, and thus these characteristics are not further included. Similarly, all species can be considered to have a co-enzyme Q system with nine isoprenologs, i.e., CoQ-9 [33], even though this has not yet been investigated for all species [17].

2.1.3.1
Malassezia pachydermatis (Weidman) Dodge (1925)

This species, isolated for the first time by Weidman in 1925 [35] from a captive Indian rhinoceros, was described as *Pityrosporum pachydermatis* and transferred to the genus *Malassezia* by Dodge in 1935 [34]. In contrast to similar organisms observed in scales of PV or pityriasis capitis, *M. pachydermatis* was able to grow on regular rich media. Unfortunately, the original strain isolated by Weidman was not preserved. In 1955 Gustafson described *Pityrosporum canis* from the ear of a healthy dog [36]. For years, its synonymy with the previously described species *P. pachydermatis* remained uncertain, but molecular approaches, such as analysis of mol% G+C, nuclear DNA/DNA reassociation experiments, and rRNA or rDNA sequence comparisons, applied to isolates from various sources, in particular, five captive Indian rhinoceros, allowed to validate the species name *M. pachydermatis* [37]. However, the species was also found to be genetically heterogeneous and in the course of evolution, probably due to adaptation to their hosts [38, 39].

2.1.3.1.1
Neotype Strain

CBS 1879, isolated from otitis externa in dog. Because the original culture of *Pityrosporum pachydermatis* [35] has not been preserved, the strain CBS 1879, which had been deposited as the type strain of *Pityrosporum canis* [36], was selected as neotype of *M. pachydermatis,* when it was demonstrated that both old names were synonyms [23, 37].

2.1.3.1.2
Morphological Characteristics

After growth at 32°C on mDA on SA for 7 days, single colonies (Fig. 2.1a) are convex, 4–5 mm in diameter, butyrous to brittle, somewhat shiny, pale yellowish-cream, and with an entire, straight or somewhat undulating margin. A few isolates from dogs were found to be lipid-dependent with smaller colonies [16], and belonged to the same sequevar d [21, 38]. An unusual pink isolate from civet (GM 439) clustered as sequevar a with strains from dogs, including neotype strain CBS 1879, and all strains from humans [21]. Cells are ellipsoidal to

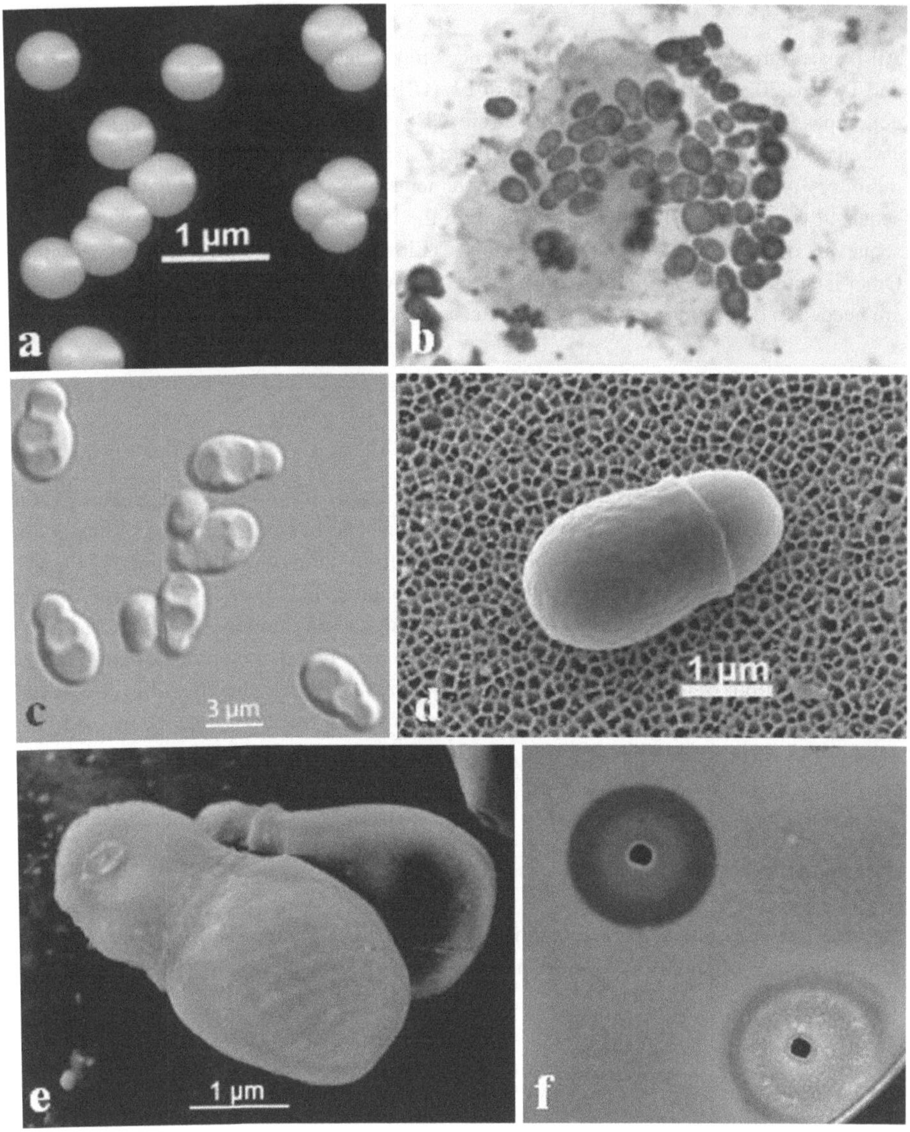

Fig. 2.1 *M. pachydermatis*. (**a**) Convex colonies with an entire margin; (**b**) ovoid Gram stained yeasts in dog ear cerumen (picture by the courtesy of Guillot); (**c**–**e**), Nomarski's and SEM micrographs showing ovoid to short cylindrical yeast cells with a broad budding site; (**e**), notice the helicoidal structure of the cell wall; (**f**), details of Cr EL and Tween 40 utilization showing secondary growth within the inhibitory areas

short cylindrical, 4–5 × 2–2.5 µm, with monopolar budding on a broad base (Fig. 2.1b–e). Filaments were never observed in this species.

2.1.3.1.3
Physiological and Biochemical Characteristics

All isolates grow at 37 and 40°C. Differences in catalase and β-glucodidase expression, and Tweens 20, 40, 60, 80 and Cremophor EL (CrEL) growth reactions occur in all rDNA genotypes. For instance, the neotype strain CBS 1879 presents growth inhibition with Tweens 20, 40 and CrEL (Fig. 2.1f and Plate 2.1-1), with secondary growth within this inhibitory area after diffusion of these supplements [40]. Strains from dogs display three sequevar types, namely sequevar a, including the type strain and all human isolates, and sequevars d and e. With the latter type, only CrEL shows an inhibitory area (Plate 2.1-1). Other isolates may be not inhibited by any of the five supplements, but more experiments are needed to determine whether or not, these differences are stable strain characteristics.

2.1.3.1.4
Ecology

Malassezia pachydermatis is a lipophilic, but not a highly lipid-dependent species [3, 41, 42]. The species is most often associated with animals, particularly ears, and/or healthy or lesional skin of canines, but has also been isolated from numerous other animals [43, 44]. Its prevalence in rhinoceroses was confirmed several times [21, 37, 45, 46], but is probably largely underestimated in animals other than cats, dogs and rhinoceroses (see Chaps. 3.3 and 10). *M. pachydermatis* can be isolated from human blood and sputum, and is implicated in infections of neonates, under parenteral alimentation enriched by lipids. However, its presence on humans is transitory and the source of these infections may be linked to pet animals [47–49].

2.1.3.2
Malassezia furfur (Robin) Baillon (1889)

Because neotype cultures corresponding to the old names *Malassezia furfur* and *P. ovale* proved to be synonymous, the former name is maintained as the generic type species [23].

2.1.3.2.1
Neotype Strain

The strain CBS 7019, isolated from PV scales on the trunk of a 15-year-old girl in Finland, is considered as the neotype strain of this species. CBS 1878, however, has also been designated as neotype, but this strain originally represented *P. ovale* (Bizzozero) Castellani and Chalmers [50], a fungus that was regularly seen in scales of pityriasis capitis.

2.1.3.2.2 Morphological Characteristics

After 7 days at 32°C on mDA, single colonies (Fig. 2.2a) are 4–5 mm in diameter, dull, umbonate or slightly folded, butyrous to friable, smooth, with a convex elevation and an entire to slightly lobate margin. Colony texture is soft and the cells are easy to emulsify.

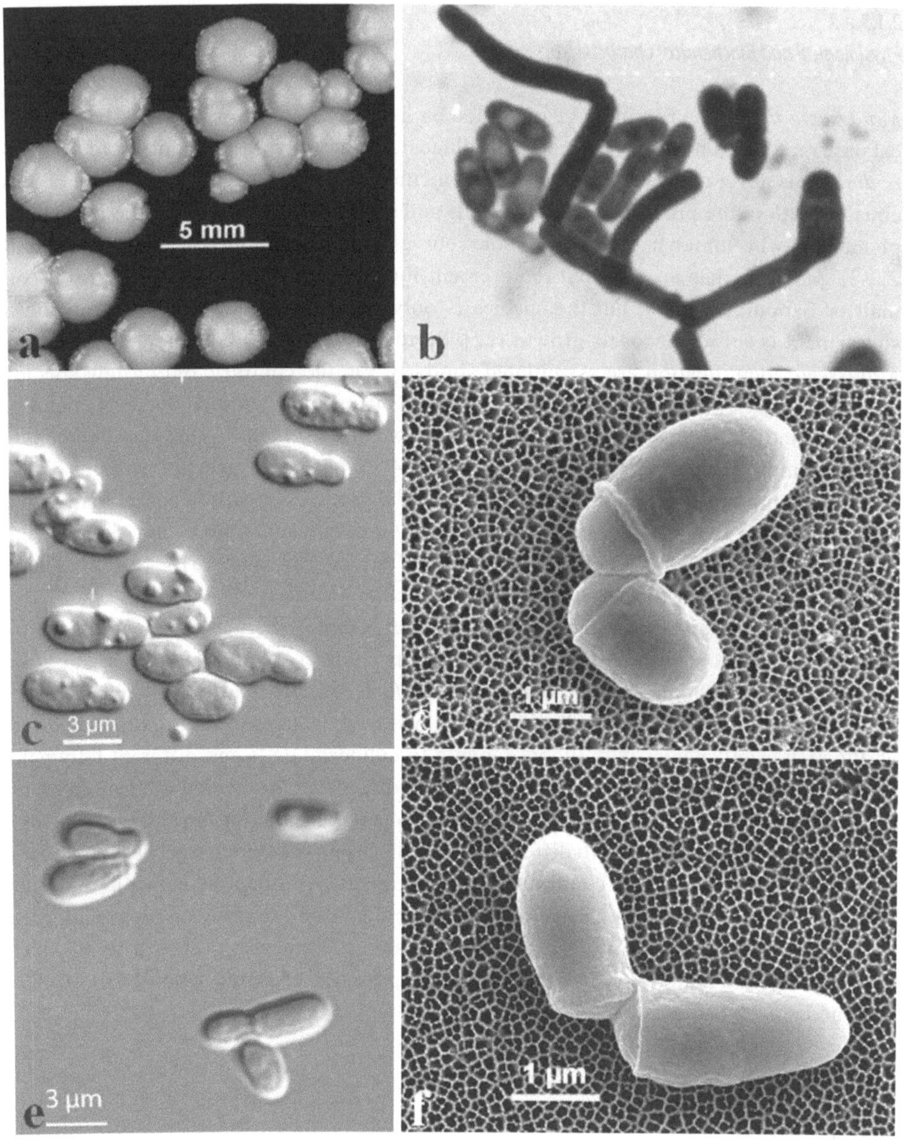

Fig. 2.2 *M. furfur.* (**a**) Umbonate colonies with a slightly lobate margin; (**b**) Gram stained ellipsoidal yeasts and pseudohyphae of a spontaneously filamentous strain; (**c–d**), Nomarski's and SEM micrographs, respectively showing the ellipsoidal yeasts of CBS 7019; (**e–f**), Nomarski's and SEM micrographs, respectively showing the cylindrical yeasts of CBS 1878

Cells are variable in size and shape, cylindrical to ovoid, 2.5–8 × 1.5–3 μm (Fig. 2.2b–f), or globose, 2.5–5 μm in diameter, with percurrent budding on a more or less broad base. Pseudohyphae may be occasionally produced in some cultures (Fig. 2.2b) and, if present, this feature seems to be a stable strain character. Pseudohyphae may also occur after culturing under appropriate medium conditions (see Chap. 2.3).

2.1.3.2.3
Physiological and Biochemical Characteristics

Malassezia furfur can routinely be identified by its capacity to grow up to 41°C, a strong catalase reaction, a more or less marked β-glucosidase activity (but in contrast to *M. sympodialis* this never turns very dark after 24 h incubation at 37°C), and shows more or less equal growth in the presence of four Tweens and CrEL (Note: growth for the latter supplement may be somewhat weaker) as sole sources of lipid (Scheme 2.1, Plates 2.1-2 and 2.2-2). *M. furfur* has an essential requirement for olive oil or oleic acid for growth on malt or Sabouraud agars, but the species is only mildly lipid-dependent, as any lipid supplement is sufficient for its growth [22]. Furthermore, all lipid supplements used for *Malassezia* identification are assimilated similarly. On the contrary, only a few species are able to grow well with oleic acid or olive oil as lipid supplementation, namely *M. furfur* and *M. pachydermatis* and, to a lesser extent, *M. japonica* and *M. yamatoensis* [17]. Optimum temperature for growth is near 34°C, but good growth occurs at 37°C, and the maximum temperature for growth is 41°C. Species-specific characteristics are rather stable for this species. However, a few atypical variants may occur, such as isolates that grow well with Tween 80 only [51] or fail to grow with CrEL [8, 52]. These atypical isolates need further studies with molecular methods. *M. furfur* is one of the most robust lipid dependent species, and its growth is merely induced by an amino nitrogen in combination with a lipid source [26]. In contrast to *M. globosa, M. obtusa, M. restricta, M. slooffiae* and *M. sympodialis, M. furfur* is able to utilize glycine as nitrogen source [53]. Salkin and Gordon [54] examined fresh isolates of *Malassezia* (reported as *Pityrosporum*) species with globose, ovoid to cylindrial cells and different fatty acid requirements, and they suggested that *P. ovale* and *P. orbiculare* were both synonymous with *M. furfur*. However, in this pre-DNA era it was impossible for the authors to recognize that they were dealing with different taxa. This proposed synonymy was maintained by Yarrow and Ahearn [55]. Unfortunately, the original type material and isolates designated as *P. orbiculare* on the basis of cell shape and inability to grow on oleic acid were not preserved. Therefore, in the taxonomic revision [23], it was proposed to consider *P. orbiculare* as a doubtful species, which may represent a probable synonym of *M. globosa* and not *M. furfur*.

2.1.3.2.4
Ecology

Malassezia furfur is known from various hosts and body sites. In humans, it has been isolated from scalp, face, dandruff, arms, legs, urine, blood, hair, nails, eyes, and the nasal cavity. The high temperature tolerance of *M. furfur*, contrary to *M. globosa*, could explain why *M. furfur* is more frequently isolated from pityriasis versicolor (PV) in warmer climates [17, 48]. The species has also been detected from a hospital floor [56], and occasionally from cats [57, 58], dogs [59], horses [60], cows, asymptomatic or with otitis [61, 62], and bats [63]. However, more veterinary surveys will be necessary to fully evaluate the prevalence of *Malassezia* lipid-dependent species on animals.

2.1.3.3 *Malassezia yamatoensis* Sugita, Tajima, Takashima, Amaya, Saito, Tsuboi & Nishikawa (2004)

2.1.3.3.1 Type Strain

CBS 9725 (JCM 12262), isolated from a nose lesion of a seborrheic patient [64].

2.1.3.3.2 Morphological Characteristics

After 7 days at 32°C on mDA, single colonies (Fig. 2.3a) are flat to convex, 3–4 mm in diameter, butyrous, shiny, pale yellowish-cream, smooth, and with an entire, somewhat

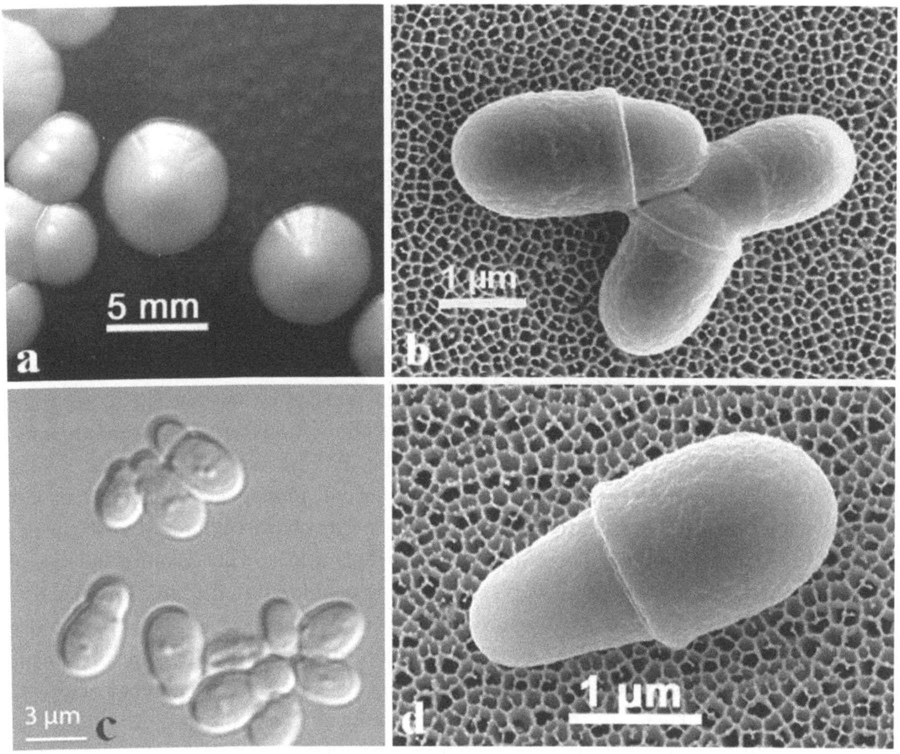

Fig. 2.3 *M. yamatoensis.* (**a**) Convex colonies with a slightly folded and undulate margin; (**b–d**) Nomarski's and SEM micrographs showing short cylindrical yeasts and their broad budding sites

undulating margin. Cells are ovoid to short cylindrical, 3–4 × 2.4–3 µm, with monopolar budding on a broad base (Fig. 2.3b–d).

2.1.3.3.3
Physiological and Biochemical Characteristics

Malassezia yamatoensis can be identified by its capacity to grow at 37°C, but not at 40°C, a strong catalase reaction and lack of β-glucosidase activity. These characteristics distinguish the species from *M. sympodialis*. Equal growth that progresses centripetally appears in the presence of all four Tweens, and a more or less marked growth occurs with CrEL (Plates 2.1-3 and 2.2-3).

2.1.3.3.4
Ecology

Malassezia yamatoensis seems to be a rare species, which has been reported from humans with atopic or seborrheic dermatitis, and more rarely from healthy individuals [64].

2.1.3.4
Malassezia sympodialis Simmons & Guého (1990)

2.1.3.4.1
Type Strain

CBS 7222, isolated from the auditory tract of a healthy 33-year-old male [65]. *M. sympodialis* corresponds to *M. furfur* serovar A as previously recognized [66].

2.1.3.4.2
Morphological Characteristics

After 7 days at 32°C, single colonies (Fig. 2.4a) are flat to somewhat elevated in the center, approximately 6–8 mm in diameter, pale cream to yellowish-brown, shiny, smooth, butyrous, and with an entire or finely folded margin. The cells are ovoid to globose, 2.5–4.0 x 1.5-3.5 µm (Fig. 2.4b–e), with enteroblastic, percurrent, monopolar budding, and with buds that may emerge sympodially from a relatively narrow base. Cultures of fresh isolates on mDA develop crystal precipitates in the agar, diffusing around the colonies and resembling eventual contaminations. This phenomenon is common to other species related to *M. sympodialis* [67], namely *M. dermatis, M. caprae, M. equina,* and *M. nana,* and also *M. globosa.* This precipitate in the culture medium is different from the white precipitate that usually surrounds Tween 40 and 60 in the testing dishes, even in the absence of growth (Plate 2.1 and 2.2). There are conflicting reports on the ability of the species to form filaments in culture. According to some workers, *M. sympodialis* is able to form filaments

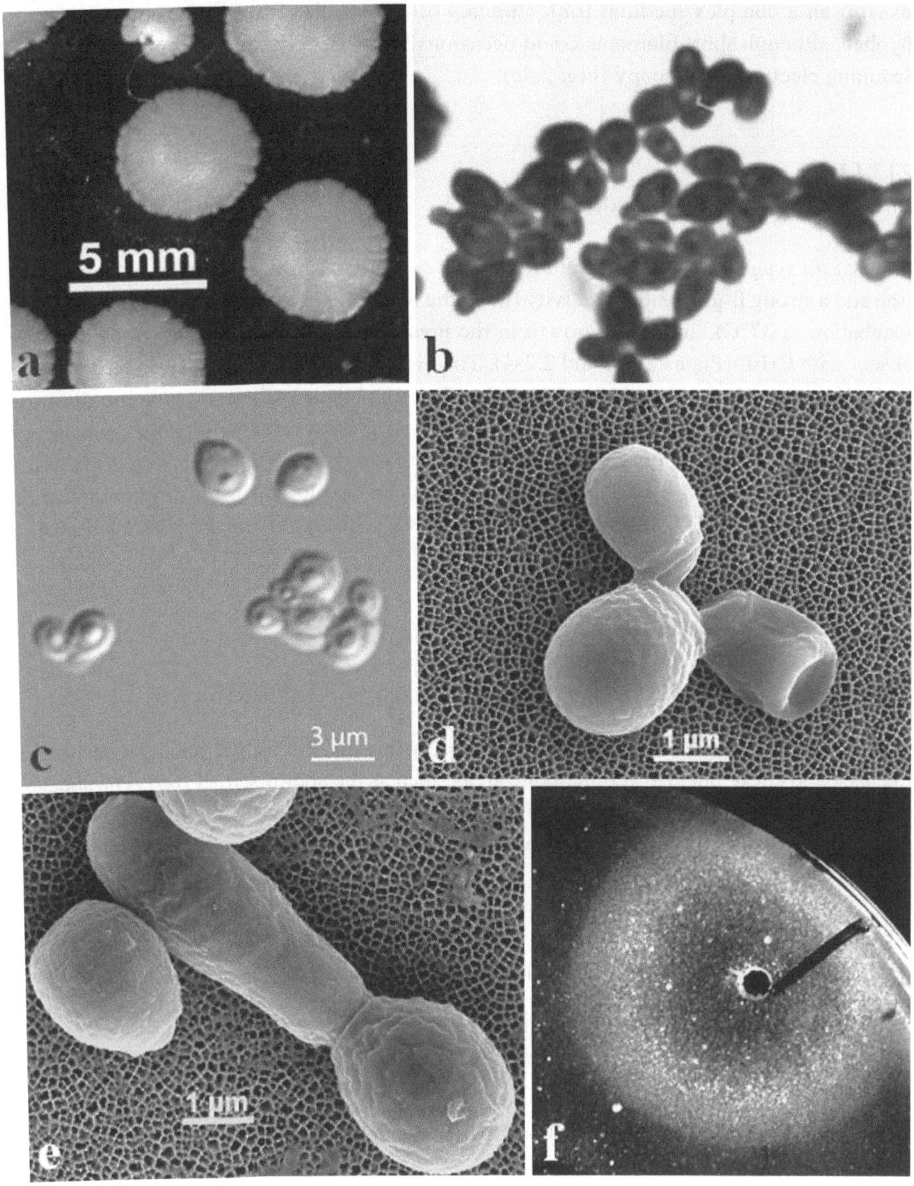

Fig. 2.4 *M. sympodialis*. (**a**) Flat colonies that are smooth and slightly elevated in the center with a finely folded margin; (**b–e**) Nomarski's and SEM micrographs showing ovoid to globose yeasts; (**d**) regular clover leaf configuration of cells, which is a typical feature of this species; (**e**) germination tube showing the helicoidal structure of the cell wall; (**f**) growth with Tween 20 that progresses centripetally, but is still incomplete towards the well after 4 days of incubation

in vitro on a complex medium [68]. Cultures of this species usually do not display any hyphae, although short filaments could occasionally be observed in the type culture using scanning electron microscopy (Fig. 2.4e).

2.1.3.4.3 Physiological and Biochemical Characteristics

Malassezia sympodialis can be identified by its capacity to grow at 40°C, a catalase reaction and a strong β-glucosidase activity (note: the reaction is getting dark already after 24 h incubation at 37°C), and good growth in the presence of all four Tweens, but growth is absent with CrEL (Plates 2.1-4 and 2.2-4). The disk of colonies around Tween 20 is often wider than that obtained with the three other Tweens, but growth starts at some distance from the well after diffusion of the compound and progresses centripetally, thus remaining after a few days' incubation as a zone of incomplete growth near the well (Fig. 2.4f). With CrEL, growth is usually absent, but fresh isolates can develop a ring of tiny colonies, or they show a white precipitate only, which occurs at some distance from the well (Plates 2.1-4 and 2.2-4).

2.1.3.4.4 Ecology

Malassezia sympodialis is an inhabitant of healthy human skin. In particular, it occurs on the back and chest, but also at other body areas, such as the auditory tract [65]. The species has also been isolated from human PV [21, 23, 69, 70] and occasionally from healthy feline skin [57, 71–73], goats [60], cows [61] and bats [63].

2.1.3.5 Malassezia caprae Cabañes & Boekhout (2007)

2.1.3.5.1 Type Strain

CBS 10434 (MA 383), isolated from healthy skin of goats, Barcelona, Spain [18].

2.1.3.5.2 Morphological Characteristics

After 7-10 days at 32°C on mDixon agar, single colonies (Fig. 2.5a) are small, 1–2 mm in diameter, whitish to cream-colored, smooth, glistening or dull, butyrous, and moderately convex with an entire to lobate margin. Cells are globose, somewhat rhomboidal or ovoid, 2.5–4 × 2.2–3.5 µm, with buds emerging from a narrow to moderately broad base (Fig. 2.5b–d). Hyphae have not been observed.

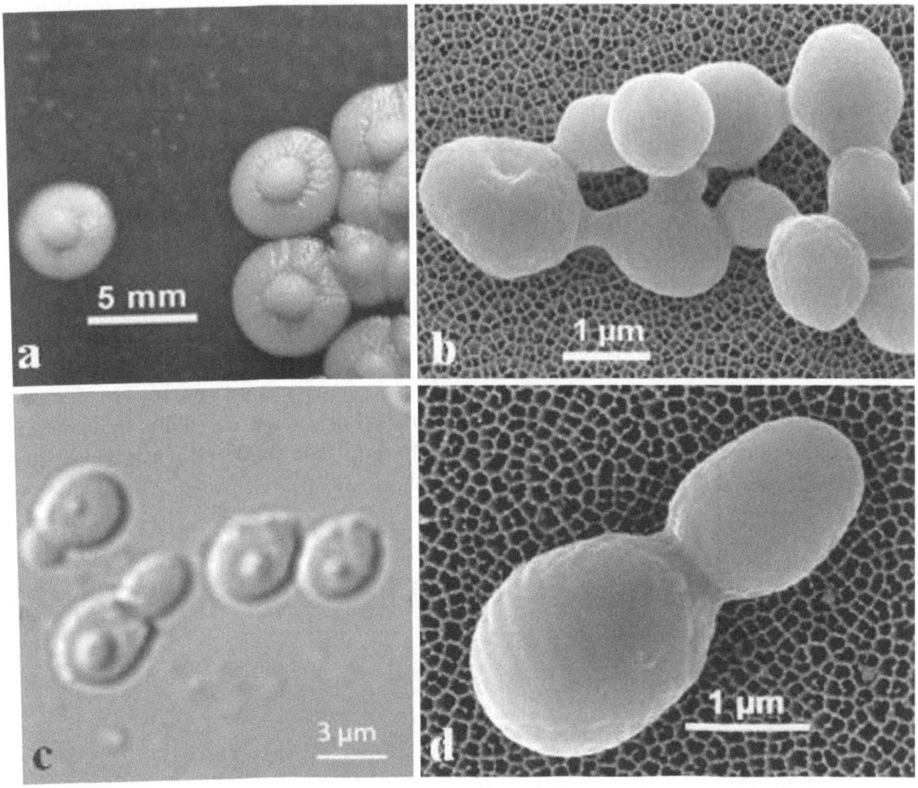

Fig. 2.5 *M. caprae.* (**a**) Convex to umbonate colonies with a finely folded to undulate margin; (**b**–**d**), Nomarski's and SEM micrographs of ovoid to slightly rhomboidal (**b**) yeasts showing a quite narrow budding site, and the corrugate helicoidal feature of the cell wall (**d**)

2.1.3.5.3 Physiological and Biochemical Characteristics

Malassezia caprae can be identified by lack of growth at 40°C, weak growth at 37°C, presence of a catalase reaction, β-glucosidase activity, and good growth in the presence of four Tweens. Growth may be somewhat weaker with Tween 80, and delayed centripetal growth with Tween 20 is observed to be similar to that of *M. sympodialis* (Plates 2.1-5 and 2.2-5). This species belongs to the cluster of species related to *M. sympodialis* [67]. It can be differentiated from the latter species by growth that is limited at 37°C. *M. slooffiae* is characterized by a weak and restricted growth with Tween 80, but the latter species, in contrast to *M. caprae,* is able to grow at 40°C and does not split esculin. *M. caprae* does not grow with CrEL, but can develop a weak precipitate with this nutriment (Plate 2.1-5).

2.1.3.5.4 Ecology

Malassezia caprae is so far only known from the healthy skin of goats and horses [18].

2.1.3.6
Malassezia dermatis Sugita, Takashima, Nishikawa & Shinoda (2002)

2.1.3.6.1
Type Strain

CBS 9169 (JCM 11348), isolated from skin lesions of a patient with atopic dermatitis, Tokyo, Japan [74].

2.1.3.6.2
Morphological Characteristics

After 7 days at 32°C on mDixon agar, single colonies (Fig. 2.6a) are 5–6 mm in diameter, flat to somewhat apiculate centrally, butyrous, shiny to dull, pale yellowish-white, and with an entire or finely folded margin. Cells are globose, ovoid or ellipsoidal, 3.8–4.8 × 2.5–

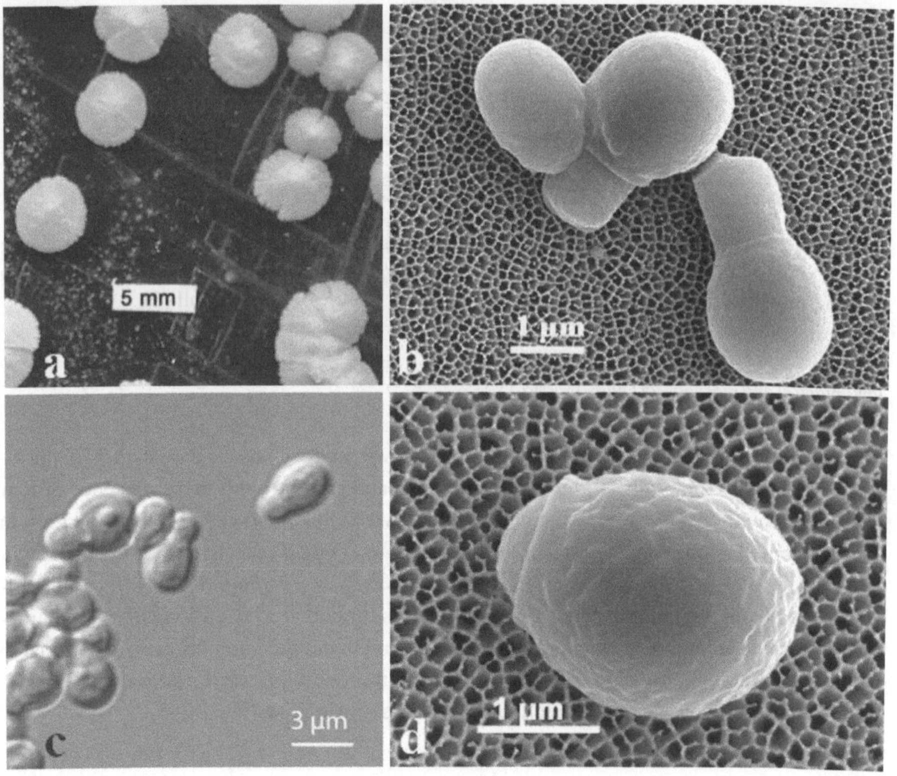

Fig. 2.6 *M. dermatis.* (**a**) Flat colonies which are slightly elevated in the center and have a folded margin (note the typical precipitate within the agar around the colonies); (**b–d**) Nomarski's and SEM micrographs of ovoid to globose yeasts resembling those of *M. sympodialis.* (**d**) The ovoid yeast shows a regular helicoidal feature of the cell wall

3.2 µm, with monopolar, eventually sympodial budding on a moderately broad base (Fig. 2.6b–d). The species is not known to produce any filaments.

2.1.3.6.3
Physiological and Biochemical Characteristics

Malassezia dermatis can be identified by lack of growth at 40°C, and lack of both catalase and β-glucosidase reactions. The former test appears contradictory with the original description in which it was given as positive [74], and the latter is absent in this original description but was confirmed by Kaneko et al. to be negative [29, 30].

Growth occurs with all four Tweens. Growth with Tween 80 may be weaker, similar to that of *M. caprae*, and is absent with CrEL (see *M caprae* in Plates 2.1-5 and 2.2-5). As this assimilation pattern may look similar to that of *M. sympodialis* or *M. caprae*, which may have the same delayed centripetal growth with Tween 20, the absence of the two enzymes clearly differentiate *M. dermatis* from the other two species. *M. restricta* is the only other lipid-dependent species that lacks both catalase and β-glucosidase activities, but this species grows only on complex media, including several lipid compounds.

2.1.3.6.4
Ecology

Malassezia dermatis was first isolated from skin lesions of a few patients with atopic dermatitis [74]. It has also been isolated from 19 out of 160 healthy volunteers in Korea [75]. Contrary to the reported catalase positive reaction in the two corresponding papers [74, 75], the three strains, including the type CBS 6169, examined for this work and the 5th edition of "The Yeasts, a taxonomic study" were found to have a catalase negative reaction [17]. The molecular identification is more trustable than the routine laboratory techniques, but it is important to control carefully all characteristics. Indeed a strain, growing with the four Tweens but unable to give a catalase reaction, can only be *M. dermatis*.

2.1.3.7
Malassezia slooffiae Guillot, Midgley & Guého (1996)

2.1.3.7.1
Type Strain

CBS 7956 (JG 554), isolated from skin of the ear of a pig [23].

2.1.3.7.2
Morphological Characteristics

After 7 days at 32°C on mDixon agar, single colonies (Fig. 2.7a) are flat or somewhat raised, 3–4 mm in diameter, shiny, pale yellowish-brown, butyrous, with a somewhat

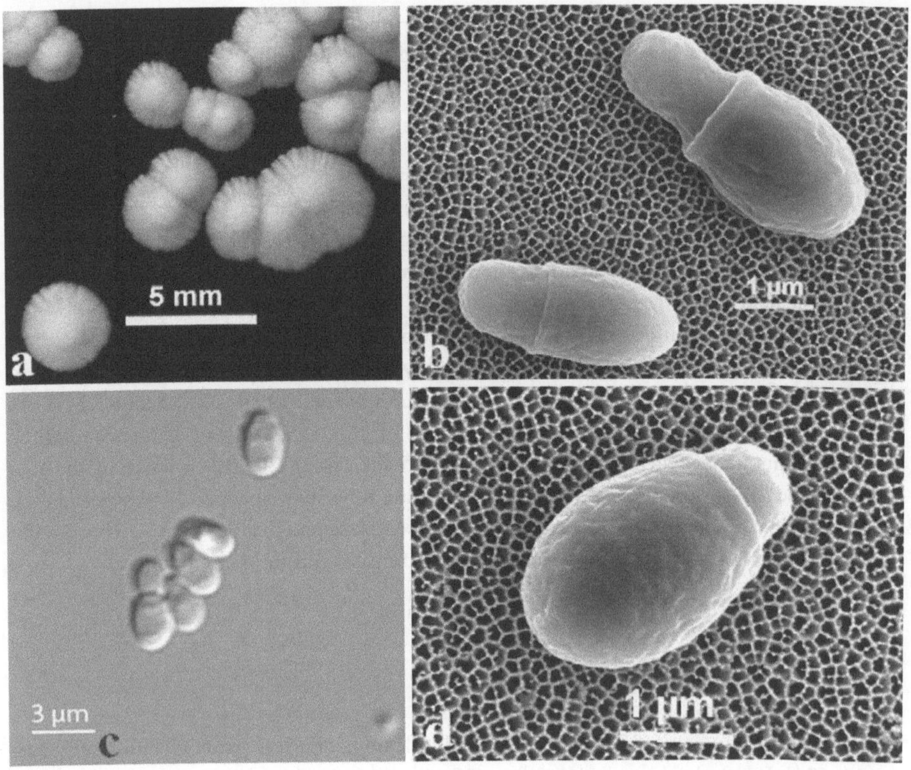

Fig. 2.7 *M. slooffiae.* (**a**) Flat to slightly apiculate colonies with a regularly folded margin; (**b**–**d**), Nomarski's and SEM micrographs of short cylindrical yeasts with a wide and more or less marked budding site

roughened surface and a finely folded margin. The cells are short, cylindrical, 1.5–4 × 1–2 µm, and budding is monopolar, percurrent and occurs on a broad base (Fig. 2.7b–d). The species is not known to produce any filaments.

2.1.3.7.3
Physiological and Biochemical Characteristics

Malassezia slooffiae can be identified by its capacity to grow at 40°C, and shows a catalase reaction, but β-glucosidase activity is absent. The key characteristic is that growth with Tween 80 is always weak and restricted in comparison to the equal growth obtained with the other three Tweens, and growth with CrEL is absent (Plates 2.1-7 and 2.2-7).

2.1.3.7.4
Ecology

The species is found in low frequency on healthy or lesioned human skin, usually in association with *M. sympodialis, M. furfur, M. globosa* or *M. restricta* [23]. *M. slooffiae* is also commonly isolated from animal skin, especially that of pigs [43], and also from cats [76, 77], cows [61] and goats [78].

2.1.3.8
Malassezia japonica Sugita, Takashima, Kodama, Tsuboi & Nishikawa (2003)

2.1.3.8.1
Type Strain

CBS 9431 (JCM 11963), isolated from a healthy Japanese woman [79].

2.1.3.8.2
Morphological Characteristics

After 7 days at 32°C on mDixon agar, single colonies (Fig. 2.8a) are 2–3 mm in diameter, flat to slightly wrinkled, dull, pale yellowish-cream, butyrous to somewhat brittle, and with

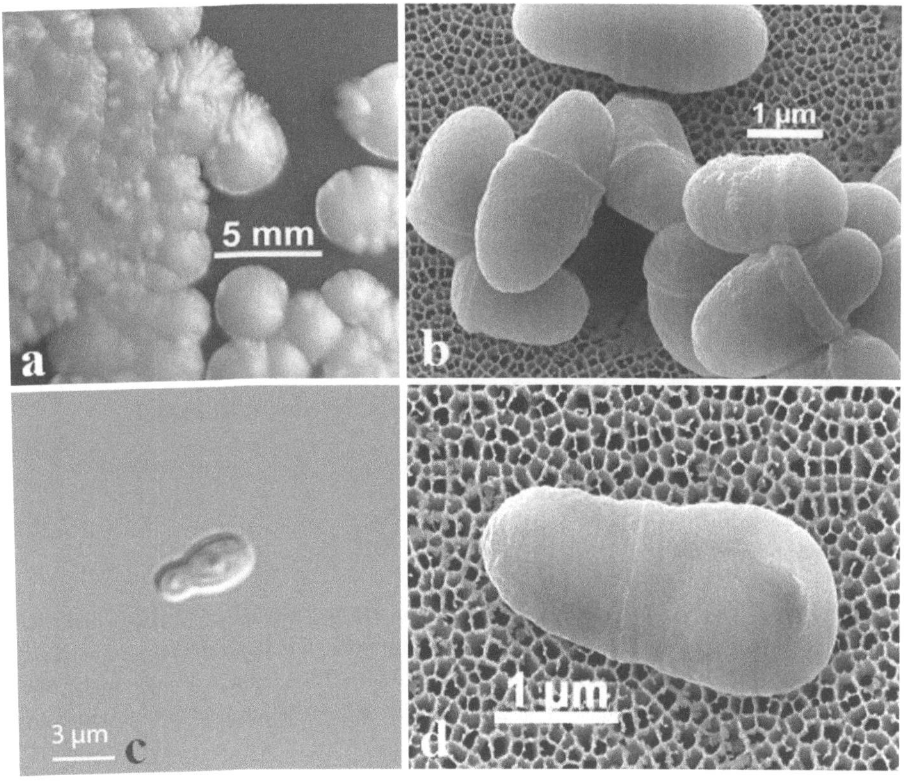

Fig. 2.8 *M. japonica.* (**a**) Flat to slightly wrinkled colonies with a slightly undulate margin; (**b–d**) Nomarski's and SEM micrographs of short cylindrical yeasts with a wide and more or less marked budding site

a straight to somewhat undulate margin. Cells are short cylindrical, 3.0–3.5 × 2.0–2.5 μm, with monopolar budding on a broad base (Fig. 2.8b–d). Filaments have not been observed.

2.1.3.8.3
Physiological and Biochemical Characteristics

This species grows at 37°C, but not at 40°C. The catalase reaction is strong, and the species is able to split esculin due to β-glucosidase activity. Only Tweens 60 and 80 are well assimilated, whereas Tween 20 gives a neat ring of colonies at some distance of the well after diffusion of the compound, and Tween 40 is very weakly assimilated at some distance from the well or within the inhibitory area after diffusion of the compound. CrEL is weakly assimilated, but more often only a reactive precipitate appears around the well containing this compound (Plates 2.1-8 and 2.2-8).

2.1.3.8.4
Ecology

Malassezia japonica has been reported from healthy human skin and from the skin of atopic dermatitis patients [79]. The species is not known from another source yet.

2.1.3.9
Malassezia nana Hirai, Kano, Makimura, Yamaguchi & Hasegawa (2004)

2.1.3.9.1
Type Strain

CBS 9557 (JCM 12085), isolated from a cat with otitis externa in Japan [80].

2.1.3.9.2
Morphological Characteristics

After 7 days at 32°C on mDixon agar, single colonies (Fig. 2.9a) are 1.5–2 mm in diameter, cream to yellow, convex, shiny to dull, smooth, butyrous, and have an entire to narrowly folded margin. The cells are ovoid to globose, 3.0-4.0 x 2.0-3.0 μm, with monopolar budding on a relatively narrow base (Fig. 2.9b–d). The species is not known to produce any filaments.

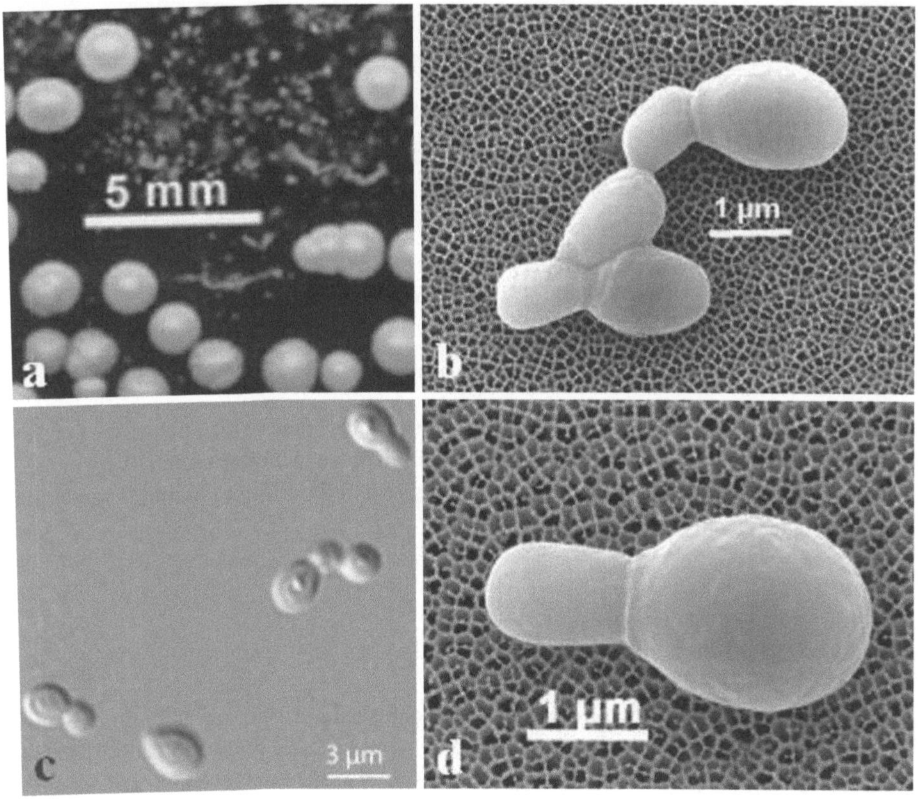

Fig. 2.9 *M. nana.* (**a**) Small, umbonate colonies which are surrounded by a precipitate within the agar; (**b–d**) Nomarski's and SEM micrographs of ovoid yeasts with a rather narrow budding site

2.1.3.9.3
Physiological Characteristics

Malassezia nana grows at 37°C, but this is also the maximum temperature for this species, and, hence, growth at 40°C is absent. Catalase and β-glucosidase activities are present. Hirai et al. [80] reported in their description the absence of this latter activity, but we believe that this might have resulted from the technical procedure applied by these authors (see above 2.1.2.3). Tweens 40, 60 and 80 are well assimilated, whereas Tween 20, gives a neat ring of colonies at some distance of the well after diffusion of the compound similarly to *M. equina* and *M. japonica* [81]. CrEL is not utilized (Plates 2.1-9 and 2.2-9). Surprisingly, strains originating from cattle give a complete disk of growth around Tween 20, making the separation with *M. caprae* impossible with the routine methods only. Filaments were not observed.

2.1.3.9.4
Ecology

Malassezia nana is known from healthy cats [77], cats with otitis externa, healthy cows or those with otitis [80].

2.1.3.10
Malassezia equina Cabañes & Boekhout (2007)

2.1.3.10.1
Type Strain

CBS 9969 (MA 146), isolated from healthy anal skin of a horse [18]. This species corresponds to the species that was isolated previously from horse, which was tentatively named *M. equi* by Nell et al. [82], but without providing a valid description, and thus rendering this latter name, invalid.

2.1.3.10.2
Morphological Characteristics

After 7 days at 32°C on mDixon agar, single colonies (Fig. 2.10a) are 2–3 mm in diameter, cream-colored, glistening to dull, butyrous, wrinkled, and with a folded to fringed margin. Crystal precipitates surround the colonies, as was also observed with *M. sympodialis* and *M. dermatis*. Cells are ovoid to ellipsoidal, 3.0–4.5 × 2.2–3.5 µm, with monopolar budding occurring at a narrow base (Fig. 2.10b–d). Filaments have not been observed.

2.1.3.10.3
Physiological and Biochemical Characteristics

Malassezia equina has a maximum temperature at 37°C. The catalase reaction is strong, but in contrast to *M. japonica* and *M. nana,* this species is unable to split esculin. Tweens 40, 60 (note: the white precipitate surrounding these two compounds makes the colony disks look brighter) and 80 are well assimilated, whereas Tween 20, similar to the type strain of *M. nana*, gives a neat ring of colonies at some distance of the well after diffusion of the compound. CrEL is not assimilated, but sometimes a weak precipitate occurs around the well containing this compound (Plates 2.1-10 and 2.2-10). *M. equina*, *M. nana* and *M. japonica* show a neat ring of colonies around Tween 20, but the esculin test discriminates among them, as does the shape of cells, which is ovoid to elongate with a narrow budding site in *M. equina*, short cylindrical with a broad budding site for *M. japonica,* and ovoid and small in size for *M. nana*.

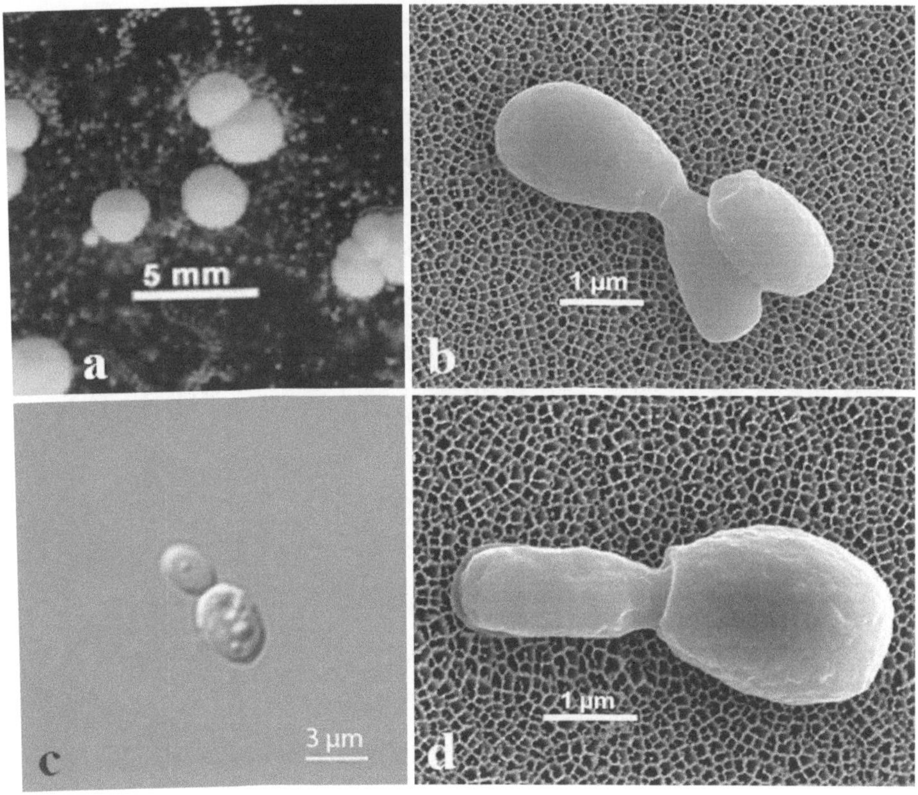

Fig. 2.10 *M. equina.* (**a**) Small, convex colonies that are surrounded by a precipitate within the agar; (**b–d**) Nomarski's and SEM micrographs of elongate yeasts with a narrow budding site

2.1.3.10.4
Ecology

So far, *M. equina* is only known from healthy anal skin of horses and from skin of cows in Spain [18].

2.1.3.11
Malassezia obtusa Midgley, Guillot & Guého (1996)

2.1.3.11.1
Type Strain

CBS 7876 (GM215), isolated from human groin [23].

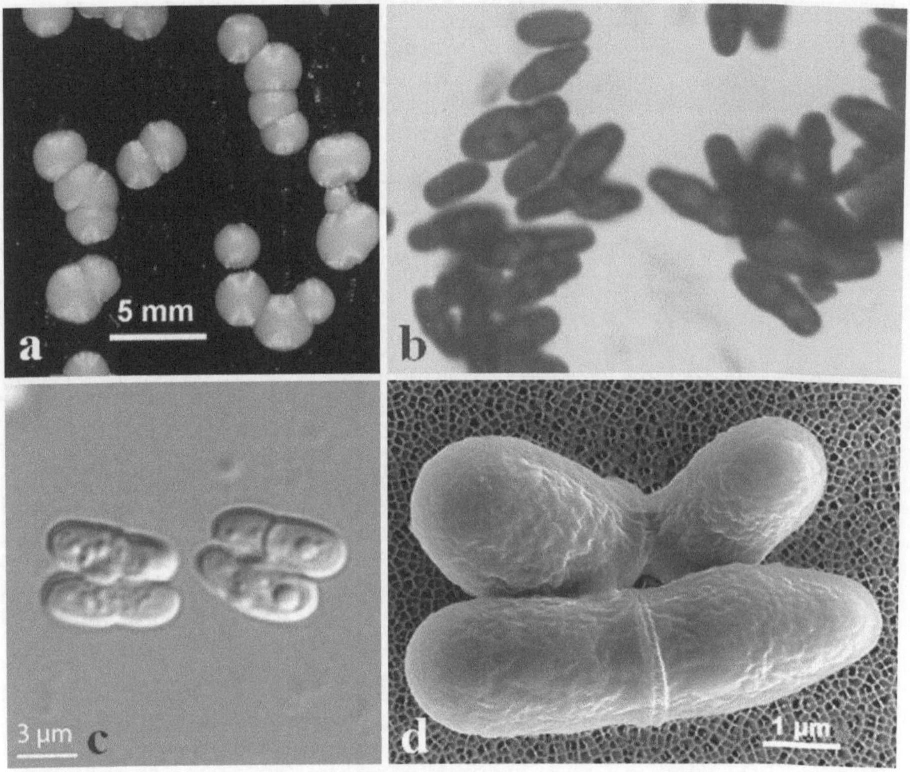

Fig. 2.11 *M. obtusa.* (**a**) Small, convex colonies with an entire to slightly lobate margin; (**b**–**d**) cylindrical to slightly rhomboidal yeasts with a broad budding site (Gram stained in b, Normaski's in c and SEM in d)

2.1.3.11.2
Morphological Characteristics

After 7 days at 32°C on mDixon agar, single colonies (Fig. 2.11a) are slightly convex, smooth, on average 1.5–2 mm in diameter, shiny or dull, butyrous to pasty, and with the margin, entire to slightly lobate. The cells are cylindrical, somewhat rhomboidal, with obtuse apices, 4.0-6.0 x 1.5-2.0 µm, showing monopolar budding on a broad base (Fig. 2.11b–d). Filaments may be present.

2.1.3.11.3
Physiological and Biochemical Characteristics

Malassezia obtusa is among the most demanding species together with *M. globosa* and *M. restricta.* None of these species grow with individual sources of lipids, although a white precipitate mimics growth around the wells containing Tweens 40 and 60 (Plates 2.1-11 and

2.2-11). Only with fresh isolates and/or very young cultures of *M. obtusa* and *M. globosa*, tiny colonies may appear at some distance from the supplements, particularly due to synergy between two of them (Plate 2.1-11, 12 and Plates 2.2-12). However, the three species can be well identified with other key characteristics. All three species have a maximum temperature at 37°C, but only *M. obtusa* combines positive reactions of catalase and β-glucosidase, whereas *M. globosa* lacks β-glucosidase activity, and *M. restricta* both activities.

2.1.3.11.4 Ecology

Malassezia obtusa is a rare species, which is mainly known from healthy human skin. Together with *M. furfur*, the species has occasionally been isolated from animals, viz., in a case of canine otitis [59] and from healthy horses and goats [60].

2.1.3.12 *Malassezia globosa* Midgley, Guého & Guillot (1996)

2.1.3.12.1 Type Strain

CBS 7966 (GM35), isolated from PV, UK [23]. *M. globosa* corresponds to the formerly recognized *M. furfur* serovar B [66].

2.1.3.12.2 Morphological Characteristics

After 7 days at 32°C on mDA, single colonies (Fig. 2.12a) are raised, wrinkled to cerebriform, 3–4 mm in diameter, rough and brittle, pale yellowish, shiny or dull, and with the margin slightly lobate. In primary cultures, colonies are surrounded by an abundant precipitate, as in species of the *M. sympodialis* complex. Cells are spherical, 2.5–8 μm in diameter, and budding is monopolar on a narrow base (Figs. 2.12b–f). In contrast to *M. furfur*, this micromorphology is a stable character in *M. globosa*. Short filaments, reminding germinative tubes of *Candida albicans*, may be present, particularly in primary cultures (Fig. 2.12e, f). On the other hand, pseudo-hyphae are almost always present in PV scales (Fig. 2.12b and see Chap. 6.1). The morphology of *M. globosa* is similar to that of the species known under the old name *Pityrosporum orbiculare* Gordon [83]. Unfortunately, the original type material and isolates designated as *P. orbiculare* on the basis of cell shape and their inability to grow on oleic acid, were not preserved. In the taxonomic revision [23], it was therefore proposed to consider *P. orbiculare* as a doubtful species, which may represent a probable synonym of *M. globosa* and not of *M. furfur*.

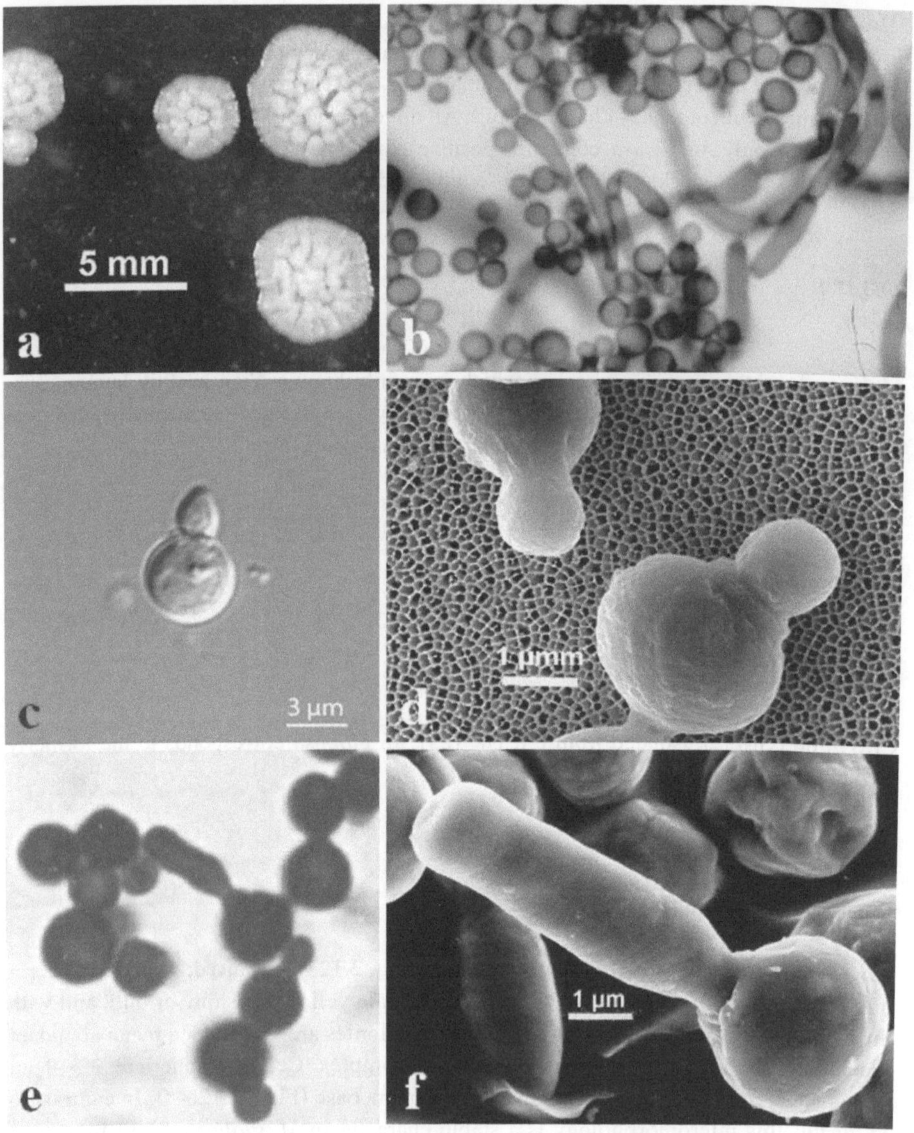

Fig. 2.12 *M. globosa.* (**a**) Large, typically wrinkled to cerebriform colonies with an undulate margin; (**b**) spherical yeasts and filaments in PV scales (i.e., typical spaghetti and meat balls feature) (picture by the courtesy of V. Crespo Erchiga); (**c–d**) Normaski's and SEM micrographs showing spherical yeasts with a narrow budding site; (**e–f**) Gram stained and SEM micrographs of a primary culture from PV showing typical spherical yeasts, some of them developing a germinative tube

2.1.3.12.3 Physiological and Biochemical Characteristics

Malassezia globosa has a strong catalase activity, similar to that of *M. obtusa*. In contrast to the latter species, *M. globosa* lacks β-glucosidase expression. Growth is limited at 37°C, and no growth occurs with individual lipid supplements, or may be very weak, with fresh cultures appearing as a ring of tiny colonies at some distance from the well containing Tween 20 (Plates 2.1-12 and 2.2-12) or Tween 80. Due to absence of good growth with individual lipid supplements, and lack of β-glucosidase activity, the species is easily recognized morphologically by its cerebriform colonies and spherical cells.

2.1.3.12.4
Ecology

Malassezia globosa is known from healthy and diseased human skin, mainly from PV [69], but also seborrheic dermatitis and even atopic dermatitis (see Chap. 6.1). One out of four genotypes may be better adapted to such pathologies [84]. The species is also known from animal skin, e.g., cats [72], horses and domestic ruminants [60]. The species has been reported to occur with high frequency in both healthy and diseased bovines with otitis in Brazil [61, 62], with the latter disease known to be associated with nematodes [62]. The species was also isolated from the acoustic meatus of bats [63]. Interestingly, DNA of *M. globosa* has been detected in European soil forest nematodes of the genus *Malenchus* [12], thus suggesting that the occurrence of the species may not be limited to warm blooded animals. Even more surprising was the detection of DNA of the species in soils from Antarctic Dry Valleys [14]. The authors postulated that the occurrence of the yeasts may be associated with nematodes, which are prevalent in Dry Valley soils. The possible interactions between nematodes and *M. globosa* need further study in order to see if nematodes represent a natural reservoir of the species. Further studies, including the characterization of the different rDNA markers and attempts to obtain cultures especially from nematodes, will be necessary to elucidate this possible ecological habitat.

2.1.3.13
Malassezia restricta Guého, Guillot & Midgley (1996)

2.1.3.13.1
Type Strain

CBS 7877 (RA 42.2C), isolated from healthy human skin [23]. *M. restricta* corresponds to the formerly recognized *M. furfur* serovar C [66]. At that time, the authors were not able to demonstrate that they were dealing with different species, but it is interesting to point out that they separated serologically *M. sympodialis* (A), *M. globosa* (B) and *M. restricta* (C), the three species that predominantly occur on human skin, either healthy or lesioned [85].

2.1.3.13.2
Morphological Characteristics

After 7 days at 32°C on mDA, single colonies (Fig. 2.13a) are small, 1–2 mm in diameter on average, flat to somewhat raised, dull, pale yellowish-brown, hard and brittle, smooth and somewhat ridged near the edge , and with a lobate margin. The cells are ovoid to glo-

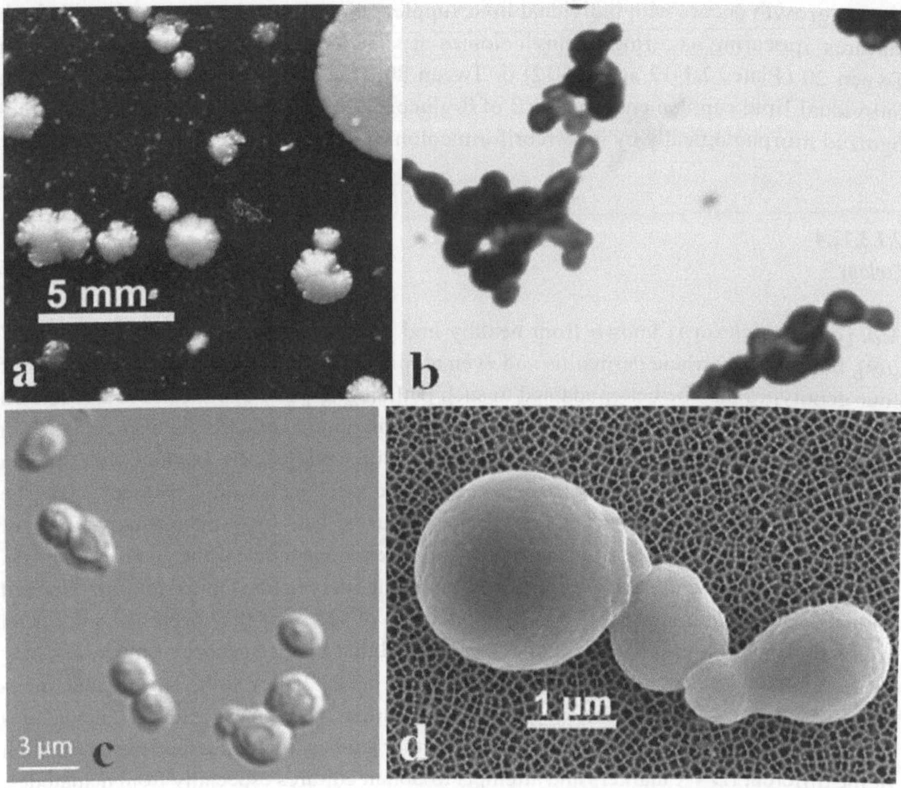

Fig. 2.13 *M. restricta.* (**a**) Very restricted colonies (if compared with *M. sympodialis* (*top right corner*)), that are somewhat raised and have a lobate margin; (**b–d**) Gram stained, Nomarski's and SEM micrographs showing ovoid and globose yeasts with a relatively narrow budding site

bose, 2.5–4 × 1.5–3 µm, with monopolar, percurrent budding on a relatively narrow base (Fig. 2.13b–d). The constant presence of ovoid and globose yeast cells in this species may erroneously suggest a mixture of *M. globosa* and *M. restricta* in the same culture (Fig. 2.13d). However, colonies of both species are sufficiently different to avoid this confusion. Filaments are not known for this species, and, therefore, it may correspond to the species that was described with the old name *P. ovale* [50].

2.1.3.13.3
Physiological and Biochemical Characteristics

Malassezia restricta lacks both, catalase and β-glucosidase activities. This species, which is the most fastidious of the genus, does not grow at 37°C or with any of the lipid supplements. Tweens 40 and 60 can show the presence of a white precipitate appearing like a white disk or a ring around the corresponding wells (Plates 2.1-13 and 2.2-13). Growth with CrEL is always absent.

2.1.3.13.4 Ecology

Malassezia restricta occurs mainly on the head, including scalp, neck, face and ears [85]. Its implication in human disease is not yet elucidated (see Chap. 6) but, as for *M. globosa,* a specific genotype may play a significant role in pathology [86]. Ribosomal DNA sequences (ITS region or D1/D2 LSU domain of the rRNA gene) of the species have been detected from indoor dust in Finland [87], but, surprisingly, also from the gut of beetles in Southern Louisiana, USA [88], forest soil nematodes in Germany [12], and rock beneath a crustose lichen in Norway [13]. These findings suggest that the human body may not be the only habitat of the species. Unfortunately, most of these studies referred to ITS1 only, and attempts to cultivate *M. restricta* were not performed. It is not easy to understand that such a fastidious species can survive without its vital nutritional lipid requirement and below its optimal temperature of growth. The species was also listed to occur in deep sediments in the South China sea [15], but this DNA comparison showed only 85 % ITS sequence identity, which is less than that observed to occur among *Malassezia* species.

2.2 *Malassezia* Phylogeny

Teun Boekhout and Dominik Begerow

Members of the genus *Malassezia*, like many other anamorphic yeasts, are difficult to assign to higher taxonomic levels based on morphological structures. Moore [89] established the name Malasseziales for the basidiomycetous yeasts without ballistospores, in contrast to the Sporobolomycetales that contained genera with actively discharged ballistospores. In the circumscription of the order Malasseziales, Moore included eight genera, namely *Cryptococcus*, *Malassezia*, *Phaffia*, *Rhodotorula*, *Sterigmatomyces*, *Trichosporon*, *Trichosporonoides*, and *Vanrija*. The lack of a unifying character and the negative definition of the Malasseziales already show the difficulties in the grouping of these yeasts in a monophyletic manner.

The availability of ribosomal DNA (rDNA) sequence data allowed a better understanding of the phylogenetic relationship of the yeasts in general. The first more detailed molecular phylogenetic study of heterobasidiomycetous yeasts [90] included *Malassezia* spp. amongst others and clearly separated the genus from species of the Pucciniomycotina (i.e., *Rhodosporidium toruloides*, *Sporodiobolus johnsonii* and *Leucosporidium scottii*) and from members of the Agaricomycotina (i.e., *Filobasidiella neoformans*, *Cystofilobasdium capitatum*, *Phaffia rhodozyma*, *Sterigmatosporium polymorphum*, and some *Trichosporon* spp.). However, due to a limited number of species included, *Malassezia* could not be clearly assigned to an order or class of the Basidiomycota.

During the late 1990s, two groups analyzed the molecular phylogeny of anamorphic basidiomycetous yeasts [91, 92]. Both groups used partial sequences of the large subunit ribosomal DNA, but while Fell et al. [92] presented an overview of more than 330 yeast strains of basidiomycetes, the second group focused on the anamorph–teleomorph relationship of Ustilaginomycotina. Both studies included members of the Malasseziales, but came up with two different systematic proposals due to different sampling of species. One of the two likely phylogenetic position of Malasseziales based on LSU rDNA sequences is illustrated in Fig. 2.14. The phylogenetic affiliation of Malasseziales within the Ustilaginomycotina (Basidiomycota) is highly supported. However, the relationship between the orders of Ustilaginomycotina is still not resolved in all clades and the support for several groupings is low (Fig. 2.14). This is also reflected in the literature, thus several phylogenetic hypotheses are published. Based on phylogenies of LSU rDNA, several analyses proposed *Malassezia* as part of Exobasidiomycetidae (Exobasidomycetes) like in Fig. 2.14 [91, 93]. Different taxon sampling and other genes could not support this placement or suggested even a grouping with Ustilaginales and Urocystales [94] or sister to all other members of the Ustilaginomycotina. The most recent treatment of fungal taxonomy followed the more cautious proposals and placed the order Malasseziales in the Ustilaginomycotina *incertae sedis* [95].

All 13 species of the genus *Malassezia* form a strongly supported monophyletic group (Figs. 2.14 and 2.15) and support the conclusions based on physiological characters like lipophily and the peculiar cell wall ultrastructure (see Sect. 2.3.1). Based on the analysis of the D1/D2 domains of the large subunit (LSU) rRNA gene, some groups of species are well supported (Figs. 2.14 and 2.15) [also ref. 18]. *M. furfur* forms a monophyletic group with *M. obtusa*, *M. japonica* and *M. yamatoensis*. *M. globosa* is related to *M. restricta*. The group of *M. sympodialis*, *M. nana*, *M. caprae*, *M. dermatis* and *M. equina* form a well supported clade as well, which has been discussed in detail recently [18]. However, *M. slooffiae* and *M. pachydermatis* do not clearly group with any other known species and more detailed studies might present more interesting results. Note that other analyses of partial LSU rRNA gene sequences yielded the same relationships for *M. obtusa* (cited as *M.* species 3) and *M. furfur* [21]. In the ITS analysis, however, the furfur and globosa clades are not separated and form a well supported clade with *M. furfur, M. japonica, M. obtusa, M. pachydermatis, M. yamatoensis* and *M. slooffiae* [18]. Partial sequences of the chitin synthase gene (CHS2) were in agreement with those based on the D1/D2 domains of the LSU rRNA gene and the ITS 1+2 regions [18, 67]. In this analysis, the sympodialis clade comprised *M. sympodialis, M. caprae, M. equina* and *M. dermatis,* whereas *M. nana* formed a basal lineage to this cluster; the furfur clade contained *M. furfur, M. japonica* and *M. obtusa*, and the globosa clade, *M. pachydermatis, M. yamatoensis, M. restricta, M. slooffiae* and *M. globosa*. Partial sequences of the RNA polymerase subunit 1 (RPB1) gene supported the sympodialis clade that also included *M. nana* [18]. All phylogenetic analyses performed so far supported all species recognized, as well as some major clades, but some species, e.g., *M. slooffiae* and *M. pachydermatis*, tend to jump between clades depending on the gene analyzed and the set of other fungi included. This implies that further phylogenetic research is needed to establish the position of these species in the tree of life. A further explanation of these incongruent results may be that hybridization may have occurred during speciation [18].

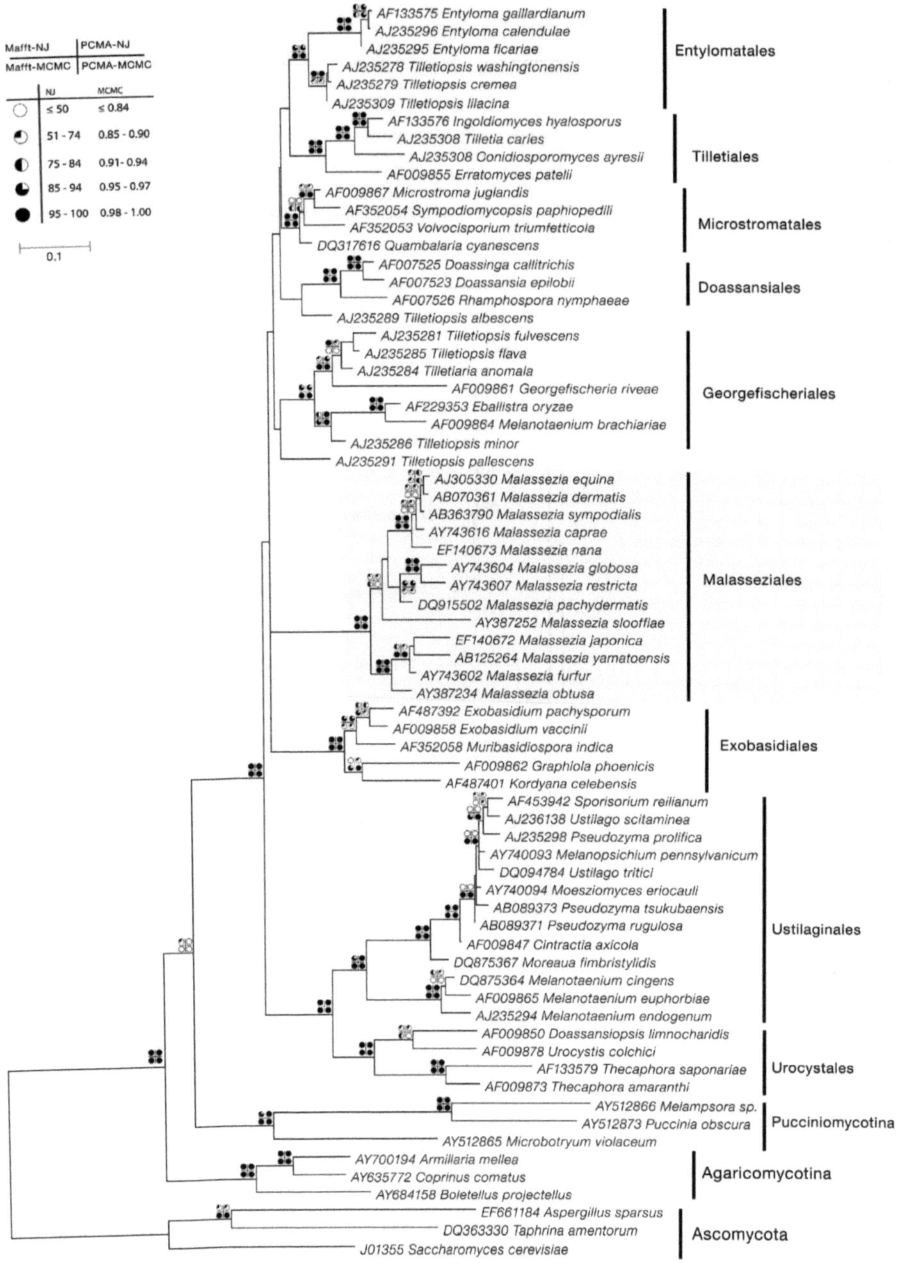

Fig. 2.14 Phylogenetic hypothesis based on partial LSU rDNA sequences of 67 basidiomycetes and three ascomycetes aligned with MAFFT (version 6.525) and analyzed using maximum likelihood in PAUP* 4b10. Support values are calculated based on two different alignment algorithms as implemented in MAFFT and PCMA and two different tree reconstruction methods, viz., neighbor-joining using PAUP* and Markov chain Monte Carlo using MrBayes. Branch lengths represent expected substitutions per base pair

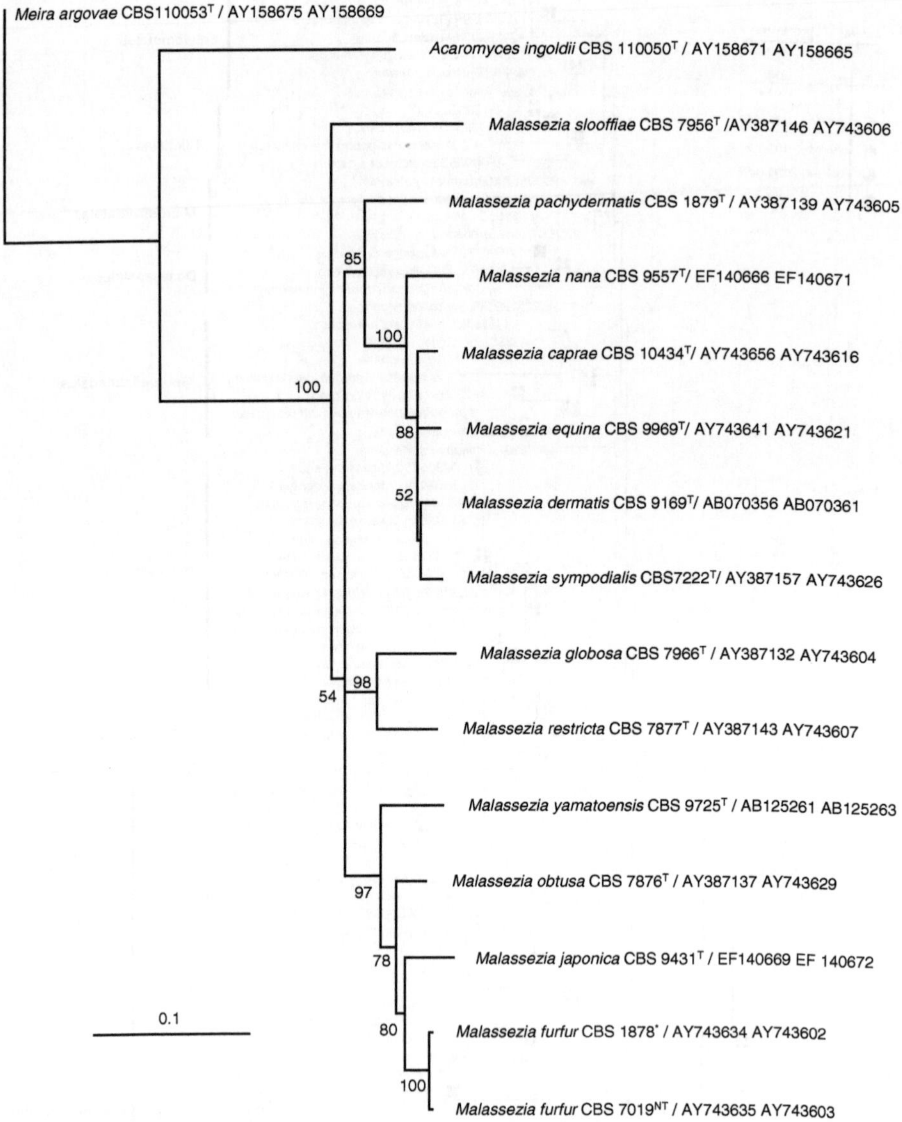

Fig. 2.15 Phylogenetic tree based on concatenated sequences of the ITS1, 5.8s and ITS2 regions of the ribosomal DNA and the D1/D1 part of the LSU rRNA gene generated in Megalign version 7.2.1 (DNAstar Inc.). The tree was generated with PAUP (version 4.0b 10) using the neighbor-joining algorithm with Kimura 2 as a distance measure and 1000 bootstrap replicates. *T* = neotype strain of *M. furfur*; *NT of *Pityrosporum ovale*

2.3
Malassezia Ultrastructure

Eveline Guého-Kellermann

The phylogenic studies, as developed above, have demonstrated the affiliation of the genus *Malassezia* to the Ustilaginomycotina of the Basidiomycetes despite the lack of a sexual state (Chap. 2.2). Some other characteristics have been used for a long time to relate these anamorphic yeasts to the Basidiomycetes. The most important of these basidiomycetous markers are the multilamellar cell wall [96, 97] and the monopolar percurrent budding. Unfortunately, due to the absence of hyphae the septal pore structure could not be used to further clarify the taxonomic position of the genus [98].

2.3.1
Cell Wall Ultra-Structure

Transmission electron microscopical (TEM) studies demonstrated that *M. furfur* and *M. pachydermatis* have a thick, electron-dense and multilayered cell wall, which is more or less coated with fibrillar material [99].

This typical basidiomycetous-type cell wall is crossed by a helicoidal translucent band, which arises from regular indentations of the plasma membrane (Fig. 2.16a, b). As far as is known, these cell wall characteristics are unique among the fungi [96–104]. Based on a similar cell wall ultra-structure of the filaments of *M. furfur*, the spherical yeasts (i.e., *Pityrosporum orbiculare*) as observed in PV, and in the oval yeasts (i.e., *Pityrosporum ovale*) as observed in lesions of pityriasis capitis, Keddie [105] suggested that *Pityrosporum* and *Malassezia* might be the same, and, hence, are synonyms. Due to nomenclatural rules, the name *Malassezia* had priority. The freeze-fracture replica technique demonstrated that the plasma-membrane indentations correspond to a regular, helicoidal and left-handed groove [102, 106, 107] which is surrounded by an electron-lucent band, visible as white lines in tangential sections (Fig. 2.16b). This specific parietal ultrastructure is also present in *M. sympodialis* [65], *M. globosa* and *M. restricta* [23] and all other species, as seen by TEM and numerous observations using scanning electron-microscopy (SEM) as well. All *Malassezia* species have a smooth cell wall, but microscopy using a mirror lighting microscope ([108] and Fig. 2.16d) or SEM, of somewhat retracted yeast cells showed, the double helicoidal system appearing on the surface as more or less marked and spaced grooves (see in Chap. 2.1, Figs. 2.1e, 2.4e, 2.5d, 2.6d, and 2.8d of *M. pachydermatis*, *M. sympodialis*, *M. caprae*, *M. dermatis* and *M. nana*, respectively). The cell wall of *M. globosa* appears to be somewhat different. By means of the freeze-fracture replication technique, Breathnach et al. [109] showed that the initial major combination of plasma membrane groove and electron-lucent band (Fig. 2.16c) was associated to a minor system of grooves and electron-lucent bands occurring at more

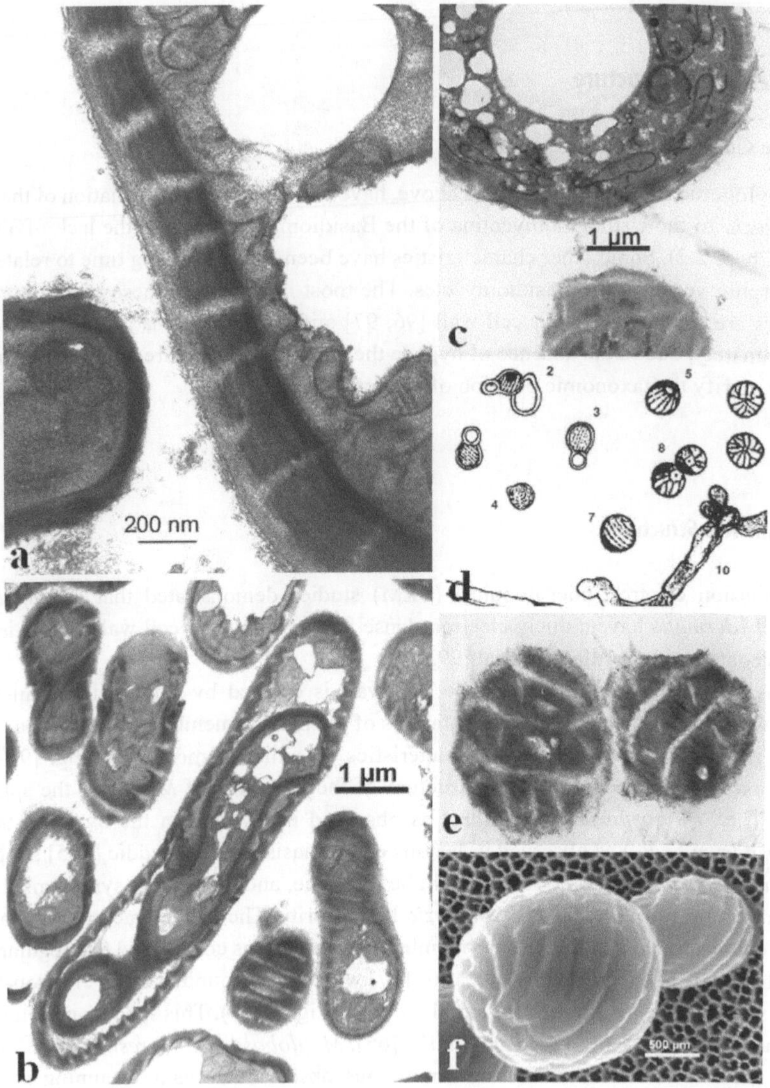

Fig. 2.16 *Malassezia* cell wall ultra-structure. (**a–b**) Micrograph of *M. furfur* made by transmission electron-microscopy (TEM). (**a**) The typical multilamellar cell wall with the corrugate invagination of the plasma membrane and its corresponding electron lucent band; (**b**) endospore within elongate yeast cells or short filaments showing the typical lamellate cell wall, and the regularly spaced electron translucent spiral, as can be clearly seen in tangential sections. (**c–f**) *M. globosa*. (**c**) Micrograph (TEM) showing the multilamellar cell wall and the electron lucent band; (**d**) drawings after Matakieff [108] of globose yeasts with an helicoidal sculpturing; (**e**) Micrograph (TEM) (courtesy of Hannelore Mittag) of a tangential cut showing the double system of electron lucent bands as described above in Fig. 2.1c; (**f**), Micrograph made by scanning electron-microscopy (SEM) of a retracted yeast and its bud showing the spaced major groove system, that is comparable with Matakieff's drawings

or less right angles. This has also been visualized by TEM analysis of tangential sections of the cell wall (Figs. 2.16c, e). The cell wall thickness appears different from one species to another, but may also be related to the age of the cells and the growth conditions.

2.3.2 Budding Process Ultra-Structure and Endosporulation

The basidiomycetous nature of the genus *Malassezia* is also revealed by its monopolar, blastic and percurrent budding process [100, 101, 104]. Buds emerge from the innermost layer of the wall, and leave a collarette on the mother cell after release (i.e., thus representing a phialidic conidiogenesis) (Fig. 2.17a, b). The scar on the mother cell becomes thicker with the increasing number of collarettes after each subsequent budding (Figs. 2.17c–e, g).

These typical scars of the genus *Malassezia* occur independently of the shape of the yeast cells, but appear more clearly under light microscopy in species having a broad bud site (i.e., *M. pachydermatis, M. furfur, M. yamatoensis, M. slooffiae, M. japonica* and *M. obtusa,* see Chap. 2.1). *M. sympodialis* differs from the other species in that a sympodial budding occurs at the monopolar budding site [97], with the buds appearing alternatively on the left and the right side, thus resulting in a clover leaf-like configuration of the parent cell and its buds (Fig. 2.17f, g). This process, which may also be present in *M. dermatis*, is difficult to visualize under the light microscope and reminds the polyphialides described in hyphomycetes such as *Chloridium* spp. [110].

2.3.3 Other Ultrastructural Characteristics

Endospores (or endoconidia) may be present in *M. furfur* (Fig. 2.16b), and it has been postulated that their presence may suggest an affiliation with teliospore-forming yeasts [103]. Furthermore, it has been suggested that the phenomenon of endosporulation might represent initial steps of basidium development [98, 103]. However, formation of endoconidia has been observed to occur during asexual reproduction in many ascomycetous and basidiomyceous yeast-like fungi. Mittag [111] considered the variable size and shape of the yeast cells within *M. furfur* (CBS 1878 and 6001) due to differences in ploidy. This was indeed confirmed by analysis of chromosomal DNAs using pulsed field gel electrophoresis (for refs. see Chap. 3).

The basidiomycetous affinities of the genus *Malasssezia* are further supported by other cell biological features, such as migration of the nucleus in the bud before mitosis [112]. The distribution of chitin, however, seems restricted to the budding site, similar to that in *Saccharomyces cerevisiae*, whereas it is distributed evenly throughout the cell wall of *Cryptococcus neoformans* [113], another basidiomycetous fungus.

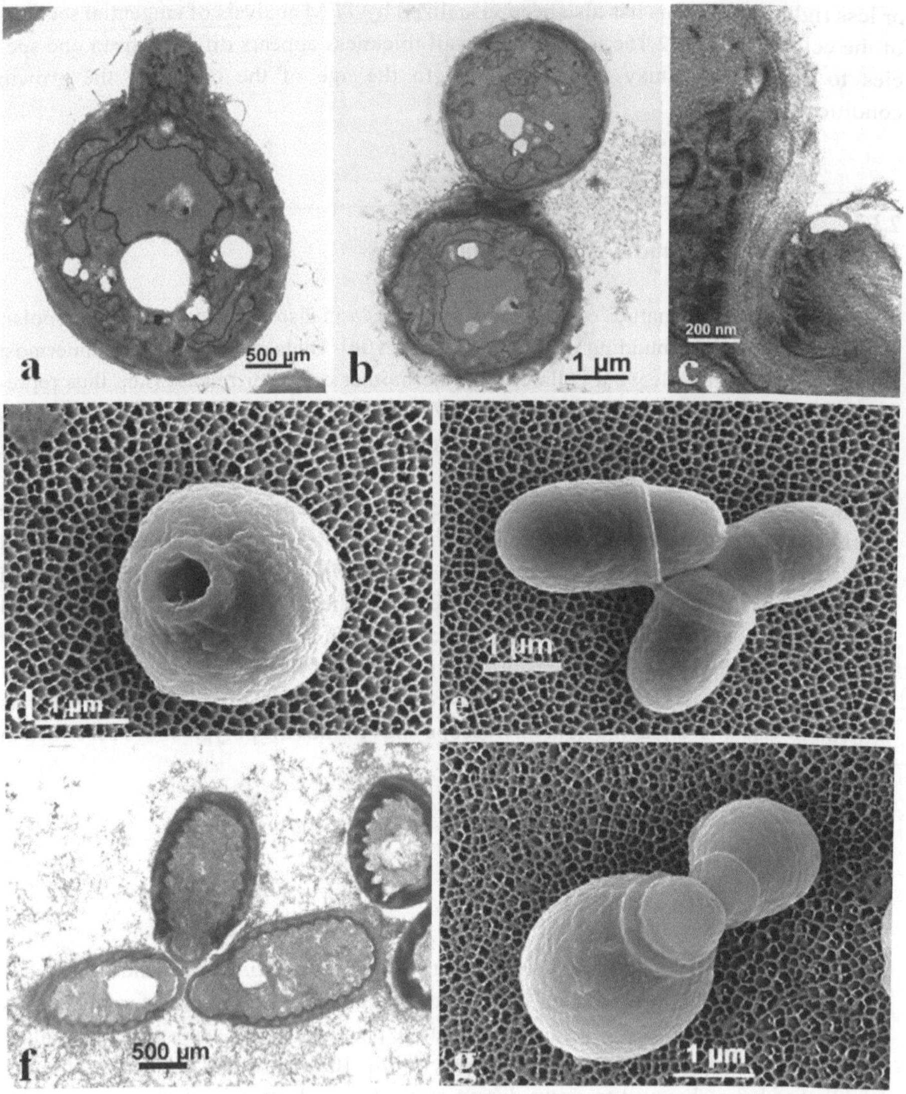

Fig. 2.17 *Malassezia* budding process. (**a–d**) *M. globosa*. (**a–b**) Micrograph (TEM) showing two states of the monopolar budding process, namely emergence and release of the bud; (**c**) Micrograph (TEM) showing several collarettes resulting from the monopolar percurrent budding; (**d**) Micrograph (SEM) of the budding site with the typical thick budding scar of *Malassezia* spp. (**e**) *Malassezia yamatoensis* (SEM), the youngest yeast cell on the right has a thinner scar. (**f–g**) *Malassezia sympodialis*; (**f**) micrographs (TEM) showing the typical clover leaf-like configuration resulting from sympodial budding; (**g**) micrograph (SEM) showing percurrent budding with the youngest bud still emerging

Acknowledgements The authors would like to warmly thank Anne-Françoise Miegeville (Laboratoire de Bactériologie-Antibiologie, Faculté de Médecine, Nantes, France), who made many scanning electron micrographs (SEM), namely all the pictures on filters presented in this chapter, and also Bart Theelen (Centraalbureau for Schimmelcultures, Utrecht, The Netherlands) for his help in preparing one of the phylogenetic trees.

References

1. Baillon H (1889) Traité de botanique médicale cryptogamique, 4th edn. Octave Doin, Paris
2. Sabouraud R (1904) Les maladies du cuir chevelu. II. Les maladies desquamatives. Masson et Cie, Paris
3. Slooff WC (1970) Genus 6 *Pityrosporum* Sabouraud. In: Lodder J (ed) The yeasts, a taxonomic study, 2nd edn. North-Holland, Amsterdam, pp 1167–1186
4. Acton HW, Panja G (1927) Seborrhoeic dermatitis or pityriasis capitis: a lesion caused by the *Malassezia ovale*. Indian Med Gaz 62:603–614
5. Panja G (1927) The *Malassezia* of the skin, their cultivation, morphology and species. Trans Th Cong Far Easthern Ass Trop Med 2:442–446
6. Shifrine M, Marr AG (1963) The requirement of fatty acids by *Pityrosporum ovale*. J Gen Microbiol 32:263–270
7. Van Abbe NJ (1964) The investigation of dandruff. J Soc Cosmetic Chem 52:173–188
8. Midgley G (1989) The diversity of *Pityrosporum* (*Malassezia*) yeasts *in vivo* and *in vitro*. Mycopathologia 106:143–153
9. Leeming JP, Notman FH (1987) Improved methods for isolation and enumeration of *Malassezia furfur* from human skin. J Clin Microbiol 25:2017–2019
10. Kaneko T, Makimura K, Onozaki M et al (2005) Vital growth factors of *Malassezia* species on modified CHROMagar Candida. Med Mycol 43:699–704
11. Lorenzini R, de Bernadis F (1987) Studies on the isolation, growth and maintenance of *Malassezia pachydermatis*. Mycopathologia 99:129–131
12. Renker C, Alphei J, Buscot F (2003) Soil nematodes associated with the mammal pathogenic fungal genus *Malassezia* (Basidiomycota: Ustilaginomycetes) in Central European forests. Biol Fertil Soils 37:70–72
13. Bjelland T, Ekman S (2005) Fungal diversity in rock beneath a crustose lichen as revealed by molecular markers. Microbiol Ecol 49:598–603
14. Fell JW, Scorzetti G, Connell L et al (2006) Biodiversity of micro-eukaryotes in Antarctic Dry Valley soils with <5% soil moisture. Soil Biol Biochem 38:3107–3115
15. Lai X, Cao L, Tan H et al (2007) Fungal communities from methane hydrate-bearing deep-sea marine sediments in South China Sea. ISME J 1:756–762
16. Bond R, Anthony RM (1995) Characterization of markedly lipid-dependent *Malassezia pachydermatis* isolates from dogs. J Appl Microbiol 78:537–542
17. Guého E, Batra R, Boekhout T (2010) The genus *Malassezia* Baillon. In: Kurtzman CP, Fell JW, Boekhout T (eds) The yeasts, a taxonomic study, 5th edn. Elsevier, Amsterdam (in press)
18. Cabañes FJ, Theelen B, Castellá G et al (2007) Two new lipid-dependent *Malassezia* species from domestic animals. FEMS Yeast Res 7:1064–1076
19. Crespo MJ, Abarca ML, Cabañes FJ (2000) Evaluation of different preservation and storage methods for *Malassezia* spp. J Clin Microbiol 38:3872–3875
20. Breierová E, Kocková-Kratochilová A, Šajbidor J et al (1991) *Malassezia pachydermatis*: properties and storage. Mycoses 34:349–352
21. Guillot J, Guého E (1995) The diversity of *Malassezia* yeasts confirmed by rRNA sequence and nuclear DNA comparisons. Antonie van Leeuwenhoek 67:297–314

22. Crespo Erchiga V, Guého E (2005) Superficial diseases caused by *Malassezia* species. In: Merz WG, Hay RJ (eds) Topley and Wilson's microbiology and microbial infections, medical mycology, 10th edn, vol 5. Hodder Arnold, London, pp 202–219
23. Guého E, Midgley G, Guillot J (1996) The genus *Malassezia* with description of four new species. Antonie van Leeuwenhoek 69:337–355
24. Guého E, Boekhout T, Ashbee HR et al (1998) The role of *Malassezia* species in the ecology of human skin and as pathogen. Med Mycol 36(suppl 1):220–229
25. Guillot J, Guého E, Lesourd M et al (1996) Identification of *Malassezia* species. A practical approach. J Mycol Méd 6:103–110
26. Mayser P, Haze P, Papavassilis C et al (1997) Differentiation of *Malassezia* species: selectivity of Cremophor EL, castor oil and ricinoleic acid for *M. furfur*. Br J Dermatol 137:208–213
27. Hägler AN, Ahearn DG (1981) Rapid diazonium blue B test to detect basidiomycetous yeasts. Int J Syst Bacteriol 31:204–208
28. Kurtzman CP (2010) Methods for the isolation, maintenance and identification of yeasts. In: Kurtzman CP, Fell JW, Boekhout T (eds) The yeasts, a taxonomic study, 5th edn. Elsevier, Amsterdam, pp (in press)
29. Kaneko T, Makimura K, Sugita T et al (2006) Tween 40-based precipitate production observed on chromogenic agar and development of biological identification kit for *Malassezia* species. Med Mycol 44:227–231
30. Kaneko T, Makimura K, Abe M et al (2007) Revised culture-based system for identification of *Malassezia* species. J Clin Microbiol 45:3737–3742
31. Boekhout T, Guého E (2003) Basidiomycetous yeasts. In: Howard DH (ed) Pathogenic fungi in humans and animals, 2nd edn. Marcel Dekker, New York, pp 535–564
32. Kurtzman CP, Fell JW, Boekhout T (2010) The Yeasts, a taxonomic study, 5th edn. Elsevier, Amsterdam
33. Hamamoto M, Uchida K, Yamaguchi H (1992) Ubiquinone system and DNA base composition of strains in the genus *Malassezia* determined by high-performance liquid chromatography. J Gen Appl Microbiol 38:79–82
34. Dodge CW (1935) Medical mycology. fungous diseases of men and other mammals. Mosby, St. Louis
35. Weidman FD (1925) Exfoliative dermatitis in the Indian rhinoceros (*Rhinoceros unicornis*), with description of a new species: *Pityrosporum pachydermatis*. In: Fox H (ed). Rep Lab Museum Comp Pathol Zoo Soc Philadelphia, Philadelphia, pp 36–43
36. Gustafson BA (1955) Otitis externa in the dog. A bacteriological and experimental study, Thesis, Stockholm
37. Guillot J, Guého E, Chermette R (1995) Confirmation of the nomenclatural status of *Malassezia pachydermatis*. Antonie van Leeuwenhoeck 67:173–176
38. Guillot J, Guého E, Chévrier G et al (1997) Epidemiological analysis of *Malassezia pachydermatis* isolates by partial sequencing of the large subunit ribosomal RNA. Res Vet Sci 6:22–25
39. Midreuil F, Guillot J, Guého E et al (1999) Genetic diversity in the yeast species *Malassezia pachydermatis* analysed by multilocus enzyme electrophoresis. Int J Syst Bacteriol 49:1287–1294
40. Guého E, Guillot J (1999) Comments on *Malassezia* species from dogs and cats. Mycoses 42:673–674
41. Ahearn DG, Simons RB (1998) *Malassezia* Baillon. In: Kurtzman CP, Fell JW (eds) The yeasts, a taxonomic study, 4th edn. Elsevier, Amsterdam, pp 782–784
42. Guillot J, Bond R (1999) *Malassezia pachydermatis*: a review. Med Mycol 37:295–306
43. Guillot J, Guého E, Mialot M et al (1998) Importance des levures du genre *Malassezia* en dermatologie vétérinaire. Point Vét 29:21–31
44. Guillot J, Petit T, Degorce-Rubiales F et al (1998) Dermatitis caused by *Malassezia pachydermatis* in a California sea lion (*Zalophus californianus*). Vet Rec 142:311–312

45. Bauwens L, De Vroey Ch, De Meurichy W (1996) A case of exfoliative dermatitis in a captive southern white rhinoceros (*Ceratotherium simum simum*). JK Zoo Wildlife Med 27:271–274
46. Wesche P, Bond R (2003) Isolation of *Malassezia pachydermatis* from the skin of captive rhinoceroses. Vet Rec 153:404–405
47. Chang HJ, Miller HL, Watkins N et al (1998) An epidemic of *Malassezia pachydermatis* in an intensive care nursery associated with colonization of health care workers' pet dogs. New Engl J Med 338:706–711
48. Midgley G (2000) The lipohilic yeasts: state of the art and prospects. Med. Mycol 38(suppl 1): 9–16
49. Morris DO, O'Shea K, Shofer FS et al (2005) *Malassezia pachydermatis* carriage in dog owners. Emerg Infect Dis 11:83–88
50. Castellani A, Chalmers AJ (1913) Manual of tropical medicine, 3rd edn. Baillière, Tindal and Cox, London, p 1747
51. González A, Sierra R, Cárdenas ME et al (2009) Physiological and molecular characterization of atypical isolates of *Malassezia furfur*. J Clin Microbiol 47:48–53
52. Duarte ER, Lachance M-A, Hamdan JS (2002) Identification of atypical strains of *Malassezia* spp. from cattle and dog. Can J Microbiol 48:749–752
53. Murai T, Nakamura Y, Kano R et al (2002) Differentiation of *Malassezia furfur* and *Malassezia sympodialis* by glycine utilization. Mycoses 45:180–183
54. Salkin IF, Gordon MA (1977) Polymorphism of *Malassezia furfur*. Can J Microbiol 23:471–475
55. Yarrow D, Ahearn D (1984) Genus 7 *Malassezia* Baillon. In: Kreger-van Rij NJW (ed) The yeasts, a taxonomic study, 3rd edn. Elsevier, Amsterdam, pp 882–885
56. Tanaka R, Nishimura K, Kamei K et al (2001) Assimilation test of *Malassezia furfur* isolated from the environment. Nippon Ishinkin Gakkai Zasshi 42:123–126
57. Colombo S, Nardoni S, Cornegliani L et al (2007) Prevalence of *Malassezia* spp. yeasts in feline nail folds: a cytological and mycological study. Vet Dermatol 18:278–283
58. Crespo MJ, Abarca ML, Cabañes FJ (1999) Isolation of *Malassezia furfur* from a cat. J Clin Microbiol 37:1573–1574
59. Crespo MJ, Abarca ML, Cabañes FJ (2000) Atypical lipid-dependent *Malassezia* species isolated from dogs with otitis externa. J Clin Microbiol 38:2383–2385
60. Crespo MJ, Abarca ML, Cabañes FJ (2002) Occurrence of *Malassezia* spp. in horses and domestic ruminants. Mycoses 45:33–337
61. Duarte ER, Melo MM, Hahn RC et al (1999) Prevalence of *Malassezia* spp. in the ears of asymptomatic cattle and cattle with otitis in Brazil. Med Mycol 37:159–162
62. Duarte ER, Resende CP, Rosa CA et al (2001) Prevalence of yeasts and mycelial fungi in bovine paratitic otitis in the State of Minas Gerals, Brazil. J Vet Med B Infect Dis Public Health 48:631–635
63. Gandra RF, Gambale W, de Cássia Garcia Simão R et al (2008) *Malassezia* spp. in acoustic meatus of bats (*Molossus molossus*) of the Amazon region, Brazil. Mycopathologia 165:21–26
64. Sugita T, Tajima M, Takashima M et al (2004) A new yeast, *Malassezia yamatoensis*, isolated from a patient with seborrheic dermatitis, and its distribution in patients and healthy subjects. Microbiol Immunol 48:579–583
65. Simmons RB, Guého E (1990) A new species of *Malassezia*. Mycol Res 94:1146–1149
66. Cunningham AC, Leeming JP, Ingham E et al (1990) Differentiation of three serovars of *Malassezia furfur*. J Appl Bacteriol 68:439–446
67. Cabañes FJ, Hernández JJ, Castellá G (2005) Molecular analysis of *Malassezia sympodialis*-related strains from domestic animals. J Clin Microbiol 43:277–283
68. Saadatzadeh MR, Ashbee HR, Holland KT et al (2001) Production of the mycelial phase of *Malassezia* in vitro. Med Mycol 39:487–493

69. Crespo Erchiga V, Ojeda Martos A, Vera Casaño A (1999) Mycology of pityriasis versicolor. J Mycol Méd 9:143–148
70. Gupta AK, Batra R, Bluhm R et al (2004) Skin diseases associated with *Malassezia* species. J Am Acad Dermatol 51:785–798
71. Bond R, Anthony RM, Dodd M et al (1996) Isolation of *Malassezia sympodialis* from feline skin. J Med Vet Mycol 34:145–147
72. Bond R, Howell SA, Haywood PJ et al (1997) Isolation of *Malassezia sympodialis* and *Malassezia globosa* from healthy pet cats. Vet Rec 141:200–201
73. Crespo MJ, Abarca ML, Cabañes FJ (2000) Otitis externa associated with *Malassezia sympodialis* in two cats. J Clin Microbiol 38:1263–1266
74. Sugita T, Takashima M, Shinoda T et al (2002) New yeast species, *Malassezia dermatis,* isolated from patients with atopic dermatitis. J Clin Microbiol 40:1363–1367
75. Lee YW, Kim SM, Oh BH et al (2008) Isolation of 19 strains of *Malassezia dermatis* from healthy human skin in Korea. J Dermatol 35:772–777
76. Ähman S, Perrins N, Bond R (2007) Carriage of *Malassezia* spp. yeasts in healthy and seborrhoeic Devon Rex cats. Med Mycol 45:449–455
77. Bond R, Stevens K, Perrins N et al (2008) Carriage of *Malassezia* spp. yeasts in Cornish Rex, Devon Rex and domestic short hair cats: a cross selectional survey. Vet Dermatol 198:299–304
78. Uzal FA, Paulson D, Eigenheer AL et al (2007) *Malassezia slooffiae*-associated dermatitis in a goat. Vet Dermatol 18:348–352
79. Sugita T, Takashima M, Kodama M et al (2003) Description of a new species, *Malassezia japonica,* and its detection in patients with atopic dermatitis and healthy subjects. J Clin Microbiol 41:4695–4699
80. Hirai A, Kano R, Makimura K et al (2004) *Malassezia nana* sp. nov., a novel lipid-dependent yeast species isolated from animals. Int J Syst Evol Microbiol 54:623–627
81. Perrins N, Gaudinio F, Bond R (2007) Carriage of *Malassezia* spp. yeasts in cats with diabetes mellitus, hyperthyroidism and neoplasia. Med Mycol 45:541–546
82. Nell A, James SA, Bond CJ et al (2002) Identification and distribution of a novel *Malassezia* species yeast on normal equine skin. Veter Rec 150:395–398
83. Gordon MA (1951) The lipophilic mycoflora of the skin. I. *In vitro* culture of *Pityrosporum orbiculare* sp. Mycologia 43:524–535
84. Sugita T, Kodama M, Saito M et al (2003) Sequence diversity of the intergenic spacer region of the rRNA gene of *Malassezia globosa* colonizing the skin of patients with atopic dermatitis and healthy individuals. J Clin Microbiol 41:3022–3027
85. Aspiroz C, Moreno LA, Rezusta A et al (1999) Differentiation of three biotypes of *Malassezia* species on normal human skin. Correspondence with *M. globosa, M. sympodialis* and *M. restricta.* Mycopathologia 145:69–74
86. Sugita T, Tajima M, Amaya M et al (2004) Genotype analysis of *Malassezia restricta* as the major cutaneous flora in patients with atopic dermatitis and healthy subjects. Microbiol Immunol 48:755–759
87. Pitkäranta M, Meklin T, Hyvärinen A et al (2008) Analysis of fungal flora in indoor dust by ribosomal DNA sequence analysis, quantitative PCR, and culture. Appl Environ Microbiol 74:233–244
88. Zhang N, Suh SO, Blackwell M (2003) Microorganisms in the gut of beetles: evidence from molecular cloning. J Invertebr Pathol 84:226–233
89. Moore RT (1980) Taxonomic proposals for the classification of marine yeasts and other yeasts-like fungi including the smuts. Bot Mar 23:361–373
90. Guého E, Kurzman CP, Peterson SW (1989) Evolutionary affinities of heterobasidiomycetous yeasts estimated from 18 S and 25 S ribosomal RNA sequence divergence. Syst Appl Microbiol 12:230–236

91. Begerow D, Bauer R, Boekhout T (2000) Phylogenetic placements of ustilaginomycetous anamorphs as deduced from LSU rDNA sequence. Mycol Res 104:53–60
92. Fell JW, Boekhout T, Fonseca A et al (2000) Biodiversity and systematics of basidiomycetous yeasts as determined by large-subunit rDNA D1/D2 domain sequence analysis. Int J Syst Evol Biol 50:1351–1371
93. Begerow D, Stoll M, Bauer R (2006) A phylogenetic hypothesis of Ustilaginomycotina based on multiple gene analyses and morphological data. Mycologia 98:906–916
94. Matheny PB, Wang Z, Binder M et al (2006) Contributions of *rpb2* and *tef1* to the phylogeny of mushrooms and allies (Basidiomycota, Fungi). Mol Phyl Evol 43:430–451
95. Hibbett DS, Binder M, Bischoff JF et al (2007) A higher-level phylogenetic classification of the Fungi. Mycol Res 111:509–547
96. Mittag H (1995) Fine structural investigation of *Malassezia furfur*. II. The envelope of the yeast cells. Mycoses 38:13–21
97. Simmons RB, Ahearn DG (1987) Cell wall ultrastructure and diazonim blue B reaction of *Sporopachydermia quercuum, Bullera tsugae*, and *Malassezia* spp. Mycologia 79:38–43
98. Guillot J, Guého E, Prévost MC (1995) Ultrastructural features of the dimorphic yeast *Malassezia furfur*. J Mycol Méd 5:86–91
99. Winiarczyk S (1992) The ultrastructure of *Pityrosporum pachydermatis*. Arch Vet Polonicum 32:5–13
100. Abou-Gabal M, Fagerland JA (1979) Electron microscopy of *Pityrosporum canis "pachydermatis"*. Mykosen 22:85–90
101. Kreger-van Rij NJ, Veenhuis M (1970) An electron microscope study of the yeast *Pityrosporum ovale*. Arch Mikrobiol 71:123–131
102. Marek D, Miroslav G, Kopecká M (2007) Microtubular and actin cytoskeletons and ultrastructural characteristics of the potentially pathogenic basidiomycetous yeast *Malassezia pachydermatis*. Cell Biol Inter 31:16–23
103. Mittag H (1994) An approach of the fine structure of *Malassezia furfur*. XII Congress of the International Society for Human and Animals Mycology, March 1994, Abst D139, Adelaide, Australie
104. Nishimura K, Asada Y, Tanaka S et al (1991) Ultrastructure of budding process of *Malassezia pachydermatis*. J Med Vet Mycol 29:387–393
105. Keddie FM (1966) Electron microscopy of *Malassezia furfur* in tinea versicolor. Sabouraudia 5:134–137
106. Takeo K, Nakai E (1980) Tilting method for distinguishing the side of freeze-fracture replicas and for describing mirror-image asymmetry. J Electron Microsc 29:91–97
107. Takeo K, Nakai E (1986) Mode of cell growth of *Malassezia (Pityrosporum)* as revealed by using plasma membrane configurations as natural markers. Can J Microbiol 32:389–394
108. Matakieff E (1899) Le pityriasis versicolor et son parasite. Thèse Fac Méd Univ Nancy
109. Breathnach AS, Gross M, Martin B (1976) Freeze-fracture replication of cultured *Pityrosporum orbiculare*. Sabouraudia 14:105–113
110. Cole GT, Samson RA (1979) Patterns of development in conidial fungi. Pitman, London, p 70
111. Mittag H (1995) Fine structural investigation of *Malassezia furfur*: I. Size and shape of the yeast cells and a consideration of their ploidy. Mycoses 37:393–399
112. David M, Gabriel M, Kopecká M (2007) Microtubular and actin cytoskeletons and ultrastructural characteristics of the potentially pathogenic basidiomycetous yeast *Malassezia pachydermatis*. Cell Biol Intern 31:16–23
113. Simmons RB (1989) Comparison of chitin localization in *Saccharomyces cerevisiae, Crytococcus neoformans*, and *Malassezia* spp. Mycol Res 94:551–553

Epidemiology of *Malassezia*-Related Skin Diseases

3

Takashi Sugita, Teun Boekhout, Aristea Velegraki, Jacques Guillot, Suzana Hađina and F. Javier Cabañes

Core Messages

> This chapter deals with the range of molecular biology methods including PCR-based assays, high-throughput DNA sequence analysis, and the use of real-time PCR in performing studies of skin and environmental community structure and dynamics. It highlights the utility of each molecular biology method in yielding *Malassezia* epidemiological data, the limitations of culture-based methods, and the possible biases that may influence *Malassezia* epidemiological studies. As in any disease or health-impacting event, the frequency and patterns of *Malassezia*-related diseases is examined by descriptive epidemiology. Thus, the chapter includes assessment of patient-related factors such as age, immunological status and gender, and environmental factors such as the geographical area and time of the year that contribute to the analysis of risks from *Malassezia*-induced or *Malassezia* -exacerbated disease. Also, a comprehensive account is given on the *Malassezia* epidemiology in animals and the factors associated with the animal hosts and their skin microenvironment.

3.1 Molecular detection and Identification of *Malassezia* yeasts in the Epidemiological Studies

Takashi Sugita and Teun Boekhout

3.1.1 Introduction

In recent years, molecular studies have considerably changed the taxonomy of the genus *Malassezia*. The molecularly defined species can still, to some extent, be discriminated

T. Sugita (✉)
Department of Microbiology, Meiji Pharmaceutical University, 2-522-1 Noshio, Kiyose, Tokyo 204-8588, Japan
e-mail: sugita@my-pharm.ac.jp

T. Boekhout et al. (eds.), *Malassezia and the Skin*
DOI: 10.1007/978-3-642-03615-6_3,

using morphological and physiological methods (see Chap. 2.1). Among the earliest molecular characteristics that were used to discriminate between species were analysis of Mol% guanine plus cytosine, and the assessment of DNA relatedness using DNA reassociation techniques [86]. Since the recognition of lipid dependent species, other than *M. furfur* [87], it became clear that molecular approaches are needed for better diagnostics, as well as our understanding of *Malassezia* community dynamics. For instance, 13.8% of isolates identified by phenotypic means were found to be misidentified after molecular re-identification using sequence analysis of the D1/D2 domains of the large subunit ribosomal rDNA (LSU rDNA) and the ITS1+2 regions [103].

Here, we present an overview of the molecular methods that have been applied to detect and identify *Malassezia* species composition on humans and animals. The methods used can be divided into four main approaches: (1) Biotyping using enzymatic methods; (2) Chromosomal analysis using pulsed field gel electrophoresis (PFGE); (3) PCR-based methods; and (4) DNA sequence based methods. It is noteworthy that the PCR- and sequence-based methods used for *Malassezia* biodiversity studies, and those employed to study *Malassezia* community structure on skin and molecular epidemiology are often similar and the distinction is not always clear (see Chap. 2.1).

3.1.2 Biotyping

In biotyping of *Malassezia* isolates, the use of enzyme activities has received little attention. In a Spanish study, 120 isolates from skin were biotyped using Api 20 NE and ApiZym (bioMérieux, Marcy l'Etoile, France) and they could be identified as *M. globosa*, *M. sympodialis*, and *M. restricta* using enzyme activity profiles. Biotype 1 (=*M. globosa*) was catalase positive, esculin negative, and lipase (C14) positive; biotype 2 (=*M. sympodialis*) was catalase positive, esculin positive, and lipase (C14) negative; biotype 3 (=*M. restricta*) was catalase negative, esculin negative, and lipase (C14) negative [7]. In another study, enzyme activities of 33 *Malassezia* isolates belonging to six species were studied using ApiZym. Species-specific differences were observed, and, in general, limited infraspecific variation was noted [130].

Phospholipase, proteinase, hyaluronidase, and chondroitin-sulfatase activities were studied in 30 isolates of *M. pachydermaris* obtained from otic secretions and skin scrapings from dogs. No significant differences in enzyme activities were observed to occur between the isolates [46]. Phospholipase activity, considered to contribute to fungal virulence, has been studied in *M. pachydermatis* and differences were observed in phospholipase activity between isolates obtained from lesional and nonlesional skin [32, 37, 38].

Multilocus enzyme electrophoresis of 13 enzymes from 52 isolates of *M. pachydermatis* yielded 27 electrophoretic types (ETs). Interestingly, a correlation was observed between ETs and host specificity. Moreover, the results indicated that genetic exchange may occur between the various ETs, thus suggesting the presence of sexual recombination in this species [138].

3.1.3 Electrophoretic Karyotyping

Electrophoretic karyotyping using PFGE demonstrated a heterogeneity in chromosomal patterns among *Malassezia* yeasts. Various techniques have been applied, e.g., LKB

Table 3.1 Genome size of some *Malassezia* species based on PFGE [13, 15]

Species	Number of chromosomes	Genome size (Mb)
M. furfur type I	7–8	8.5
M. furfur type II	10–11	14
M. globosa	8	8.5–8.9
M. obtusa	6	7.4
M. pachydermatis	5	6.7–9.3
M. restricta	9	8.0
M. slooffiae	7	8.6
M. sympodialis	7	6.4–7.8

Pulsaphor and CHEF Dr-II (Biorad), with the latter technologically being more reliable due to the absence of mechanically moving parts. Using Pulsaphor, isolates of *M. furfur* were found to have seven chromosomes, which could separate them in three groups that correlated with the three morphological types that were distinguished before [115, 136].

CHEF analysis of *M. pachydermatis* isolates and lipophilic *Malassezia* spp. [13, 14] demonstrated that the karyotypes of *M. pachydermatis* were similar to each other and contained five chromosomes, whereas among the lipophilic isolates four different patterns could be discerned. In the latter study [15], the lipophilic species with different karyotypes could be linked to the species described by Guého et al. [87]. One of the main conclusions was that *Malassezia* species showed limited intraspecific chromosomal length polymorphism (CLP). The observed heterogeneity in *M. furfur* could be related to the presence of putative hybrids between different AFLP genotypes of that species (T. Boekhout, unpublished observation). Another remarkable feature of the *Malassezia* genomes is their small size ranging from 6.4 to 14 Mb (Table 3.1) [15, 208]. The genome size of *M. globosa* of 8.5–8.9 Mb as estimated by PFGE analysis agreed with that based on whole genome sequencing, namely 9 Mb [208]. The latter analysis also suggested that the genome is haploid as the level of polymorphism present was found to be very low (>0.0004%). PFGE has also been used to confirm phenotypically identified *Malassezia* isolates. In a study of 220 isolates from affected skin of animals and humans, 217 were identified as *M. pachydermatis* and three as *M. sympodialis*. The majority of the *M. pachydermatis* isolates contained six chromosomes and 17 had seven. Interestingly, two isolates obtained from both ears of a single dog revealed 6 and 7 chromosomes, respectively [182]. The *M. sympodialis* isolates also showed some CLP. On the basis of these studies, PFGE may be used to identify *Malassezia* species, but the method is time consuming and methodologically demanding. Moreover, karyotypes of the recently described species (e.g., *M. caprae, M. dermatis, M. equina, M. japonica, M. nana* and *M. yamatoenis*) have not yet been analyzed.

3.1.4
PCR-Based Methods

Many genotypic methods used to discriminate between isolates of *Malassezia*, or to identify them, are based on PCR. Among these are genotypic methods that generate a banding pattern per strain that can be compared with other such banding patterns available (e.g., RAPD, AFLP, PCR-RFLP). Alternatively, amplicon lengths may be compared (e.g., *t*FLP), or the migration behavior of single stranded DNA is used to infer relationships between isolates in DGGE or SSCP analyses (see below). Species-specific PCR probes have been developed for diagnostics. These can be used in individual PCRs or, technically more advanced, they can be hybridized to probes bound to latex beads (e.g., Luminex technique) thus allowing further, high-throughput automation for use in large-scale epidemiological studies. Furthermore, real-time PCR has been used to study *Malassezia* community structure and dynamics.

3.1.4.1 Random Amplification of Polymorphic DNA (RAPD) and DNA Fingerprinting

RAPD typing of *Malassezia* isolates has been used in various studies. The epidemiological analysis of an outbreak at a neonatal ward caused by both *M. furfur* and *M. pachydermatis* [201] was among the first. This study used prokaryotic repeat consensus primers based on enterobacterial repetitive intergenic consensus (ERIC) or repetitive extragenic palindromic (REP) motifs. The outbreak isolates showed identical banding patterns if compared with the reference strains. Hence, it was concluded that the outbreak was caused by a single genotype and most probably a single strain, despite occasional subtle differences that were observed by PFGE [13]. Apparently, ERIC and REP PCR-typing showed better epidemiological resolution than PFGE. It is important to realize that PFGE is not directly comparable with ERIC and REP PCR-typing, as each method targets different elements of the *Malassezia* yeast genome. In another study of an outbreak in a neonatal ward caused by *M. pachydermatis*, RAPD analysis showed the presence of genetically distinct isolates involved in the outbreak [44], and similar observations were made in a *M. pachydermatis* outbreak in a US-based hospital [199]. The use of RAPD typing for taxonomic and henceforth epidemiological purposes has also been investigated [15]. Representatives of all known species at that time were investigated by 20 decamer primers. Primers OPA 02, OPA 04, OPA 05, and OPA 13 (Operon Technologies, Alameda, CA, USA) gave best resolution. All species could be discriminated, but straightforward analysis was hampered by the presence of considerable intraspecific variation [15]. Similar results were obtained in an epidemiological study of isolates from skin, where *M. sympodialis* showed the greatest homogeneity [41]. Within 55 isolates of *M. pachydermatis* obtained from various domestic animals and body sites, RAPD discriminated four genotypes; one of them occurred in various animals (cats, horse, goat, pig) and the other three were observed only in the external ear canal of dogs; different genotypes could occur on a single animal or even at a single body site [40]. Other studies on *M. pachydermatis* also demonstrated genetic heterogeneity [114, 126]. In an extensive study of 210 isolates using RAPD, morphology, catalase and β-glucosidase activities, lipid and carbohydrate growth requirements, and tryptophan-dependent pigment synthesis [114], it was concluded that extensive phenotypic and genotypic diversity occurs within the species. A single genotype was found to be correlated with pigment production and the authors proposed to use this observation as a possible screening to search for this character in a larger number of isolates. In a study on the presence of genetic diversity in *M. furfur*, 47 isolates from either pityriasis versicolor (PV) patients, seborrheic dermatitis (SD) patients, or SD in AIDS patients, RAPD patterns were found to correlate with these three patient categories. On the basis of this observation, the authors suggested that RAPD is a useful epidemiological tool [76]. M13-DNA fingerprinting of 42 globally collected isolates of *M. furfur* yielded six genotypic clusters that correlated with the underlying skin condition and the hosts' geographic origin. Isolates from Scandinavia, Greece, Bulgaria, and China showed a propensity for geographic clustering. The authors concluded that *M. furfur* may be a good candidate species to investigate human phylogeography patterns [74].

3.1.4.2 Restriction Fragment Length Polymorphism (RFLP) Analysis

Digestion of PCR amplicons has shown to be useful for the separation of *Malassezia* species [71, 72, 82, 96, 114]. Sequence polymorphisms in the target gene or DNA region of choice can be made visible by the use of restriction enzymes. The ribosomal DNA, including the large subunit (LSU) rDNA and internally transcribed spacer (ITS) regions have been DNA targets explored for diagnostics [96, 100, 103]. Using the *Malassezia* specific primers Malup (5'-AGC GGA GGA AAA GAA ACT-3') and Maldown (5'-GCG CGA AGG TGT CCG AAG-3'), the LSU rRNA gene could be amplified resulting in a fragment between 541 and 579 bp [96]. All seven species recognized at that time could be discriminated by digestion of the amplicon using the restriction enzymes BanI, HaeII, and MspI. Moreover, no intraspecific variation was observed, rendering this method a *Malassezia*-specific identification tool. In addition, species in mixed samples of *M. globosa* and *M. sympodialis*, or *M. furfur* and *M. sympodialis*, respectively, could be discriminated [96]. Later, this method was extended to separate 11 species [159] (Table 3.2). In another RFLP study, five out of seven species could be discriminated using AvaI digestion of the LSU rDNA and digestion of the ITS regions by EcoRI and NcoI [100]. However, *M. globosa* and *M. restricta* could not be discriminated using the latter enzymes, and, therefore, an alternative hot start touch-down PCR of the

Table 3.2 RFLP profile of eleven *Malassezia* species based on three restriction endonucleases

Species	ITS 3/4 PCR product (bp)	AluI	BanI	MspAI
Malassezia dermatis	416	NRS	186, 230	192, 85, 109, 30
Malassezia furfur	557	306, 251	389, 168	525, 32
Malassezia globosa	477	221, 16, 240	NRS	447, 30
Malassezia japonica	528	394, 134	183, 199, 146	498, 30
Malassezia nana	428	NRS	188, 240	286, 110, 32
Malassezia obtusa	554	NRS	396, 158	NRS
Malassezia pachydermatis	529	412, 117	NRS	499, 30
Malassezia restricta	463	NRS	186, 277	432, 31
Malassezia slooffiae	505	385, 120	NRS	472, 33
Malassezia sympodialis	420	NRS	NRS	281, 109, 30
Malassezia yamatoensis	470	NRS	NRS	NRS

NRS no restriction site

β-tubulin gene was needed for a proper identification of these two species. Gaitanis et al. [71] developed a PCR-RFLP aiming to identify *Malassezia* species present in skin scales. For DNA extraction, a cetyltrimethylammonium bromide (CTAB) protocol was employed. Panfungal primers targeting the ITS, namely ITS 1 (5'-TCC GTA GGT GAA CCT GCG G-3'), ITS 3 (5'- GCA TCG ATG AAG AAC GCA GC-3') and ITS 4 (5'-TCC TCC GCT TAT TGA TAT GC-3'), were used to amplify the ITS regions. In case of a weak amplification result using ITS 3 and ITS 4 primers, the primer combination ITS 1 and ITS 4 was used, followed by a nested PCR using primers ITS 3 and ITS 4. Length estimates of the amplicons generated by primers ITS 3/ITS 4 discriminated between *M. furfur, M. sympodialis, M. globosa*, and *M. pachydermatis*. However, using standard agarose conditions, *M. restricta* could not always be separated from *M. slooffiae* or *M. pachydermatis.* Digestion of the amplicons by AluI and HinfI discriminated between all seven species included in the study, namely *M. furfur*, *M. obtusa*, *M. slooffiae*, *M. restricta*, *M. sympodialis*, *M. globosa*, and *M. pachydermatis*. The relative sensitivity was found to be 10±5 cfu. Subsequently, 11 *Malassezia* species using pure culture were discriminated (Table 3.2) through restriction of PCR amplicons that were generated with the ITS 3/4 primers [72]. PCR-RFLP of the LSU-rDNA using CfoI and BstF51 discriminated among 11 species, and was successfully applied to clinical isolates [139]. PCR-RFLP has also been used to identify larger numbers of clinical isolates [39, 212] or isolates from animals [160]. Importantly, tracing an outbreak of *M. pachydermatis* in an intensive care unit was possible using RFLP analysis. Analysis of isolates obtained from neonates, health care workers and their pet dogs demonstrated that transmission occurred from a pet dog via the health care worker to the IC patients [42] (see Chaps. 3.3 and 8).

3.1.4.3 Amplified Fragment Length Polymorphism (AFLP™) analysis

AFLP, a DNA fingerprinting technique based on the selective amplification of restriction fragments from a digest of genomic DNA [202], has been successfully applied to understand the genetic relationships among isolates of *Malassezia* [31, 103, 199]. In AFLP analysis, all lanes contain a size standard and, therefore, the method is, at least theoretically, less prone to experimental variation than, for instance, RAPD analysis. Practically, variation has been noted to occur between separate runs, especially when a time interval occurs between runs, thus rendering an easy comparison sometimes difficult. The same may be true for interlaboratory comparisons, although the variation observed may be less than in RAPD analyses. In case of critical studies, we therefore recommend running the experiment(s) in a single run, in order to avoid experimental bias as much as possible.

Using AFLP all species recognized could be separated [31], the genetic diversity within species was illustrated, and in some species distinct genotypes could be discerned [103, 199] (Fig. 3.1). In *M. furfur*, four [199] and in a more extended study eight [103] AFLP groups could be recognized. Probably the most significant observation was that 80% of the isolates of AFLP genotype 4 came from deep or mucosal sources, such as urine, blood, tracheal secretion, nasal smear, and feces [199]. In a more extensive sampling, genotype 1 isolates came from chest and back, genotypes 2 and 8 contained animal isolates, genotypes 4 and 5 originated from deep body sites, and isolates from genotypes 1 and 6 seemed to have a preference for

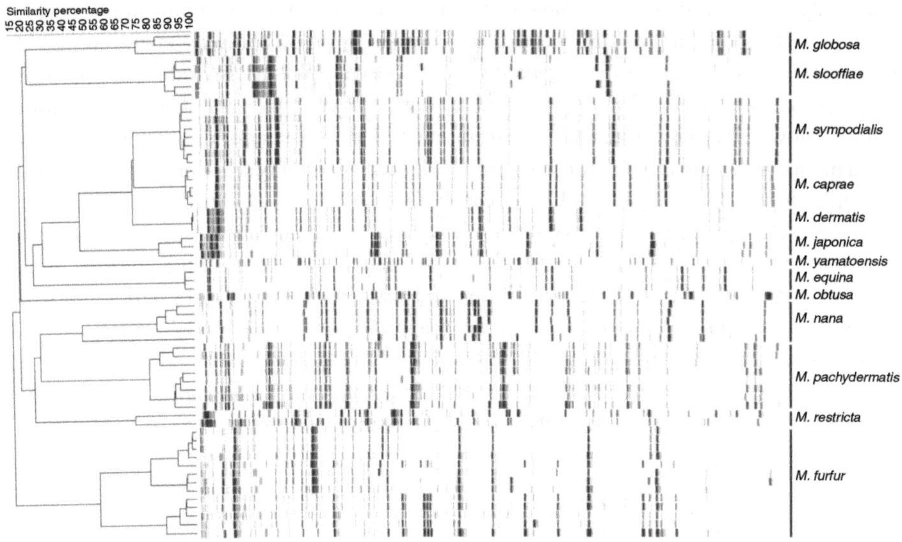

Fig. 3.1 Genotypic diversity among *Malassezia* species determined by amplified fragment length polymorphism (AFLP). The AFLP patterns were generated using the selective primers EcoR1A and Mse1G as described [199], and the tree was created with Bionumerics (Applied Math, Belgium) using curve-based Pearson correlation in combination with single linkage

neonates. The distribution of AFLP genotypes between North America and Europe was found to be significantly different [103]. AFLP seems to be the tool of choice for genotypic analyses when detailed genotypic information is needed and many isolates are available.

3.1.4.4 Terminal Fragment Length Polymorphism (*t*FLP) Analysis

The size of ITS 1 and ITS 2 amplicons can potentially be used to discriminate between species (see also above under RFLP [71]). Similar approaches have been described to identify other clinically important yeasts, such as *Candida* spp. [56]. This method was further developed using capillary detection of the amplicon size, making the method suitable for automation. This so-called terminal fragment length polymorphism (*t*FLP) used the amplicon lengths of both the ITS 1 and ITS 2 for discrimination of *Malassezia* species [80]. The method can also be used on noninvasively acquired swab samples and uses only three ITS-based primer sets, and therefore, minimizes the potential bias related to amplification efficiency. It is sensitive enough to detect *Malassezia* with as few as 100 cells per sample, either dosed directly onto the swab for extraction control or for data collection from 1 cm^2 of skin surface sample. This method involves isolation of fungal DNA, followed by nested PCR of the ribosomal gene cluster and amplification with ITS 1 and ITS 2 specific fluorescently labeled primers. The resulting terminally-labeled products are analyzed for fragment length on a fluorescent DNA sequencer. The amplifications are carried out with universal fungal primers [80], and therefore the methodology should be broadly applicable to other fungal species. The ITS 1 and ITS 2 amplifications are carried out individually and combined for final analysis. The primary advantage of this method is that the resultant sample contains only two labeled fragments per species, allowing analysis of complex communities. All *Malassezia* species can be differentiated by length polymorphisms, including multiple genotypes per known species. Results obtained for standards and mixtures of standards showed that all known *Malassezia* genotypes can be identified in a single amplification reaction based on unique fragment lengths, thus eliminating the need for restriction analysis. Results from this study also showed that *t*FLP is capable of reproducibly assessing the *Malassezia* species present in complex mixtures and clinical samples. In an analysis of subjects assigned for different grades of dandruff, i.e., a comparison between persons with composite adherent flaking score (ASFS)<10 (low dandruff) and those with a composite ASFS>24 (high dandruff) showed that in both human populations *M. globosa* and *M. restricta* were predominant species in samples from scalp, with both species being more prominent in high ASFS subjects. Importantly, *M. furfur* was not detected [80]. The authors concluded that both *M. globosa* and *M. restricta* are involved in causing dandruff.

3.1.4.5 Denaturing gradient gel electrophoresis (DGGE) and single strand conformation polymorphism (SSCP) analysis

DGGE and SSCP are methods that can discriminate among PCR amplicons that have a similar size, and, hence cannot be separated using standard electrophoretic techniques. The principle is that the tertiary conformation of single stranded DNA, which determines the running distance in a gel, depends on the nucleotide sequence.

DGGE uses a denaturing gel, e.g., containing a urea gradient, to separate both strands of the PCR amplicon [143, 144], whereas SSCP separates, in a nondenaturing gel, heat separated single stranded DNA that has been cooled down quickly [79]. Both methods require dedicated equipment, and as the running behavior in the acrylamide gels depends on the experimental conditions, results are not always straightforward to interpret. It seems that results based on SSCP are less prone to experimental variation, and, in addition, running conditions do not need to be optimized for each new PCR. For these reasons, DGGE analysis has only rarely been applied to differentiate *Malassezia* strains [199]. All seven species included in the study could be distinguished, and two isolates of each *M. furfur* and *M. pachydermatis* were found to have identical banding patterns.

PCR-based SSCP allows detection of point mutations in PCR amplicons, e.g., those of the widely used ITS regions. Commonly used primers (e.g., ITS 1 and ITS 2) allow efficient amplification. Separation of single stranded DNA that is obtained after denaturation of the amplicons at e.g., 94°C followed by immediate snap-cooling on ice or a freezing block is performed on a nondenaturing acrylamide/bis-acrylamide gel at low temperatures (e.g., 11°C) (for a detailed protocol see ref. [79]). Five genetic subtypes were identified in *M. globosa* isolates obtained from PV and SD by SSCP using the ITS 1 spacer [73]. SSCP subgroup A was found to be correlated with the presence of more extensive lesions. Twelve isolates of *M. sympodialis* did not show any SSCP ITS subtypes [73]. Among 185 isolates of *M. pachydermatis* obtained from diseased and nondiseased skin of 30 dogs and from various body sites, eight different ITS 1 genotypes could be discriminated by SSCP that were also confirmed by sequencing analysis [38]. One genotype (i.e., genotype B) was observed to occur mainly on healthy skin. Fifty percent of the dogs contained multiple ITS subgenotypes at different body sites [38]. Next to SSCP separation of ITS amplicons, the method has also been used to separate amplicons based on the chitin-synthase-2 (CHS-2) gene [35].

3.1.4.6
Detection by Luminex Technology

Sensitive detection of *Malassezia* species, either alone or in mixed populations, has been made possible using a bead suspension array that combines the specificity and reliability of DNA-based probes and the speed and sensitivity of the Luminex analyzer [57]. Probes based on sequence divergence in the D1/D2 domains of the LSU rDNA and the ITS spacers are covalently bound to unique sets of fluorescent beads allowing detection of the specific probes using a red laser (636 nm), and, hence, identification of the species present. After hybridization, the biotinylated PCR amplicon is detected using a green laser (532 nm) by addition of a fluorochrome coupled to a reporter molecule (e.g., streptavidin). Probes covered either a narrow or a broader taxonomic range, and discriminated between species, groups of species, or all representatives of the genus. This allowed detection of species in a multiplex and high-throughput format [57]. The hybridization assay could discriminate between sequences that only differed in 1 or 2 bp, but specificity depends on the position of the mismatch as well as the probe design. Accurate identification of species is achieved using a multiplex format, thus allowing the analysis of mixed clinical samples. Interestingly, the method could even use direct amplification from a pin-head-sized portion of cells of *M. pachydermatis*, thus without the need to isolate DNA. If this is true for the other species as well, this method may be used to analyze the *Malassezia* community structure fast and reliably [57]. However, one may expect considerable differences in the efficiency of the PCR performance between the various species when omitting DNA extraction, thus complicating the analysis.

3.1.4.7
Epidemiological Investigations by Real-time PCR

PCR approaches, such as real-time PCR as well as *t*FLP (see Sect. 3.1.4.4 above) have been used to investigate the presence of *Malassezia* species without the need to culture isolates. Detection of *Malassezia* species from clinical samples has been performed by real-time PCR using a nested PCR approach targeting the intergenic spacer (IGS) or the internal transcribed spacer (ITS) [5, 188]. In a series of studies, the number of *Malassezia* cells present on healthy and diseased skin was quantified using Taq-Man probes based on D1/D2 LSU rRNA gene sequences (Fig. 3.2). LSU rDNA was amplified using primers pITS (5'-GTCGTAACAAGGTTAACCTGCGG-3') and NL4 (5'-GGTCCGTGTTTCAA GACGG-3'). After insertion in a plasmid, the plasmid DNA was quantified using a NanoDrop ND-1000 spectrophotometer. For real-time PCR, 200 nm of each primer and 250 mM of the Taq-Man primer were used (Table 3.3).

Ten to the power of two to ten to the power of nine (10^2–10^9) copies of plasmid DNA of all 11 species tested could be detected using the primers Mala-F, Mala-R, and Mala-MGB. Importantly, cross reactivity was neither observed against DNA from a variety of clinically relevant yeast species, e.g., belonging to *Candida*, *Cryptococcus*, and *Rhodotorula*, nor from

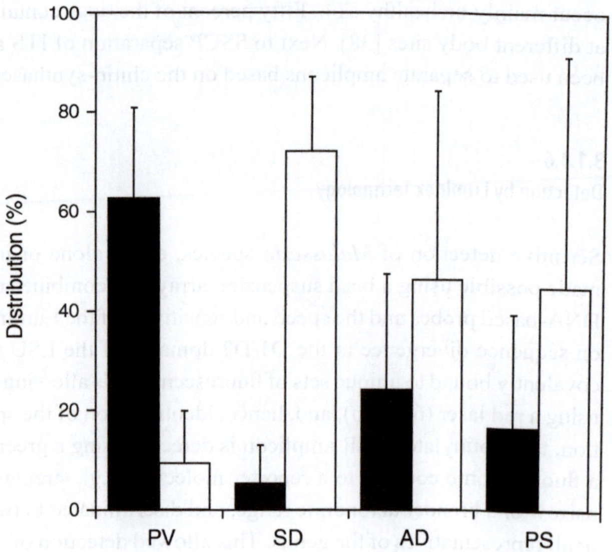

Fig. 3.2 Relative occurrence of *Malassezia globosa* (black bar) and *M. restricta* (*empty bar*) as detected by real-time PCR in pityriasis versicolor (PV), seborrheic dermatitis (SD), atopic dermatitis (AD) and psoriasis (PS)

Table 3.3 Primers used in real-time PCR [193]

Universal *Malassezia* primers	
Mala-F (forward)	5'-CTAAATATCGGGGAGAGACCGA-3'
Mala-R (reverse)	5'-GTACTTTTAACTCTCTTTCCAAAGTGCTT-3'
Mala-MGB (probe)	FAM-TTCATCTTTCCCTCACGGTAC-MGB
***M. globosa* primers**	
M.glob-F (forward)	5'-GGCCAAGCGCGCTCT-3'
M.glob-R (reverse)	5'-CCACAACCAAATGCTCTCCTACAG-3'
M.glob.MGB (probe)	FAM-ATCATCAGGCATAGCATG-MGB
***M. restricta* primers**	
M.rest-F (forward)	5'-GGCGGCCAAGCAGTGTTT-3'
M.rest-R (reverse)	5'-AACCAAACATTCCTCCTTTAGGTGA-3'
M.rest-MGB (probe)	FAM-TTCTCCTGGCATGGCAT-MGB

bacteria. Using real-time PCR, it could be demonstrated that the number of *Malassezia* cells present in the head and neck of 34 AD patients was 12.4 times higher than that on the trunk and 6.8 times higher than that present on the limbs [193]. Both *M. globosa* and *M. restricta* occurred on all AD patients investigated, and these species caused approximately 80% of the colonization by *Malassezia*. In another study of AD patients, it was observed that *M. restricta* and *M. globosa* were the dominant species in adults [188], whereas *M. restricta* occurred dominantly in children with AD [197]. *M. dermatis*, *M. sympodialis*, *M. furfur*, *M. obtusa*, and *M. yamatoensis* were detected in 2–8% of the subjects, and *M. japonica* and *M. slooffiae* were not detected. Using real-time PCR, the colonization rate by *Malassezia* was found to be 2.4-fold higher in adults with AD than in children with this disease. In children and adults, colonization by *M. restricta* was, respectively, 3.5 and 1.5 times higher than that of *M. globosa* [197].

In a similar study on the presence of *Malassezia* species occurring on the skin of 31 SD patients (i.e., 31 lesional and 27 nonlesional samples), *M. globosa* and *M. restricta* were observed in 93.5 and 61.3% of the SD patients, respectively [195]. *M. dermatis*, *M. slooffiae*, and *M. sympodialis* were detected in 20–30% of the patients, and the remaining species occurred in less than 10%. In both lesional and nonlesional skin, *M. globosa* and *M. restricta* occurred at 70.4 and 55.6%, respectively, but both types of skin did not differ in species composition. In a comparison between AD, SD, PV, and healthy skin, the detection rate of all species was the highest in AD patients with 4.1 ± 1.9 species occurring [195]. Using real-time PCR, lesional skin was found to harbor three times as many *Malassezia* cells when compared with nonlesional skin. *M. restricta* was by far the dominant species in both lesional- and nonlesional skin with an occurrence of 72.3 ± 15% in lesional skin and 63.6 ± 16.4% in nonlesional skin, whereas the occurrence of *M. globosa* was 5.1 ± 7.9% and 6.5 ± 6.3 %, respectively (Fig. 3.2). In conclusion, these analyses suggested that *M. restricta* is the dominant species occurring on SD patients. Genotypically, eight IGS sequence types occurred among the isolates of *M. globosa*, and five of them (i.e., IGS types II, III, IV, V, VI) occurred on SD patients only, whereas type VII occurred on healthy skin, and types I and VIII occurred on both skin from SD and healthy individuals. In *M. restricta*, two main IGS genotypes were apparent, and the isolates of cluster II originated from SD patients and those of cluster I came from both SD and healthy individuals [195]. The authors concluded that SD is a *M. restricta* dominated disease.

In a study of 20 psoriatic patients, real-time PCR analysis indicated that *Malassezia* spp. colonized the trunk and head more heavily than the limbs [196]. *M. globosa* and *M. restricta* occurred on almost all patients with psoriasis, namely 98.0% and 91.8% of lesional skin, and 100% and 91.7% of nonlesional skin, respectively. *M. restricta* was found to be the dominant species in both lesional and nonlesional skin, with a presence of 46.9% and 50.3%, respectively, whereas *M. globosa* occurred only in 12.6% and 8.2% of these samples, respectively [196]. Interestingly, psoriatic patients with normolipidaemia showed greater colonization by *Malassezia* yeasts, including *M. restricta* and *M. globosa*, than patients with hyperlipidaemia. The reason for this is not known, but the authors suggested a possible relationship with the observed presence of atrophic sebaceous glands and reduced lipid production as observed in hyperlipidaemic transgenic mice. Second, it was suggested that antifungal agents might be more effective in patients with normolipidaemia [196].

Paulino et al. [158] investigated the dynamics of the *Malassezia* community structure in time on six healthy skin and two psoriatic skin samples using real-time PCR with dual-labeled probes based on the 5.8 S rRNA gene and the ITS2 spacer that targeted either each of six species or the entire genus. A good correlation was observed between data obtained by real-time PCR and those obtained in a clone library study [157, 158]. *M. restricta* occurred in 57–100% and 7–99% of skin samples from healthy patients and psoriasis patients, respectively, and *M. globosa* was the dominant species, but no differences were observed between lesional and nonlesional skin. *M. sympodialis* was detected exclusively on the upper back of healthy individuals, but only in low amounts (i.e., 10^2–10^5-fold less than *M. restricta*). When comparing the community structure during a 4-month sampling period in 1-month intervals, little variation in the *Malassezia* communities was observed [158].

Lesional skin and nonlesional skin of psoriasis patients revealed that all nine species, for which *Taq*-Man probes were available, could be detected in 22 patients. Detection rates of *M. globosa, M. restricta*, and *M. sympodialis* were 82, 96, and 64%, respectively. No differences occurred between lesional and nonlesional skin, and no species-specific colonization dependent on age, gender, body site, severity of psoriasis, or treatment was observed [5]. The skin of both psoriasis and AD patients showed a greater diversity in *Malassezia* species as compared with normal skin, with an average 3.7±1.6, 4.1±1.9, and 2.8±0.8 species occurring, respectively.

In a study on PV, all PV lesions from 49 patients were found to be positive for *Malassezia* spp. and *M. globosa* and *M. restricta* predominated in 93.9% of the cases [142]. Other species as *M. sympodialis* (34.6%), *M. dermatis* (24.4%), *M. furfur* (10.2%), *M. obtusa* (8.1%), *M. japonica* (6.1%), and *M. slooffiae* (4.1%) were less frequently detected. In lesions that showed hyphal growth of *Malassezia* cells, which is considered an important factor contributing to PV, only *M. globosa* could be detected. Therefore, the authors concluded that PV is mainly caused by *M. globosa* [141].

These molecular studies suggest that *M. globosa* and *M. restricta* are the most important species involved in the different skin disorders, AD, SD, PV, and probably psoriasis as well. Therefore, emphasis to understand the pathophysiology of *Malassezia*-mediated skin disorders should focus on these two species.

3.1.5 DNA Sequence Analysis

Nucleotide sequence analysis of the ribosomal DNA locus has fundamentally contributed to our current understanding on the biodiversity of *Malassezia* yeasts. More recently, other genes have also been explored for their ability to contribute in understanding the species boundaries within the genus.

3.1.5.1 Ribosomal DNA Sequences

The ribosomal RNA (rRNA) genes (also referred to as rDNA, but this also includes the IGS regions) has been widely used for the identification of microorganisms, including clinically relevant yeasts. Fig. 3.3 shows the primary structure of the *Malassezia globosa* rRNA genes and spacer regions. Fungal rRNA genes occur in tandem repeats, with each repeat encoding 18S (or small subunit, SSU), 5.8S, 26S (or large subunit, LSU), and 5S rRNA genes. Two

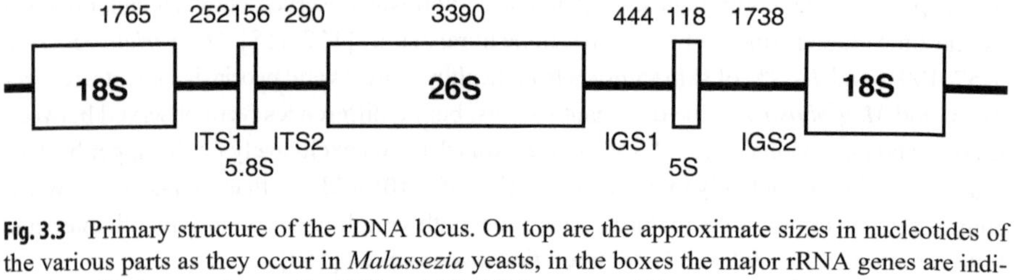

Fig. 3.3 Primary structure of the rDNA locus. On top are the approximate sizes in nucleotides of the various parts as they occur in *Malassezia* yeasts, in the boxes the major rRNA genes are indicated, and below are the smaller 5.8S and 5S rRNA genes and the spacer regions

spacer regions exist in each repeat, namely the Internal Transcribed Spacer (ITS) regions located between SSU and LSU, and the InterGenic Spacer (IGS) region located between LSU and SSU. The four subunits have more or less constant lengths regardless the species concerned as follows: SSU 1800 bp, 5.8S 160 bp, LSU 3,500bp, and 5S 120 bp. The rRNA genes can be compared among phylogenetically distant species because the sequences are highly conserved, but phylogenetically closely related species or isolates can be compared as well because of the presence of more variable stretches of DNA, such as the ITS and IGS spacers (Table 3.4). At present, sequences of the rRNA genes and ITS regions of almost all human- and animal pathogenic fungi have been deposited in GenBank.

To be useful as a species-specific diagnostic tool, DNA sequences must be evaluated and compared with other criteria, such as phenotype. In general, conspecific strains have identical (or highly similar) sequences, whereas different species have divergent sequences [186, 187]. The domains 1 and 2 (i.e., the D1/D2 domains) of the LSU rRNA gene, which is the most variable region of LSU, and the ITS 1+2 regions are useful for the identification of unknown fungal isolates. Peterson and Kurtzman [161] and Sugita et al. [186] suggested that conspecific strains have a similarity of 99% or higher in both these regions. In practice, almost all pathogenic fungi can be identified following this approach. Sequencing of rDNA is relatively straightforward because universal PCR primers are available. PCR primers used to amplify the ITS region, including the 5.8S rRNA gene, and the D1/D2 domains of the LSU rRNA gene are listed in Table 3.5. Other primer sequences have also been used for amplification, but those designated by White et al. [206] and O'Donnell [154] are the most commonly used. Primers ITS1, ITS4, NL1, and NL4 are useful to amplify rDNA of all fungi. Usually, a nested approach is followed, in which approximately 1.2–1.3 kb of the ITS and D1/D2 LSU is amplified using primers ITS1 (forward) and NL4 (reverse). Subsequently, a sequencing PCR is performed targeting either the ITS regions or the D1/D2 domains of the LSU rRNA gene, or both, using primers ITS1 and ITS4 or NL1 and NL4, respectively.

Guillot and Guého [90] were the first to investigate the usefulness of LSU rRNA sequences to address species diversity in *Malassezia*. This study clearly revealed the

Table 3.4 Taxonomic resolution of each subunit or spacer region in the ribosomal DNA unit

Subunit of spacer region	Class	Family	Genus	Species	Strain
SSU	—	—	—	—	
D1/D2	—	—	—	—	
ITS			·····	—	
IGS				—	—

Table 3.5 PCR primers used for D1/D2 LSU rRNA gene and ITS spacer sequence analysis

Subunit or spacer region	Names of primers	Sequences	Annealing temperature
D1/D2 LSU	NL1 (forward)	GCATATCAATAAGCGGAGGAAAAG	65.3 °C
	NL4 (reverse)	GGTCCGTGTTTCAAGACGG	65.5 °C
ITS region	ITS1 (forward)	TCCGTAGGTGAACCTGCGG	68.4 °C
	ITS4 (reverse)	TCCTCCGCTTATTGATATG	57.6 °C

presence of additional species of lipid-dependent *Malassezia* species, next to *M. furfur*, which were subsequently described as *M. globosa*, *M. obtusa*, *M. restricta*, and *M. slooffiae* [87]. Subsequent studies demonstrated the presence of additional species, such as *M. caprae*, *M. dermatis*, *M. equina*, *M. japonica*, *M. nana* and *M. yamatoensis* (see Chap. 2.1). The D1/D2 regions of the LSU rRNA gene, which is approximately 600 bp in length, is the most variable region in the LSU, and generally useful to identify *Malassezia* species [103]. The amplicon length is identical across all species, and the within-species sequence similarity is above 99% for these *Malassezia* species (Table 3.6).

The length of the ITS region differs between species, whereas its length among strains of a species is usually identical, and this can be used as an identification tool (see also RFLP and *t*FLP, above). Some *Malassezia* species, however, are an exception to this. For instance, the *M. globosa* ITS1 region ranges from 237 to 266 bp in length and the within-species DNA sequence similarity is 88% (Table 3.6). The *M. restricta* ITS1 region ranges from 210 to 261 bp in length showing 73% within-species similarity (Table 3.6). Of the 13 *Malassezia* species, *M. caprae*, *M. dermatis*, *M. japonica*, *M. obtusa*, and *M. yamatoensis* do not show infraspecific ITS diversity. The nucleotide order of the ITS regions turned out to be useful for the identification of *Malassezia* strains as well as to infer their phylogenetic relationships [129].

In general, most *Malassezia* species can be identified using the D1/D2 LSU rRNA gene sequence only. Caution is needed, however, as some *Malassezia* species, namely *M. sympodialis* and the phylogenetically closely related species *M. caprae*, *M. dermatis*, and *M. equina*, could not easily be identified following this approach (Table 3.7). *M. caprae* and *M. equina* were separated from *M. sympodialis sensu lato* based on phylogenetic analyses [30, 31], but they also differ phenotypically. D1/D2 LSU sequences of these four species show

Table 3.6 Summary of the ITS1, ITS2, and D1/D2 LSU rDNA sequences of *Malassezia* spp

Species	ITS1		ITS2		D1/D2 LSU
	Length (bp)	Similarity (%)	Length (bp)	Similarity (%)	Similarity (%)
Malassezia caprae	164	100	231	100	100
Malassezia dermatis	161	100	233	100	100
Malassezia equina	162–164	>99	232	100	100
Malassezia furfur	209–210	>95	370	100	>99
Malassezia globosa	237–266	>88	278–290	>91	>99
Malassezia japonica	208	100	337	100	100
Malassezia nana	182–186	>97	241–242	>98	>99
Malassezia obtusa	213	100	367	100	100
Malassezia pachydermatis	186–187	>94	328–344	>94	>99
Malassezia restricta	210–261	>73	276	>99	>99
Malassezia slooffiae	195–196	>99	313–316	>99	100
Malassezia sympodialis	162–163	100	234–235	100	100
Malassezia yamatoensis	176	100	298	100	100

Table 3.7 DNA sequence similarity of rDNA units among the related species *M. sympodialis*, *M. caprae*, *M. dermatis*, and *M. equina*

Subunit or spacer region	Species	*M. caprae*	*M. dermatis*	*M. equina*
D1/D2 LSU	*M. dermatis*	98	X	98
	M. equina	99	98	X
	M. sympodialis	99	99	99
ITS1	*M. dermatis*	93	X	85
	M. equina	82	85	X
	M. sympodialis	88	90	85
ITS2	*M. dermatis*	95	X	91
	M. equina	91	91	X
	M. sympodialis	91	91	88

more than 98–99% sequence similarity and they share 82–93% and 88–95% sequence similarity in the ITS1 and ITS2 regions, respectively (Table 3.7). Therefore, DNA sequencing of both the ITS regions and the D1/D2 domains of the LSU rRNA gene is recommended to identify isolates that may belong to *M. sympodialis* and related species. In practice, it may, therefore, be best to use both these rDNA regions to identify isolates of unknown identity.

The more variable IGS regions are less suitable as an identification tool of fungal species. Sequence analysis of the IGS regions has, however, been applied to assess species diversity within *Malassezia* and this turned out to be particularly useful to study strains genotypically as an epidemiological tool [189–192].

Recently pyrosequencing, yielding c. 30 bp fragments of the ITS1+2 regions and the 5.8S rDNA, has been used to identify *M. slooffiae*, *M. pachydermatis*, and *M. sympodialis* [140]. However, species of *Trichosporon* and some members of *Cryptococcus,* including members of the *C. neoformans* complex, and some related *Bullera* species could not be identified. *Malassezia* yeasts have also been identified with the MicroSeq D2 LSU ribosomal DNA sequence kit [110]. Both these latter methods need further evaluation before they can be implemented in routine diagnostics of *Malassezia* yeasts in the clinical laboratory.

3.1.5.2
Other Genes

Analysis of partial sequences of the mitochondrial Large subunit ribosomal RNA (mtLrRNA) that were generated with primers ML7 (5'-GACCCTATGCAGCTTCTACTG-3') and ML8 (5'-TTATCCCTAGCGTAACTTTTATC-3') discriminated between all seven species studied [209]. Two loci of the chitin-synthase 2 (*CHS2*) gene and the ITS 1 region have been used to genotype isolates of *M. pachydermatis* obtained from dogs and cats [2, 3, 38, 117]. In a sample of 185 dogs, three *CHS2* genotypes and eight ITS1 genotypes were distinguished. Two ITS1 types (namely A_{11} and B_{11}) occurred on all skin sites investigated, whereas one type (C_{12}) was associated with specific body sites [38]. Analysis of partial sequences of the RNA polymerase subunit 1 (*RPB1*) has been used to assess biodiversity within the *M. sympodialis* clade that also included *M. nana* [31]. The authors, however, were unable to easily amplify the *RPB1* gene in all species and, therefore, the amplification method of this locus needs further optimization.

3.1.6 Skin Community Analysis and Environmental Samples

One of the major challenges in the epidemiology of *Malassezia*-related diseases is the analysis of the (skin) microbial community structure in order to establish the role of *Malassezia* species in the various diseases, and to study the *Malassezia* community dynamics in time or as a consequence of treatment with antifungals. Multiple methods have been reported to identify *Malassezia* species from skin without culture of clinical material [71, 80, 185, 188, 190], but most of these methods require either separate amplification with specific primer sets or digestion of the amplification products (see RFLP above). Some real-time PCR protocols have been useful to study and quantify the presence of individual species (see PCR methods above). In microbial ecology, the study of community structure using the analysis of clone libraries obtained with taxon specific (e.g., panfungal) primers that target the rDNA locus has received considerable attention. This approach was applied to the analysis of psoriatic skin of five healthy humans and three patients with psoriasis [157]. The panfungal primers used targeted the SSU rDNA or were *Malassezia* specific based on the 5.8S rDNA and ITS 2 region (viz. Mal1F 5'-TCTTTGAACGCACCTTGC-3' and Mal1R 3'-CATGATACGTCATTTGCT-HT-5'). After cloning, the SSU rDNA amplicons were digested with HaeIII and the nucleotide order was determined [157]. The amplicons based on the Mal1F and Mal1R primers were also analyzed by digestion, in this case with DraI and XbaI, in order to specifically detect *M. globosa* and *M. restricta*. Amplicons that did not digest were sequenced to assess the species identity. Interestingly, this analysis yielded two unknown sequence types A and B (referred to as phylotypes A and B) that were the most prevalent sequences observed and that were closely related to, but not identical with *M. furfur*. On healthy skin, five *Malassezia* species were observed, with *M. restricta* dominating the samples from two subjects and *M. globosa* those from one other individual. *M. sympodialis, M. pachydermatis*, and *M. furfur* were observed as well. Healthy skin yielded four new phylotypes that may represent a yet undescribed species. Phylotype 1 occurred on all humans samples, whereas the other three phylotypes were found more rarely. In most of the healthy human beings, the *Malassezia* communities present on the right and left forearms were found to be similar. Three *Malassezia* species, namely *M. restricta, M. globosa*, and *M. sympodialis*, and an unknown phylotype occurred in psoriatic patients. Different samples obtained from one subject were similar, despite their site of isolation, from healthy or diseased skin, but those from the two other subjects showed differences in species composition. Similar to most other studies, and based on the number of clones detected, *M. restricta* and *M. globosa* were the most dominant species present in healthy and psoriatic skin [157].

In ecological studies, in which environmental samples were tested, DNA of *Malassezia* species has been detected in expected, but also in some unexpected habitats. Fourteen percent of 1,339 ITS clones made from indoor dust in Finland were related to various *Malassezia* species, and five out of 234 basidiomycetous sequences were related to *M. restricta* [164], thus indicating that *Malassezia* spp. may also occur abundantly in indoor environments. DNA of *M. globosa* was detected in low moisture soil from Antarctic dry valley [68], DNA of *M. restricta* occurred in the gut of beetles originating from Southern Louisiana [211], as well as in nematodes from German soils [175], and *Malassezia* DNA was detected in methane hydrate-bearing deep-sea sediments [124]. These observations conflict with the supposed and, so far, exclusive occurrence of *Malassezia* yeasts on warm-blooded animals. It may well be that these observations imply that *Malassezia* yeasts occur in yet unknown environmental habitats. An alternative explanation may be that during sampling or processing of the environmental samples contamination by *Malassezia* DNA originating from other sources has occurred. Until the presence of *Malassezia* DNA and cells has been demonstrated unequivocally and undisputedly in such habitats, e.g., by using detection techniques combining both molecular and morphological characteristics, such as fluorescent *in situ* hybridization (FISH), their presence in such habitats remains putative.

3.1.7 Other Methods

Fourier transform infrared spectroscopy (FT-IRS) has been applied to identify *Malassezia* isolates [116]. This rapid method separated *M. furfur* from the remaining species. Isolates obtained from PV lesions were identified as *M. globosa* (62%), and those from dandruff as *M. furfur* (60%). As this latter observation conflicts with other published data, e.g., those from Gemmer et al. [80], FT-IRS needs further evaluation before its clinical application in *Malassezia* diagnostics.

3.2 Human Epidemiology

Takashi Sugita and Aristea Velegraki

3.2.1 Introduction

Although *Malassezia* species colonize the skin of healthy individuals, they cause pityriasis versicolor (PV), seborrheic dermatitis (SD), and *Malassezia* folliculitis (MF), and can exacerbate atopic dermatitis (AD). *Malassezia* species are also thought to be involved in psoriasis and acne vulgaris, although the detailed role of the microorganisms is unknown. *Malassezia* requires lipids for its growth; thus, it colonizes the sebaceous areas of the skin such as those on the face, scalp, and upper trunk, which are colonized to a greater extent than the limbs. It has also been shown that *Malassezia* species are heterogeneous. Although *M. furfur* has long been thought to cause or exacerbate several skin diseases, Guého et al. showed in 1996 [87] that *M. furfur* actually consists of five distinct species: *M. furfur*, *M. globosa*, *M. obtusa*, *M. restricta*, and *M. slooffiae* (see Chap. 2.1 for details). Since then, several studies were conducted, which aimed at elucidating the diseases caused by specific *Malassezia* species.

Initially, most of these studies involved qualitative comparisons of the *Malassezia* microbiota between lesion and nonlesion sites in patients, or between patients and healthy individuals. More recently, quantitative analyses have been performed, as well as investigations into the relationships of the microbiota with aging or gender in healthy individuals, between children and adult patients, and among disease types. The *Malassezia* microbiota associated with different skin diseases are described in the following sections of this chapter. The apparent variability in the data may be explained by differences in race or region; it may also be the result of differences in sampling methods or culture media among the studies. Contact plates, swabs, and tape strips have been used to sample scale. Dixon media or Leeming and Notman agar (LNA) are used for *Malassezia* cultures. One member of the genus *Malassezia*, *M. restricta,* is often absent in cultures of clinical samples because its growth on plates is slower and therefore the species is easily overgrown by other isolated related species. All of the *Malassezia* species in a sample must be recovered from the medium to obtain an accurate quantitative analysis. Unfortunately, an optimized growth medium for all *Malassezia* species is not yet available. Culture-dependent methods can yield significant numbers of viable cells but have limited ability to produce accurate analytical results. Thus, methods that do not require an incubation period (i.e., molecular-based, culture-independent methods) have been developed to analyze the distribution of *Malassezia* microbiota.

In the molecular-based, culture-independent methods, sampling is performed using tape or a swab, and fungal DNA is subsequently extracted directly from the collected samples and analyzed by PCR. This method is not influenced by composition of the isolation medium, and thus slow-growing species, such as *M. restricta* and *M. obtusa*, may be detected. Even so, the results may be influenced by the DNA extraction method or the PCR detection sensitivity. In addition to *Malassezia*, fungi such as *Candida* and many bacteria colonize the skin. Therefore, high primer/probe specificity is needed to detect specific microbiota. In 2001, Sugita et al. introduced a DNA-based, culture-independent method for use in *Malassezia* research [188]. They used a transparent dressing to sample scale and nested PCR to increase the detection sensitivity for *Malassezia* DNA. In 2006, they also developed a quantitative real-time PCR method using TaqMan probes that allowed detection of as few as 10 copies of *Malassezia* DNA [193]. Retrospective studies also can be conducted using this procedure, as template DNA can be stored. This method is currently the most reliable and appropriate for analyzing *Malassezia* microbiota.

The number of accepted *Malassezia* species has doubled over the last decade, and many more species may exist. A comprehensive analysis using a clone library method is needed to identify all species of *Malassezia*. In this section, the epidemiology of each *Malassezia*-related skin disease and changes in the microbiota related to aging or gender in healthy individuals are described.

3.2.2 *Malassezia* in Healthy Individuals

The cutaneous *Malassezia* microbiota of an individual is formed immediately after birth. In a British study of 245 neonates (<28 days of age), skin swabs from 78 patients (31.8%) were positive for *Malassezia* on LNA, and 41 of 42 neonates were positive at follow-up [6]. In a second study examining 195 neonates in Iran [212], *M. furfur* and *M. globosa* were recovered from the medium in 60.5% and 7.2% of the patients, respectively; 31.3% of the cultures were negative for *Malassezia*. Recently, *M. dermatis* has been reported for the first time in 19 healthy individuals aged 17–55 years of age in Korea [125] by culture-dependent methods. The identity of *M. dermatis* isolates was corroborated by sequencing of the ITS1 and 26S rDNA. However, quantitative studies to associate *M. dermatis* colonization and exacerbation or causation of skin disorders are pending.

Malassezia colonizes the sebaceous areas of the skin on the face, scalp, and upper trunk; the extent of colonization on the limbs is comparatively lower. Sebaceous gland activity is minimal in children, begins to increase at mid- to late childhood in response to androgen secretion, continues to increase until 16–19 years of age, and then shows no further significant change until old age [166]. Thus, given the dependency of sebaceous gland activity on age and gender, the *Malassezia* microbiota of healthy individuals should be assessed at every age and in both genders. Several studies, including two large-scale studies, have examined the *Malassezia* microbiota of healthy individuals. In 2004, Gupta et al. examined 245 healthy Canadian subjects [99]. The microorganisms were sampled by pressing contact plates filled with LNA to the skin, and positive colonies were identified based on their physiological features and molecular characteristics. The subjects were classified into groups (I–VI, respectively) by age: 0–3, 4–14, 15–25, 26–40, 41–60, and >60-years old. The rate of positive cultures in children ≤14 years of age was only 36%, whereas that in subjects ≥15 years of age was approximately 90%. The number of colonies recovered from the medium increased markedly beginning at age 15 (Fig. 3.4). The distributions of *M. furfur*, *M. globosa*, and *M. sympodialis* differed significantly among the age groups. In group I, *M. globosa* and *M. furfur* were the predominant species, accounting for 90% of the colonies; no *M. sympodialis* was found in group I. In groups II–VI, *M. globosa* and *M. sympodialis* accounted for 90% of the colonies. They found no significant difference between the genders in any age group.

In the second large-scale study, 770 healthy Japanese individuals aged 0–82 years were examined using a molecular-based, culture-independent method [194]. In males, the total amount of *Malassezia* DNA remained essentially constant between 0 and 9 years of age and then increased markedly up to 16–18 years of age (Fig. 3.5). In females, the amount of *Malassezia* DNA increased until 10–12 years of age, then decreased until 19–22 years of age,

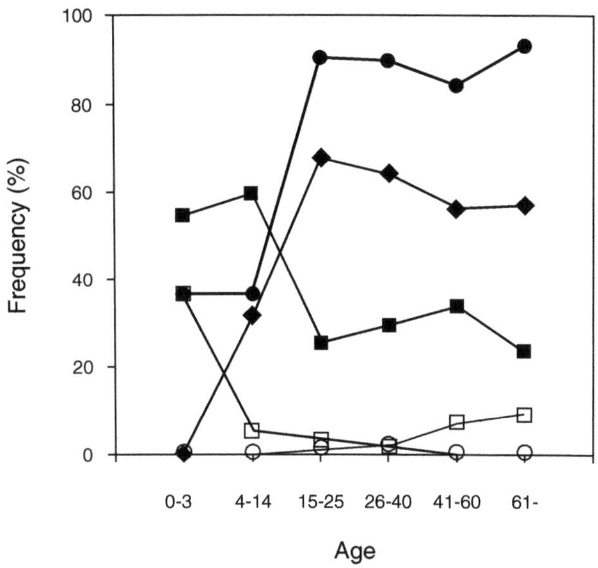

Fig. 3.4 Distribution of *Malassezia* species among individuals in different age groups as determined by culture-dependent analysis. The figure is based on original data from Gupta et al. [99]. Symbols indicate the number of individuals positive for *Malassezia* (closed circle), *M. sympodialis* (closed diamond), *M. globosa* (closed square), *M. furfur* (open square), and *M. restricta* (open circle)

and then increased again until 30–39 years of age before finally decreasing gradually with age beyond 39 years. Overall, males contained more *Malassezia* DNA than females. The ratio of total *Malassezia* colonization between males and females ranged from 0.7 to 3.4 until 15 years of age, after which it increased rapidly to 34 at 19–22 years of age (Fig. 3.5). *M. globosa* and *M. restricta* accounted for more than 70% of the total *Malassezia* DNA in both males and females at all ages. In males, *M. restricta* predominated at all ages and was overwhelmingly predominant after 16 years of age (Fig. 3.6). In females, *M. globosa* was dominant at 10–18 years of age, and *M. restricta* predominated over *M. globosa* after 23 years of age (Fig. 3.6).

Given that *Malassezia* requires lipids for growth, it is thought that changes in the *Malassezia* microbiota occur with changes in the composition of sebum, which consists of ceramides, fatty acids, cholesterol, cholesterol esters, squalene, triglycerides, and wax esters. Lipases secreted from cutaneous microorganisms hydrolyze the triglycerides into glycerin and free fatty acids, which serve as nutrients for the microorganisms. Yamamoto et al. [210] investigated the relationship between aging and the composition of saturated and unsaturated fatty acids (C_{14}–C_{18}), including straight and branched chains, in sebum. The proportion of C_{16} iso-branched fatty acids decreased markedly from infancy until the age of twenty and then increased slightly until senescence. The $C_{18:1}$ straight-chain component decreased during the twentieth year of age and did not change significantly thereafter. The percentages of $C_{15:1}$ and $C_{16:1}$ straight-chain components increased with age from infancy to the age of twenty and gradually decreased until the age of fifty. In contrast, the proportion of $C_{16:1}$ iso-branched fatty acids decreased from infancy until maturity, with its lowest point occurring at the age of twenty, and gradually increased until the age of fifty. Overall, the proportion of each fatty acid changed significantly at the age of twenty, as did the amount of sebum. Thus, age-related changes in the extent of colonization and the composition of the *Malassezia* microbiota are closely associated with age-related changes in sebaceous gland activity and the fatty acid composition of sebum. In addition, cosmetics may have an effect on the microbiota, as preservatives included in the cosmetics can inhibit the growth of *Malassezia*.

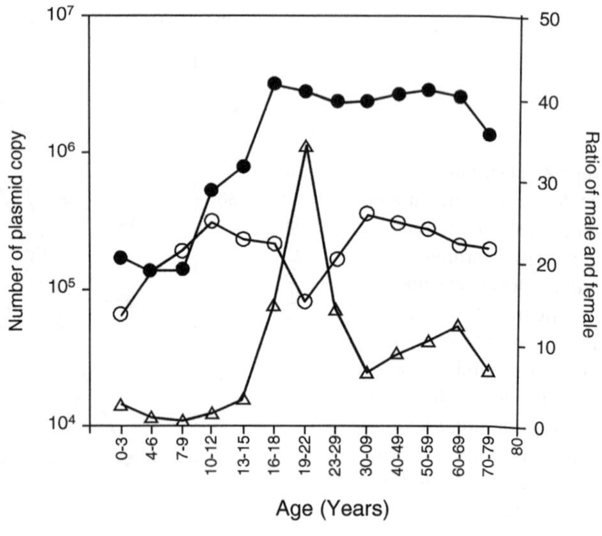

Fig. 3.5 Change in the amount of *Malassezia* DNA present in humans of different age. Males (closed circle), females (open circle), ratio of the amount of *Malassezia* DNA in males vs. females (open triangle)

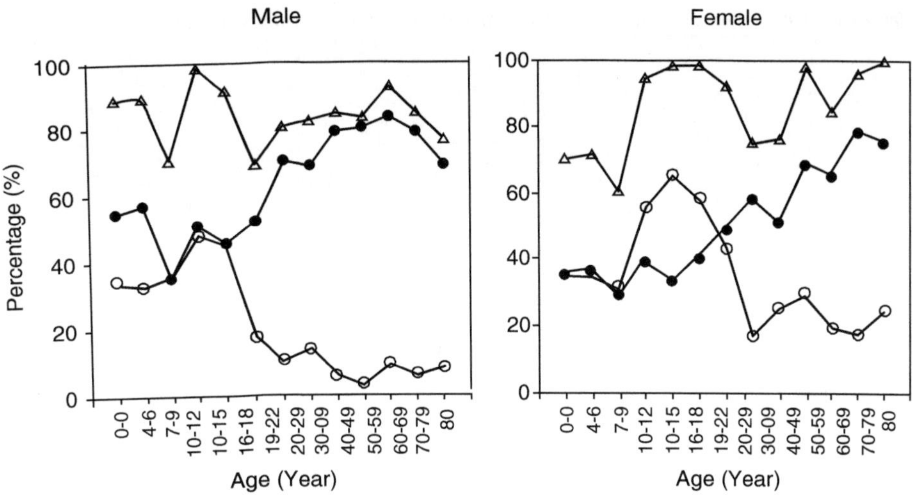

Fig. 3.6 Proportions of the predominant species *Malassezia globosa* and *M. restricta* in males and females, respectively. *M. globosa* (closed circle), *M. restricta* (open circle), *M. globosa* plus *M. restricta* (triangle)

In summary, the *Malassezia* microbiota of healthy individuals is gender-dependent and changes with age. Nevertheless, *M. globosa* and *M. restricta* predominate in both genders and in all age groups.

3.2.3
Malassezia-Related Skin Diseases

3.2.3.1
Pityriasis Versicolor

Pityriasis versicolor (PV) is a superficial infection of the stratum corneum caused by *Malassezia*. It is most commonly found on the trunk and upper arms of people in their twenties and thirties, occurring often during the summer but not during the winter. Direct microscopic examinations using methylene blue stain have revealed hyphae and yeast cells among the scales collected from the lesions of PV patients; however, hyphae are rare in other *Malassezia*-related diseases.

Studies of the *Malassezia* microbiota in PV have been conducted in several counties; six representative studies are summarized in Table 3.8. *M. globosa* was isolated at a rate of 60% by culture-dependent analysis in a Spanish study (Southwest Europe) [53] (Table 3.8). A Greek study (Southeast Europe) also suggested that *M. globosa* was a major *Malassezia* species in PV patients [73]; *M. globosa* was identified alone or in combination with other *Malassezia* species in 77 or 12.7% of the *Malassezia*-positive then patients, respectively. In a Canadian study, *M. sympodialis* (68.8%) was the predominant species, followed by *M. furfur* (12.3%) and *M. globosa* (6.3%) [101]. Also, in studies conducted in Iran (southwest

Table 3.8 Distribution of *Malassezia* species in patients with pityriasis versicolor

Analytical methods	Culture-dependent methods					Culture-independent method
Countries References	Spain [53]	Canada [101]	Iran [198]	Greece [73]	Bosnia and H* [171]	Japan [141]
Number of patients	96	23	94	71	90	49
M. globosa[a]	60[b]	6[b]	53[b]	77[b]	63[b]	94[b]
M. restricta	NI	NI	NI	2[c]	NI	94[b]
M. sympodialis	33[c]	69[b]	9[b]	13[c]	14[b]	35[b]
M. dermatis	NI	NI	NI	NI	NI	24[b]
M. furfur	NI	12[b]	25[b]	5[c]	10[b]	10[b]
M. obtusa	NI	6[b]	8[b]	NI	8[b]	8[b]
M. japonica	NI	NI	NI	NI	NI	6[b]
M. slooffiae	7[c]	6[b]	4[b]	3[c]	5[b]	4[b]
M. yamatoensis	NI	NI	NI	NI	NI	4[b]
Medium used for isolation	mDA	LNA	DA	mDA	mDA	–

DA Dixon agar; *mDA* modified Dixon agar; *LNA* Leeming and Notman agar; *NI* not identified; *H** Herzegovina

[a]Number indicates the percentage (%) of the total number of patients with PV

[b]Samples collected from lesional sites

[c]Percentage refers to co-isolation with *M. globosa*

Asia) and in Bosnia and Herzegovina (the Balkan peninsula) *M. globosa* was the predominant species isolated from lesional sites [171, 198, 213] followed by *M. sympodialis* and *M. furfur*. Little or no *M. restricta* (Table 3.8) was isolated from patients in these studies using culture-dependent analysis. However, *M. globosa* and *M. restricta* were detected in almost all PV patients by nested PCR using species-specific primers in a Japanese study [141]. Real-time PCR revealed that *M. globosa* DNA accounted for $62.8 \pm 18.1\%$ of all *Malassezia* DNA (Fig. 3.2), and the extent of colonization of *M. globosa* was six times that of *M. restricta* [184]. The detection frequency of *M. sympodialis* was 34.6%, and that of the other species ranged from 4.1 to 24.4%. Thus, DNA-based analyses suggest that *M. globosa* is qualitatively and quantitatively the predominant species in the scale of PV patients.

Hyphae are frequently observed in the scale of PV patients; therefore, the presence of hyphae is important in elucidating the causative agent of PV. Following treatment with antifungal agents, yeast cells were detected in the scale of PV patients, whereas hyphae were not observed, in accordance with the reported improvement of the patients' symptoms. Only *M. globosa* DNA was detected by PCR when scale samples containing only hyphae were harvested [141]. Furthermore, no report has shown that *M. restricta* develops hyphae in vitro. Taken together, these data suggest that *M. globosa* plays a significant role in the pathogenesis of PV.

3.2.3.2 Seborrheic Dermatitis

Seborrheic dermatitis (SD) is a common inflammatory skin disease associated with seborrhea. It occurs most often on areas of the body that are rich in sebaceous glands, such as the face, scalp, and upper trunk. The disorder is present in 1–3% of the general population and is more prevalent in men than in women; it is frequently observed in patients with acquired immunodeficiency syndrome (AIDS), with an incidence of 30–33% [8, 98]. SD is also observed in patients with Parkinson's disease. In addition, SD is among the new secondary effects appearing recently with the use of antitumor agents, such as cetuximab and erlotinib, an epidermal growth factor (EGF) receptor inhibitor [176]. The condition often has a seasonal aspect, being more common and, in chronic cases, more severe in the winter months. The cause of SD has not been clarified; however, several possible contributors to the development of the disorder, including exogenous and endogenous factors, have been described. Of these, *Malassezia* has generated a large body of data, indicating that it plays a major etiological role in SD. The clinical condition is improved in accordance with a reduction in the number of *Malassezia* cells in the lesions following antifungal treatment with itraconazole or ketoconazole, suggesting that *Malassezia* is strongly associated with the development of SD [102]. The free fatty acids produced by the hydrolysis of sebum by lipases serve as nutrients for cutaneous microorganisms and also induce inflammation of the skin.

The distribution of *Malassezia* in patients with SD from various countries is summarized in Table 3.9. A culture-dependent analysis from Canada showed that *M. globosa* was the predominant species (45.0%) in lesions, followed by *M. sympodialis* (30.8%) and *M. slooffiae* (10%) [101]. Moreover, the trunk was found to be the most highly colonized site on the body, and the overall recovery rate was 82.1%. However, the recovery rate in studies from Sweden and Japan was much lower. *M. globosa*, *M. furfur*, and *M. sympodialis* were recovered in 20.8, 20.8, and 6.3% of the samples, respectively, in a Japanese study [145], and *M. sympodialis*, *M. obtusa*, and *M. slooffiae* were recovered in 25, 25, and 12.5% of the samples, respectively, in a Swedish study [180]. A common feature among these three studies is that no *M. restricta* was recovered from the samples. In a Greek study of SD patients, *M. globosa* was isolated alone (33.3%), in combination with either *M. sympodialis* or *M. restricta* (13.3%), and together with both *M. sympodialis* and *M. restricta* (1 patient 2.2%). It was concluded in that study that *M. globosa* was the prevalent species in SD patients and occurred either alone or in combination with other *Malassezia* species [73]. However, by culture-independent analysis, a Japanese study showed that *M. globosa* and *M. restricta* were the predominant species at both lesion and nonlesion sites in SD, accounting for 5.2±7.9% and 72.3±15.0%, respectively, of the *Malassezia* DNA detected by real-time PCR [195] (Fig. 3.2). *Malassezia sympodialis*, *M. dermatis*, and *M. slooffiae* were present at rates of 25.8–35.5% at lesion sites and at rates of 14.8–22.2% at nonlesion sites; other species, including *M. furfur*, *M. obtusa*, and *M. yamatoensis*, were detected at rates below 10% at lesion and nonlesion sites.

As for colonization by *Malassezia* cells, the results varied as well. One study indicated a correlation between the density of *Malassezia* cells and the clinical severity of SD [111], yet another study indicated that the number of *Malassezia* colonies at lesion sites was significantly lower than that at nonlesion sites [101]. A DNA-based study revealed that the *Malassezia* microbiota at lesion sites was more diverse than at nonlesion sites [195], that

Table 3.9 Distribution of *Malassezia* species in patients with seborrhoeic dermatitis

Analytical methods	Culture-dependent methods				Culture-independent methods			
Countries References	Canada [101]	Japan [145]	Sweden [180]	Greece [73]	Japan [195]		USA [80]	
Number of patients	28	48	16	38	31	46	24	
M. globosa[a]	45[b]	21[b]	6[b]	58[b,c]	93[b]	70[d]	46[e]	33[f]
M. restricta	38	NI	NI	48[b,c]	61[b]	56[d]	72[e]	50[f]
M. sympodialis	37[b]	6[b]	25[b]	8[b,c]	26[b]	15[d]	7[e]	8[f]
M. dermatis	8[b]	NI	NI	NI	35[b]	19[d]	NI	NI
M. furfur	NI	21[b]	6[b]	2[b,c]	6[b]	NI	NI	NI
M. obtusa	NI	NI	25[b]	NI	10[b]	NI	2[e]	NI
M. japonica	NI	NI	13[b]	NI	13[b]	4[d]	NI	NI
M. slooffiae	10[b]	NI	12[b]	5[b,c]	32[b]	22[d]	4[e]	NI
M. yamatoensis	NI	NI	NI	NI	10[b]	4[d]	4[e]	NI
Medium used for isolation	LNA	DA	LNA	mDA	–	–	–	–

DA Dixon agar; *mDA* modified Dixon agar; *LNA* Leeming and Notman agar; *NI* not identified
[a]Number indicates the percentage for total number of patients
[b]Samples collected from lesional sites
[c]Co-isolation of mixed *Malassezia* species from lesions
[d]Samples collected from nonlesional sites
[e]Samples collected from patients at >24 of adherent scalp flaking scale (ASFS)
[f]Samples collected from patients at <10 of ASFS

the detection frequency of each species in lesion skin was higher than that in nonlesion skin, that the average numbers of species per individual in lesion and nonlesion skin were 2.8 ± 1.6 and 1.9 ± 1.2, respectively, and that the overall colonization by *Malassezia* was higher in lesion skin than in nonlesion skin.

Dandruff is an exceptionally common scalp condition characterized by flaking and scaling. Dandruff can be thought of as either a mild variant of SD or as a separate condition that is capable of worsening into true SD. The key distinction between dandruff and SD is that there is no visible inflammation associated with dandruff. Gemmer et al. [80] first attempted to detect *Malassezia* from dandruff samples using a DNA-based approach (see Sect. 3.1.4.4). The subjects were separated into two groups (<10 and >24) based on scoring according to an adherent scalp flaking scale (ASFS). Both *M. globosa* and *M. restricta* were predominant in the dandruff from each group. Those subjects with a high composite ASFS score were more likely to be positive for *Malassezia* (*M. restricta*, 71.7% for a high composite score vs. 50.0% for a low composite score; *M. globosa*, 45.7% for a high composite score vs. 33.3% for a low composite score). *M. sympodialis*, *M. slooffiae*, *M. obtusa*, and *M. furfur* were detected in a very small percentage of the subjects from each group. A significant percentage of subjects in each group (28.3% with ASFS score <10; 28% with

ASFS score >24) were positive for non-*Malassezia* species. To date, no quantitative analysis of dandruff has been reported.

3.2.3.3
Atopic Dermatitis

Atopic dermatitis (AD), which is defined as dermatitis with pruritus, is a chronic disease involving cycles of remission and deterioration. AD is characterized by hypersensitivity to dry skin, which may involve IgE antibodies and various environmental factors. As the barrier function of the cutaneous surface is low in AD patients, their skin is easily stimulated by normal cutaneous microorganisms. A relationship between AD and cutaneous *Malassezia* has been suggested based on the fact that anti-*Malassezia* IgE antibodies have been detected in patients with AD but not in healthy subjects. In particular, in head and neck AD, there is a significant correlation between the level of anti-*Malassezia* IgE and clinical severity [6]. In addition, the symptoms of AD patients may be improved by antifungal agents such as ketoconazole and itraconazole, according to a reduction in the number of *Malassezia* colonies in their lesions [55, 151]. Currently, *Malassezia* is considered to be an exacerbating factor in AD.

A study conducted by Nakabayashi et al. [145] detected *M. furfur*, *M. globosa*, *M. sympodialis*, and *M. slooffiae* in 21.4, 14.3, 7.1, and 3.6% of samples from Japanese AD patients, respectively (Table 3.10). A study performed in Sweden [180] produced similar data, while a Canadian study by Gupta et al. [101] showed that *M. sympodialis* was the predominant species in patients with AD (detection rate, 51.3%). Common among the three culture-dependent studies is that *M. restricta* was rarely or never isolated from the medium (0–7.7%). With the detection of *Malassezia* by a molecular-based, culture-independent method, which may be considered to be superior for the detection of *Malassezia* spp. by traditional methods, Tajima et al. [195] found both *M. globosa* and *M. restricta* to be the predominant species in AD, while the detection rate of *M. sympodialis* was 58.3% and the detection rates for other species ranged from 13.9 to 33.3%. The serum levels of IgE antibodies against the two major species were also found to be higher than those against other species [118].

The onset of AD usually occurs during the first 6 months of life, and the prevalence of AD, which is similar in the United States, Europe, and Japan, is increasing. AD is classified into three sequential phases: infantile, childhood, and adult. Although the incidence of AD decreases with age, the severity of AD tends to be greater in adults. The species of the *Malassezia* microbiota were similar between children with AD (9.7±1.5-years old) and adults with AD (36.1±9.5-years old). In addition, *M. globosa* and *M. restricta* were detected in 100% of child and adult AD patients, whereas the other species were detected in no more than 40% of the cases. Totals of 3.2±1.3 and 2.8±1.1 species were detected in samples from children and adults with AD, respectively [197]. Quantitatively, the *Malassezia* microbiota differed significantly between children and adults with AD. Colonization by *Malassezia* in adults was 2.4-fold of that in children. The percentages of *M. globosa* and *M. restricta* were similar in adults, while *M. restricta* was the predominant species in children. The serum level of *Malassezia*-specific IgE antibodies was lower in

Table 3.10 Distribution of *Malassezia* species in patients with atopic dermatitis

Analytical methods	Culture-dependent methods			Culture-independent methods
Countries References	Japan [145]	Canada [101]	Sweden [180]	Japan [195]
Number of patients	28	31	124	31
M. globosa[a]	14[b]	18[b,c]	11[b]	100[b]
M. restricta	NI	8[b,c]	3[b]	97[b]
M. sympodialis	7[b]	51[b,c]	NI	58[b]
M. dermatis	NI	NI	NI	31[b]
M. furfur	21[b]	10[b,c]	NI	33[b].
M. obtusa	NI	10[b,c]	5[b]	28[b]
M. japonica	NI	NI	NI	33[b]
M. slooffiae	4[b]	3[b,c]	3[b]	31[b]
M. yamatoensis	NI	NI	NI	14[b]
Medium used for isolation	DA	LNA	LNA	–

DA Dixon agar; *mDA* modified Dixon agar; *LNA* Leeming and Notman agar; *NI* not identified
[a]Number indicates the percentage (%) for total number of patients
[b]Samples collected from lesional sites
[c]Samples collected from nonlesional sites

children than in adults. Antifungal agents have been given as an alternative therapy to AD patients; however, clinical improvement of the patients' symptoms has been observed only in adults. It appears that antifungal therapy (e.g., ketoconazole or itraconazole) may be effective in adult cases of AD affecting the head and neck [55].

3.2.3.4 *Malassezia* Folliculitis

Malassezia folliculitis (MF) involves pruritic papules and pustules that occur mainly on the trunk and upper arms and may also be seen in association with pregnancy, leukemia, and Hodgkin's disease in immunocompromised patients [6]. The development of MF may be related to temperature and humidity, as MF appears to be more common in tropical countries.

Follicular occlusion is a primary event in the development of folliculitis, with *Malassezia* overgrowth as a secondary occurrence. Propionibacteria and staphylococci are found distributed within affected follicles, although few studies of MF have attempted to culture the follicular contents. A recent culture-dependent study of 32 Japanese patients demonstrated that the predominant species recovered from folliculitic lesions were *M. globosa* and *M. sympodialis*, while *M. restricta*, *M. globosa*, and *M. sympodialis* were isolated using culture-independent methods [4].

3.2.3.5 Psoriasis

Psoriasis (PS) is a multifactorial immune skin disease whose etiology involves microbial environmental factors in combination with a strong genetic component, including several genes encoding proteins involved in epidermal differentiation and immune, inflammatory, and pathogen responses. Various microorganisms are considered to be associated with the provocation and exacerbation of PS; these include *Streptococcus pyogenes*, *Staphylococcus aureus*, *Malassezia* spp., and *Candida albicans* [70]. The association between PS and *S. pyogenes* was first noted 50 years ago. However, our knowledge of *Malassezia* and PS is not as extensive as our knowledge of PV, SD, and AD. *Malassezia* upregulates keratinocyte expression of transforming growth factor-beta1 and heat shock protein 70, which are associated with hyperproliferation and cell migration in the epidermis and which are more highly expressed in psoriatic skin colonized with *Malassezia* than in noncolonized psoriatic skin [70]. Antifungal treatment has been shown to improve psoriatic scalp lesions [67]. In a Canadian study using culture-dependent methods, *M. globosa* (57.7%) and *M. sympodialis* (30.8%) were found to be the predominant species in the lesions of patients with PS (Table 3.11) [101]. In 40 psoriatic patients with scalp involvement studied in Bosnia and Herzegovina [170] on the Balkan Peninsula, *M. globosa* was also the predominant (55%) lesional isolate, followed by *M. slooffiae* (17.5%) and *M. restricta* (10%) but, surprisingly, *M. sympodialis* was absent. In that culture-based study, *M. restricta* was the predominant isolate form healthy scalp specimens of 40 clinically healthy individuals and significant difference in the distribution of *Malassezia* species was recorded between psoriatic and healthy scalp. An Iranian study of 110 PS patients and 123 healthy individuals reported that the frequency of positive cultures among the PS patients was lower than that among the healthy controls (62.7 vs. 87.0%, respectively) [213]. *Malassezia globosa*, *M. furfur*, and *M. restricta* were predominantly recovered in psoriatic scalp cultures (Table 3.11).

Using a PCR-based, culture-independent method, Amaya et al. [5] showed that *M. globosa* (68.2%) and *M. restricta* (90.9%) were the major components, followed by *M. sympodialis* (50.0%), in the lesions of 22 PS patients (Table 3.11). Colonization by *M. restricta* was five times the colonization by *M. globosa* at all body sites. In addition, *Malassezia* colonization was significantly lower in patients with hyperlipidemia than in patients with normolipidemia [196]. No correlation has been demonstrated between the psoriasis area and severity index (PASI) and the amount of colonization by *Malassezia*.

A comprehensive analysis of PS patients using a clone library was performed by Paulino et al. [157], who analyzed 25 skin samples from five healthy subjects and three PS patients using broad-range 18S rDNA and 5.8S rDNA/ITS2 *Malassezia*-specific primers. An analysis of 1,374 clones identified five species (*M. globosa*, *M. restricta*, *M. sympodialis*, *M. furfur*, and *M. pachydermatis*) and four unknown phylotypes, potentially representing new species. The *Malassezia* microbiota did not change significantly over 6 months. Samples of the *Malassezia* microbiota from healthy skin and psoriatic lesions were similar in one patient but substantially different in two others. No consistent pattern differentiating psoriatic skin from healthy skin was found.

Table 3.11 Distribution of *Malassezia* species in patients with psoriasis

Analytical methods	Culture-dependent methods			Culture-independent methods	
Countries References	Canada [101]	Bosnia and H* [170]	Iran [213]	Japan [5]	Japan [196]
Number of patients	28	40*	110	22	20
M. globosa[a]	58[b]	55[b]	47[c,d]	68[b]	98[b]
M. restricta	NI	10[b]	11[d]	91[b]	92[b]
M. sympodialis	31[b]	2[b]	3[d]	50[b]	NI
M. dermatis	NI	NI	NI	14[b]	NI
M. furfur	NI	8[b]	39[d]	18[b]	NI
M. obtusa	NI	NI	NI	18[b]	NI
M. japonica	NI	NI	NI	13[b]	NI
M. slooffiae	11[b]	17[b]	NI	14[b]	NI
M. yamatoensis	NI	NI	NI	10[b]	NI
Medium used for isolation[e]	LNA	mDA	LNA	–	–

mDA modified Dixon agar; *LNA* Leeming and Notman agar; *NI* not identified; *H** Herzegovina
*Three out of forty patients with negative cultures
[a]Number indicates the percentage (%) for total number of patients
[b]Samples collected from lesional site
[c]Samples collected from scalp
[d]Percentage refers to co-isolation with other *Malassezia* species

3.2.3.6
Genotype analysis of the two predominant species, *M. globosa* and *M. restricta*

In the rRNA gene, the IGS region, which is located between the small and large subunits, shows a remarkable diversity in fungi. The IGS region is further divided into two subregions, IGS1 and IGS2. A phylogenetic tree constructed from the IGS1 sequences of *M. globosa* consisted of eight clusters (Fig. 3.7), each of which corresponded to a different subject or patient group (i.e., healthy subjects, patients with AD, and patients with SD) [190, 195]. Cluster VII corresponded to the sequences derived from healthy subjects only, whereas clusters I and VIII corresponded to sequences from both AD/SD patients and healthy subjects. The remaining five clusters were specific to patients. *M. globosa* IGS1 has four short sequence repeats (SSRs): one $(GT)_n$ and three $(CT)_n$. The number of (CT) repeats in the $(CT)_n$ SSRs was more variable in the healthy subjects than in the AD patients, who exhibited 4–11 (CT) repeats at CT1; 3–10, at CT2; and 3–11, at CT3. This corresponds to four (CT) repeats in 50% of the samples from AD patients, eight repeats in 60%, and 9–11 repeats in 80%. For the $(GT)_n$ SSR of IGS1, the numbers of (GT) repeats in 70–80% of the samples from AD patients and healthy subjects were 9–11 and 15–19, respectively. For the *M. restricta* IGS1 sequences, the tree consisted of two major clusters: cluster I corresponding to patients and healthy subjects, and cluster II corresponding to patients only. Genotypic

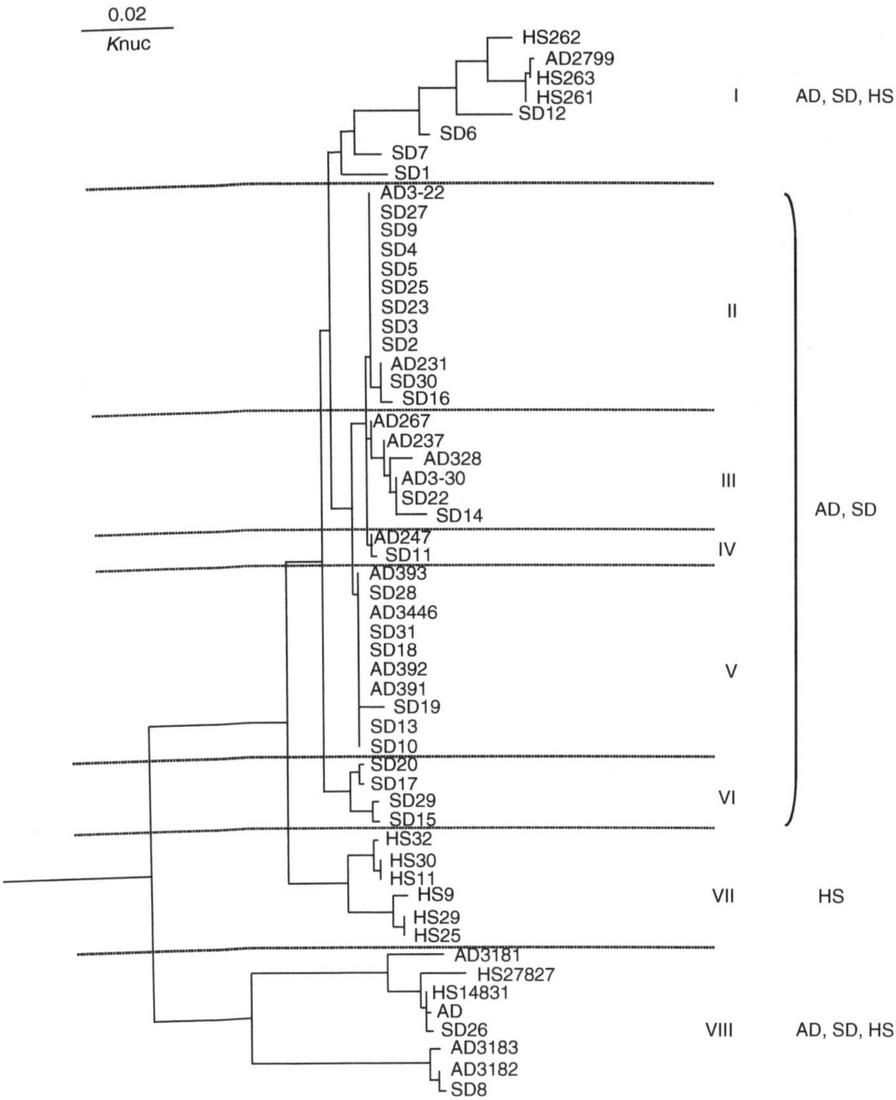

Fig. 3.7 Phylogenetic tree of *M. globosa* based on its IGS sequence. AD atopic dermatis; SD seborrheic dermatitis; HS healthy subject.

analysis of the IGS1 sequences could not separate AD patient-derived *M. globosa*/*M. restricta* from SD patient-derived *M. globosa*/*M. restricta*. Gaitanis et al. [73] reported that sequences of *M. globosa* isolated from PV patients were divided into five groups by single-strand conformational polymorphism analysis of the ITS1 region. One genotype of *M. globosa* appeared to be associated with extensive clinical disease.

3.3 Animal epidemiology

Jacques Guillot, Suzana Hađina and F. Javier Cabañes

3.3.1 Introduction

Malassezia yeasts belong to normal cutaneous or mucosal microbiota of many warm-blooded vertebrates. These yeasts are now recognized as opportunistic pathogens that play a significant role in the development of different animal diseases such as otitis externa (OE) or seborrhoeic dermatitis (SD).

Malassezia species have an affinity for lipids as substrates and the term "lipophilic yeasts" is frequently used to characterize the genus. In fact, most of the species show an absolute requirement for long fatty acid chains and they are, therefore, rarely isolated in the laboratory unless specific nutrients are provided in the medium. In 2006, lipid-dependent yeasts included ten species: *M. sympodialis, M. globosa, M. restricta, M. furfur, M. obtusa, M. slooffiae, M. dermatis, M. japonica, M. yamatoensis*, and *M. nana.* Of these, three species (*M. dermatis, M. japonica*, and *M. yamatoensis*) have so far been isolated exclusively from human skin. Recently, two new species that inhabit animals have been described: *M. equina* from healthy skin of the anus from horses, and *M. caprae* from the ear skin of healthy goat [31]. *M. pachydermatis* is the only non lipid-dependent species; it grows on Sabouraud glucose agar without any special additional requirements.

Epidemiology is traditionally defined as the study of the occurrence, distribution, and control of diseases in populations. In the present chapter, the occurrence and distribution of *Malassezia* yeasts among animal hosts will be described. The second part of this chapter is dedicated to the description of predisposing factors leading to *Malassezia* overgrowth on the skin of animals.

3.3.2 Occurrence and Distribution of *Malassezia* Yeasts in Animals

Malassezia yeasts have been isolated from almost all domestic animals, different wild animals held in captivity, but also from wildlife. Tables 3.12 and 3.13 show the animal species from which *Malassezia* yeasts have been isolated.

3.3.2.1 Animals in Zoological Parks or in Wildlife

A few epidemiological studies aimed at detecting the presence of *Malassezia* yeasts from the skin of different animal species. In 1925, Weidman was the first to observe *Malassezia* yeasts from an animal [204]. He described bottle-shaped yeast-like cells in scales from an Indian rhinoceros (*Rhinoceros unicornis*) with exfoliative dermatitis (Fig. 3.8). This microorganism

Table 3.12 *Malassezia* yeasts isolated from wild animals

Animals	Non lipid dependent species (*M. pachydermatis*) isolated from					Lipid-dependent yeasts[a] isolated from					References
	A	B	C	D	E	A	B	C	D	E	
Bear (different species)	+	+	–	–	–	–	–	–	-	–	[178]
	+	–	–	+	–	–	–	–	–	–	[107]
	+	–	–	–	–	–	–	–	–	–	[91, 120, 123]
Wolf (*Canis lupus*)	+	–	–	–	–	–	–	–	–	–	[107]
Coyote (*Canis latrans*)	–	+	–	–	–	–	–	–	–	–	[179]
Fox (*Vulpes vulpes*)	–	+	–	–	–	–	–	–	–	–	[179]
	+	–	–	+	–	–	–	–	–	–	[107]
	+	–	–	–	–	–	–	–	–	–	[54]
Fennec fox (*Vulpes zerda*)	+	–	–	–	–	–	–	–	–	–	[91]
Ferret (*Mustela putorius*)	+	–	–	–	–	–	–	–	–	–	[58, 91]
Large felids (different species)	–	–	–	–	–	–	2	–	–	–	[91]
	–	–	–	–	–	2	–	–	–	–	[47]
Small felids (different species)	+	–	–	–	–	–	–	–	–	–	[47]
Gray seal (*Halichoerus gripus*)	+	–	–	–	–	–	–	–	–	–	[91]
Sea lion (*Zalophus californianus*)	+	+	+	–	–	–	–	–	–	–	[94]
(*Otaria byronia*)	–	+	–	–	–	–	–	–	–	–	[146]
Porcupine (*Erethizon dorsatum*)	–	+	–	–	–	–	–	–	–	–	[179]
Badger (*Meles meles*)	+	–	–	–	–	–	–	–	–	–	[123]
Bat (*Molossus molossus*)	+	–	–	–	–	2,3,4	–	–	–	–	[77]
Big anteater (*M. tridactyla*)	+	–	–	–	–	–	–	–	–	–	[123]
Doe (*Capreolus capreolus*)	–	+	–	–	–	–	–	–	–	–	[106]
Camel (*Camelus dromedarius)*	–	–	–	–	–	–	–	–	–	–	[122]
Okapia (*Okapia johnstoni*)	–	–	–	–	–	1	–	–	–	–	[91]
Rhinoceros (*Rh. unicornis*)	+	–	–	–	–	–	–	–	–	–	[123]
(*C. simum simum)*	–	+	–	–	–	–	–	–	–	–	[9, 204, 205]
(*D. s. bicornis michaeli*)	+	+	–	–	–	1	–	–	–	–	[91]
Elephant (*Elephas maximus)*	+	+	–	–	–	1	1	–	–	–	[91]
	+	–	–	–	–	–	–	–	–	–	[123]
Wombat (*Vombatus ursinus*)	+	–	–	–	–	–	–	–	–	–	[123]
	–	+	–	–	–	–	–	–	–	–	[91]
Non human primates (different species)	–	–	–	–	–	–	1	–	–	–	[137]
	+	+	–	–	–	–	1	–	–	–	[91]
Birds (different species)	–	+	–	–	–	–	–	–	–	–	[66]
	–	+	–	–	–	–	1	–	–	–	[89, 137, 177]
	–	–	+	–	–	–	–	–	–	–	[29]

Localization of sampling is indicated with a capital letter: *A* external ear canal; *B* body skin (lips, axilla, groin, dorsum, pinna, precrural area, ventral midline, udder), mucosa, hair, nails, fur, feathers, and comb; *C* mouth; *D* perianal area, anus; *E* eye, conjunctival sac

[a]Key for lipid-dependent yeasts: *1* not identified or not specified; *2 M. sympodialis*; *3 M. globosa*; *4 M. furfur*

Table 3.13 *Malassezia* yeasts isolated from domestic animals

Animals	Non lipid-dependent yeasts (*M. pachydermatis*)					Lipid-dependent yeasts[a] isolated from					References
	A	B	C	D	E	A	B	C	D	E	
Dog (*Canis familiaris*)	+	–	–	–	–	–	–	–	–	–	[33, 81, 104, 121, 128, 133, 152]
	+	+	–	–	–	–	–	–	–	–	[88, 91, 95, 120, 132, 174, 182]
	–	+	–	–	–	–	–	–	–	–	[20, 21, 24, 27, 28, 36, 66, 119, 154]
	+	+	–	+	–	–	–	–	–	–	[17]
	+	+	+	+	–	–	–	–	–	–	[22, 108]
	+	+	–	+	–	1	1	–	1	–	[34]
	+	+	+	+	+	–	–	–	–	–	[127, 148]
	–	+	+	+	–	–	–	–	–	–	[18]
	+	+	–	–	+	–	–	–	–	–	[168]
	+	–	–	+	–	–	–	–	–	–	[40]
	–	–	–	–	–	4	–	–	–	–	[63]
	+	–	–	–	–	4, 6	–	–	–	–	[50]
	+	–	+	+	–	–	–	–	–	–	[162]
	–	+	+	–	–	–	–	–	–	–	[19]
	–	–	–	–	+	–	–	–	–		[167]
Cat (*Felis catus*)	+	+	–	–	–	–	–	–	–	–	[10, 40, 88, 150, 174, 182]
	+	+	–	+	–	2	2, 3	–	–	–	[25]
	+	–	–	–	–	2	–	–	–	–	[59]
	+	–	–	–	–	2, 3, 4	–	–	–	–	[147]
	+	–	–	–	–	3	–	–	–	–	[33]
	–	+	–	–	–	–	–	–	–	–	[66]
	–	–	–	–	–	2	–	–	–	–	[49]
	–	–	–	–	–	8	–	–	–	–	[112, 113]
	+	–	–	–	–	4	–	–	–	–	[48]
	–	+	–	–	–	–	2, 4	–	–	–	[45]
	–	–	–	+	–	–	–	–	–	–	[109]
	+	+	–	+	–	–	5	–	–	–	[1]
	+	+	+	+	–	8	5, 8	–	5	–	[159]
	–	–	–	+	–	2	2	–	2	–	[23]
	–	+	–	–	–	–	1	–	–	–	[155]
Swine (*Sus scrofa*)	–	–	–	–	–	1	–	–	–	–	[106, 122]
	–	–	–	–	–	–	1	–	–	–	[105]
	–	+	–	–	–	–	–	–	–	–	[66]
	+	–	–	–	–	–	–	–	–	–	[40, 123, 163]
	–	–	–	–	–	2, 5	–	–	–	–	[78]
	–	–	–	–	–	1	–	–	–	–	[91]

Table 3.13 (continued)

Animals	Non lipid-dependent yeasts (*M. pachydermatis*)					Lipid-dependent yeasts[a] isolated from					References
	A	B	C	D	E	A	B	C	D	E	
Horse *(Equus caballus)*	–	–	–	–	–	1	–	–	–	–	[106, 183]
	–	+	–	–	–	–	–	–	–	–	[66]
	–	–	–	–	–	–	2	–	–	–	[182]
	+	+	–	+	–	4,5,6,7	2,3,4,5,7	–	3,4,5,6	–	[182]
	–	–	–	–	–	–	1	–	1	–	[149]
	–	+	–	+	–	–	–	–	–	–	[40]
	–	–	–	–	–	–	5, 9	–	–	–	[207]
	–	–	–	–	–	–	–	–	9	–	[31]
Bovine *(Bos taurus)*	–	+	–	–	–	–	–	–	–	–	[66]
	–	–	–	–	–	–	1	–	–	–	[106]
	+	–	–	–	–	2,3,4,5,6	–	–	–	–	[61, 63]
	–	–	–	–	–	2, 4	–	–	–	–	[62]
	–	–	–	–	–	3, 4	2, 4, 7	–	–	–	[51]
	1	–	–	–	–	1	–	–	–	–	[65]
	–	–	–	–	–	8	–	–	–	–	[113]
	–	–	–	–	–	–	5	–	–	–	[91]
Goat *(Capra hircus)*	–	–	–	–	–	5	1	–	–	–	[89]
	+	–	–	–	–	2, 6	3, 7	–	4	–	[51]
	–	+	–	–	–	–	–	–	–	–	[161]
	+	–	–	–	–	–	–	–	–	–	[40]
	–	–	–	–	–	–	5	–	–	–	[200]
	–	–	–	–	–	–	1	–	–	–	[12]
	–	–	–	–	–	–	10	–	–	–	[31]
Sheep *(Ovis aries)*	–	–	–	–	–	–	1	–	–	–	[91]
	–	–	–	–	–	–	2, 3, 7	–	–	–	[51]
Chicken *(Gallus gallus)*	–	–	–	–	–	–	2	–	–	–	[84]
Guinea pig *(Cavia porcellus)*	–	+	–	–	–	–	–	–	–	–	[66]
Rabbit *(Oryctolagus cuniculi)*	–	+	–	–	–	–	–	–	–	–	[66]

Localization of sampling is indicated with a capital letter: *A* external ear canal; *B* body skin (lips, axilla, groin, dorsum, pinna, precrural area, ventral midline, udder), mucosa, hair, nails, fur, feathers, and comb; *C* mouth; *D* perianal area, anus; *E* eye, conjunctival sac

[a]Key for lipid-dependent yeasts: *1* not identified or not specified; *2 M. sympodialis*; *3 M. globosa*; *4 M. furfur*; *5 M. slooffiae*; *6 M. obtusa*; *7 M. restricta*; *8 M. nana*; *9 M. "equi"* and *M. equina*; *10 M. caprae*

Fig. 3.8 An exfoliative dermatitis caused by non lipid-dependant *Malassezia* yeasts was firstly reported in 1925 by Weidman on an Indian rhinoceros from Philadelphia [204]. In 1996, similar lesions were observed in a Southern white rhinoceros [9]

Fig. 3.9 The skin and mucosa of birds may be colonized by lipid-dependent or nonlipid-dependent *Malassezia* yeasts. In the present case, *Malassezia* overgrowth was responsible for the loss of small feathers around the eyes (Service de Parasitologie, ENVA)

was able to grow on ordinary media apparently without any lipid supplementation. Weidman classified this yeast in the genus *Pityrosporum* because of its resemblance to *P. ovale*, the purported agent of human dandruff. However, the regularity in form and the smaller size of the yeast recovered from the Indian rhinoceros led Weidman to describe the species *Pityrosporum pachydermatis*, which was later changed to *Malassezia pachydermatis* [87, 89]. The skin of rhinoceroses seems to be frequently colonized by *Malassezia* yeasts [91, 205].

In 1969, Midgley and Clayton [137] investigated the mycobiota of 90 mammals and 258 birds settled in a zoological park in London. They identified *Malassezia* yeasts in 32.2% of mammals (20% of primates and 56.7% of canines), 30% of water fowls, 2% of tropical birds, and 4.7% of parrots. Dufait [66] isolated *M. pachydermatis* from feathers of 15 birds out of 18 sampled. Another study reported the presence of *M. pachydermatis* in the throat of infected scarlet macaw [29]. Guillot et al. [91] isolated *Malassezia* yeasts from cutaneous samples of six birds (Fig. 3.9). However, a study performed on 103 psittacine birds, both healthy and with manifested feather-destructive behavior, did not recover any *Malassezia* yeasts using cytology examination, nor in fungal cultures of samples taken from the skin surface [169].

Wild animals have always been an interesting source of different microorganisms, including *Malassezia* yeasts [47, 91, 107, 123]. The examination of 165 foxes, one wolf and one bear resulted in the isolation of *M. pachydermatis* from the ears of 13, and anuses of 7 foxes, while one fox harbored the species on both locations. In addition, the same yeast was isolated from the external ear canal of wolf, and the ear and anus of a brown bear [107]. In 1994, Guillot et al. [91] investigated the prevalence of *Malassezia* yeasts on different animal species, including 356 domestic and wild animals that represented 40 different species. Most of the isolated yeasts belonged to *M. pachydermatis* and originated from carnivores. Lipid-dependent yeasts, which could not be precisely identified at that time, were isolated from domestic animals (viz. cow, sheep, and pigs) and from wild animals (viz. elephant, chimpanzee, rhinoceros, cheetah, and okapi). This study covered also Mammalian orders (Rodents, Lagomorphs, and Insectivores) that yielded negative results. An epidemiological study on 132 wild felids resulted in a the recovery of *M. pachydermatis* in 43.1% of small felids, while *M. sympodialis* was recovered only in large felids with 56.9% positive animals [47]. In Brazil, ear swabs from 30 bats allowed the isolation of *M. pachydermatis* in 62.5%, *M. furfur* in 20.8%, *M. globosa* in 12.5%, and *M. sympodialis* in 4.2% of sampled animals [77]. Despite these studies, the presence of *Malassezia* yeasts on the skin of many wild animals remains largely unexplored.

3.3.2.2 Domestic Carnivores

Malassezia yeasts are frequently isolated from carnivores, especially dogs. In 1955, Gustafson was the first one to isolate lipophilic yeasts from otitis externa (OE) in dogs [104]. He related the isolates to the genus *Pityrosporum* on the basis of their cellular shape and the type of budding, but distinguished them from already described species. In contrast to the other lipid-dependent species, Gustafson's isolates did not require fatty acid supplementation for growth. Gustafson created the new species *Pityrosporum canis* that was made a synonym under *M. pachydermatis*, a combination already proposed by Dodge in 1935 [60] and Gordon in 1979 [83]. Gustafson was able to induce OE by applying *M. pachydermatis* to the external ear canal, whereas lipophilic yeasts isolated from humans had no effect. Fraser [69] isolated *M. pachydermatis* in comparable frequencies from healthy dogs and those with OE, but found that affected dogs tended to have larger numbers of the yeast on smears of cerumen. Over the last 10 years, in addition to *M. pachydermatis*, some lipid-dependent species were recovered from dog skin, but, unfortunately, the identification method used was not sufficiently clear [85, 172, 173] or lipid-dependent yeasts were not identified [34]. Cafarchia et al. [34] isolated lipid-dependent species from healthy dogs and Crespo et al. [50] sampled atypical lipid-dependent species from dogs that suffered from chronic OE. However, in these last studies no molecular identification was performed. Two years later, Duarte et al. [63] isolated atypical lipid-dependent yeasts from the ear of a healthy dog.

Among the different *Malassezia* species, the non lipid-dependent *M. pachydermatis* can be isolated from the external ear canal and mucosa of healthy cats and those with OE and dermatitis [91]. Generally, *M. pachydermatis* is more frequently isolated from dogs than from cats. In healthy cats, the reported percentages of isolation of this species range from less than 10% [25, 109] to approximately 20% [52, 59, 91, 147] and up to 40% [33]. However, *M. pachydermatis* has been identified in 41–71% of cats with OE [33, 52, 59, 147].

Although lipid-dependent species have also been reported from canine specimens, they seem to occur more frequently in cats than in dogs. The presence of lipid-dependent *Malassezia* yeasts in healthy carnivores was initially demonstrated in felines. *M. globosa* was isolated from normal skin of a cheetah (*Acinonix jubatus*) [87, 91]. Lipid-

Fig. 3.10 The skin of Devon Rex cats represents a favorable biotope for the development of *Malassezia* yeasts. In this breed, high numbers of *Malassezia* yeasts were detected in different anatomic sites, including axilla, groin, ventral neck, and nail folds (Service de Zootechnie, ENVA)

dependent species, such as *M. furfur* [48, 147], *M. globosa* [25, 33, 147], and *M. sympodialis* [23, 47, 52, 59], have also been reported from the skin or the external auditory canal of healthy pet cats or other felids. On the other hand, Colombo et al. [45] recently found that *M. pachydermatis*, *M. furfur*, and *M. sympodialis* can be frequently isolated from the nail folds of healthy cats of various breeds, especially in Devon Rex cats (Fig. 3.10). These authors pointed out that the presence of a high number of yeasts in cytological investigations correlated with the clinical observation of the presence of brown, greasy material in the nail folds. *Malassezia* overgrowth was also reported in allergic cats, mainly with AD [155].

Most of these isolates, however, have been identified on the basis of phenotypic characteristics and misidentification might occur. The speciation of lipid-dependent isolates from animals by means of physiological tests presents some difficulties. Crespo et al. [49] reported for the first time lipid-dependent yeasts associated with OE in cats. These yeasts had similar morphological characteristics and some shared physiological characteristics with the type strain of *M. sympodialis*. More recently, Hirai et al. [113] described *M. nana*, a novel species from otic discharges of a cat and cows. The type strain of *M. nana* showed identical physiological characteristics to those of *M. sympodialis*, but with the only exception being the presence of ring-like growth with Tween 20 [112]. In fact, the frequently reported isolates from feline skin that showed a similar phenotype to *M. sympodialis* may be conspecific with *M. nana*. Some lipid-dependent strains isolated from cats that were found to be similar to the type strain of *M. sympodialis* were studied using DNA sequence analysis and it was found that they grouped together with the *M. nana* AB075224 LSU rRNA gene sequence [30]. *M. nana* seems to be the most common lipid-dependent species isolated from cats. This species and *M. slooffiae* have recently been isolated from cats with different disorders [159], and *M. slooffiae* has been found in the claw folds of healthy and seborrhoeic Devon Rex cats [1].

3.3.2.3 Domestic Herbivores

Lipid-dependent species constitute the major population of the lipophilic mycobiota in herbivores, such as horses, goats, sheep, and cows [51] and their occurrence in these animals was much greater than that of *M. pachydermatis*. Different species, such as *M. furfur, M. globosa, M. obtusa, M. slooffiae, M. sympodialis*, and *M. restricta*, were phenotypically identified, but they were not confirmed by rDNA sequence analysis. Moreover, a large number of isolates could not be identified by means of phenotypic techniques.

A lipid-dependent isolate identified as *M. furfur sensu lato* was recovered from the healthy skin of a cow [91]. A strain from the normal ear of a cow and a strain of the normal skin of a cow were included in the description of *M. furfur sensu stricto* and *M. globosa,* respectively [87]. Different lipid-dependent species have also been recovered from healthy cattle from Minas Gerais, Brazil [61, 63, 64], namely *M. furfur, M. globosa, M. obtusa, M. sympodialis, M. slooffiae*, and *M. globosa*, which were phenotypically identified from cerumen of the external ear of these animals. The most common species isolated from these animals was *M. sympodialis*. However, the identity of these isolates was not confirmed by rDNA sequence analysis. Four lipid-dependent strains from cows were considered to be genetically identical to the type strain of *M. nana* [113]. Molecular analysis showed, however, that some differences occurred between the strain from cat and those from cow, respectively.

Nell et al. [149] reported the presence of a novel *Malassezia* species from normal equine skin, which they tentatively named "*Malassezia equi*," but without including a valid description. The species was identified as a member of the genus *Malassezia* by sequence analysis of the D1/D2 domains of the LSU rRNA gene, and was found to be closely related to *M. sympodialis*. In a study of the lipophilic mycobiota of the intermammary and preputial fossa areas of horses "*Malassezia equi*" and *M. slooffiae* were identified by PCR sequencing analysis [207]. In another study, several isolates from horses that phylogenetically were related to *M. sympodialis* grouped together by D1/D2 and ITS sequence analysis and were described as *M. equina* [30, 31].

Few reports are available on *Malassezia* yeasts that affect the skin of goats [12, 200]. Lipid-dependent isolates identified as *M. furfur sensu lato* were recovered from the healthy skin of two sheep as well [91]. In the last revision of the genus, two strains from the normal skin of a goat and a sheep were included in the description of *M. slooffiae* [87]. Lipid-dependent *Malassezia* species, such as *M. furfur*, *M. globosa*, *M. obtusa*, *M. sympodialis*, and *M. restricta*, have also been isolated from the skin of healthy goats, and *M. globosa, M. restricta*, and *M. sympodialis* were obtained from the skin of healthy sheep. These isolates were phenotypically identified, but they were not confirmed by rDNA sequence analysis [51]. From these isolates, the sequence of the D1/D2 domains of the 26 rRNA of a lipid-dependent strain from sheep was found to be identical to that of the type culture of *M. sympodialis* [30]. *Malassezia caprae*, a lipid-dependent species isolated mainly from healthy goats, has been recently described as well [31].

3.3.2.4
Swine

In 1959, Gustafson isolated *Malassezia* yeasts from three out of 32 swine [105]. In 1985, Dufait [66] successfully cultured *M. pachydermatis* from six pigs. The isolation of *Malassezia* lipid-dependent yeasts from 23 out of 40 [91] and 73 out of 100 swine [78] suggested that this animal species is regularly colonized by lipophilic yeasts. Garau et al. [78] reported colonization of swine ears with *M. sympodialis* and *M. slooffiae*, while swab and carpet did not reveal any *Malassezia* yeasts from anus, forehead, and back [78]. Another study [163] reported the presence of nonlipid-dependent strains in swine ears. Some of these isolates were randomly chosen and their identification as *M. pachydermatis* was confirmed by karyotyping and restriction fragment length polymorphism analysis.

3.3.2.5
Rodents and Lagomorphs

Dufait isolated nonlipid-dependent *Malassezia* yeasts from guinea pigs and rabbits [66]. However, further investigations failed to demonstrate the presence of lipophilic yeasts in rodents or lagomorphs [75, 91]. Recently, Galuppi et al. [75] examined 147 laboratory rats and did not isolate any yeast.

3.3.2.6
Distribution Within a Single Animal Host

Because of the importance of *Malassezia* dermatitis in dogs, many investigations have been conducted in order to describe the distribution of *M. pachydermatis* on the body of dogs. Lipophilic yeasts can be isolated from skin, auditory canal, anus, oral cavity, lips, and mucosae of male and female genitalia (vagina, preputium) [181]. Colonization of the canine skin probably occurs in the very first days after birth, but the way by which colonization occurs is still not understood. Wagner and Schadler [203] isolated *Malassezia* yeasts in 39.4% of all samples collected from the lips, nail beds, and ears of 3-day old Rottweilers. The percentage increased to 41.2% in 1-week old puppies, and decreased to 37% when the animals were 5 weeks old. Most of the samples recovered from dams yielded negative results, thus excluding the possibility that *Malassezia* was transferred through cleaning and feeding processes [203]. Oral carriage was investigated by Bond and Lloyd [19] who compared two different antifungal shampoos against SD in basset hounds. Application of the miconazole-chlorhexidine shampoo throughout a 3-week period resulted in a significant reduction of the *Malassezia* population on both the skin and in the mouth. However, the importance and significance of oral transfer remained unclear. Another study pointed out the possible transfer of *Malassezia* yeasts from pruritic skin lesions of the inguinal area to the perioral area as a consequence of frequent licking. In addition, dogs with lesions in the perioral area yielded a greater number of *Malassezia* yeasts in the interdigital region without visible skin damage as a probable result of persistent scratching [34].

Different isolation techniques were used to assess the number of *Malassezia* and to determine a possible threshold value that could trigger *Malassezia* dermatitis. It is well known that sampling techniques determine the detection limit of *Malassezia* yeasts. Kennis et al. [119] reported that at least one positive *Malassezia* sample could be detected in 95% of examined dogs when using the culture method, while cytological examination of samples collected by the direct impression method using the glass slide, cotton swabs, and skin scraping techniques yielded positive results in 74% of dogs. Evaluation of these three techniques separately yielded positive results in only 37–57% of the dogs [119].

The results of epidemiological surveys also rely on the choice of the culture conditions. *M. pachydermatis* is considered a nonlipid-dependent species, which grows well on Sabouraud glucose agar. However, the use of specific media, such as modified Dixon medium, showed a significant higher recovery rate of lipophilic yeast in comparison with that on Sabouraud glucose agar [95]. Until 1995, it was believed that dogs were carriers of *M. pachydermatis*, a species that does not require lipids in media for their growth. Transfer of some *M. pachydermatis* colonies from modified Dixon agar to Sabouraud's glucose agar, however, showed fastidious growth and the need for the addition of lipids. Karyotyping of those isolates confirmed a typical *M. pachydermatis* pattern [16], and two years later partial sequencing of the LSU rRNA defined them as sequence type Id [92]. In 1999, Senczek et al. [182] distinguished two different types of colonies of *M. pachydermatis* isolated from inflamed external ear canals of dogs. Small colonies showing poor expansion growth on culture media reached only 1 mm in diameter after 72 h of incubation, whereas bigger ones showed faster growth and colonies of 4 mm in diameter. Ahman et al. [1] isolated two different types of *M. pachydermatis* in Devon Rex cats. One type showed good growth on Sabouraud glucose agar, while colonies of the other type remained significantly smaller, and were harder to grow on Sabouraud glucose agar.

Several investigations explored *Malassezia* colonization of various anatomical regions in different breeds of healthy dogs. Results differed from 10 to 100% and depended on the dog breed, sampled anatomic location, and identification method used [93]. Most frequently sampled sites were the external ear canal, anus, followed by mouth, buccal mucosa, haired skin of lips, pinna, dorsum, axilla, groin, interdigital area, nail beds, prepuce, and vulva [22, 33, 34, 81]. *M. pachydermatis* has been identified in 57.8% of clinically healthy dogs [108]. Lukman [127] reported that 56.6% of anal sacs were colonized with *M. pachydermatis*, followed by the auditory canal (48.6%), rectum (46%), anus (40.6%), and vagina (38.2%). A study performed on 20 beagles and 20 pet dogs of various breeds and crossbreeds showed significant differences in the number and occurrence of *Malassezia* isolation between beagles and the other breeds of dogs [22]. In 55% of examined beagles *Malassezia* prevailed in the anus and ears, while in pet dogs it prevailed in anus (50%) and mouth (30%). In pet dogs, the nose, prepuce, and vulva were colonized with lower frequency and were positive in 8.5–10% of the animals investigated. A similar situation was observed in beagles where frequencies ranged from 0 to 20%. Beagles had statistically a significant higher number of *Malassezia* yeasts than pet dogs (55 vs. 10%) in terms of population size and frequency of isolation from the ear. It is notable that isolation attempts from dorsum samples of both pet dogs and beagles remained negative. Moreover, the lower lip region was colonized with *Malassezia* in the range of 75–80%, and the interdigital area from 60 to 70% of the samples from healthy dogs [22]. A similar study was carried out on healthy basset hounds and resulted in the recovery of *M. pachydermatis* from the mouth (92.3%) and vulva (80%). The axilla was colonized in all dogs, the nose in 69.2%, and this was significantly higher than the values obtained from healthy mixed-breed dogs. By comparison, *Malassezia* yeasts were recovered from the axilla in 15%, from the mouth in 30%, the nose in 10%, and from vulva in 8.5% of mixed-breeds [18]. Comparable results were reported when colonization of *M. pachydermatis* was investigated of the hair of the lower lip and the hair follicles of healthy beagles. *M. pachydermatis* was most frequently isolated in hair samples from the lips (75% of beagles), whereas isolations from the fore foot and dorsum were found to be positive in 33% of animals [95]. An investigation conducted in France reported that 91% of the samples recovered from dogs' lips and 43% from their ears were positive in culture for *M. pachydermatis* after cytological examinations yielding 3.1 and 2.2 *Malassezia* cells per field [95]. In the study by Bensignor et al. [11], pinna was the area most frequently colonized with *Malassezia* yeasts. Contrary to previous studies, lower detection and isolation frequency were observed in the perianal area when compared with the umbilical and axilla areas [11]. Kennis et al. [119] reported the highest number of *Malassezia* yeasts in the chin region, while the inguinal and axillary zones presented the lowest number. A study on 99 healthy dogs revealed the prevalence of *M. pachydermatis* in the ear of 39.4% of cytologically examined samples and 45.5% of samples that were identified by culture [121]. Cafarchia et al. [34] isolated *Malassezia* yeasts in the perianal area of 60.6% from healthy dogs, followed by 36.4% of the perioral area from healthy dogs. The interdigital area, ear canal, and periorbital sites were found to be less frequently colonized with 18.2, 12.1, and 9.1% of the animals investigated, respectively. The inguinal area was positive in only 3% of dogs [34]. Girão et al. [81] found *M. pachydermatis* in 30% of the ears of healthy dogs after microscopic and culturing assessment. Cytological quantification of *Malassezia* in the anal sac content of healthy dogs revealed a low occurrence of yeasts, with only 12.5% of the examined dogs and 10% of the examined anal sacs demonstrating the presence of *Malassezia* yeasts [156]. Although there is no consistent pattern in the results obtained from several cases for which the anal and perianal regions were examined, the perioral area, including lips and mouths, ear skin and ear canal, were most frequently colonized with *Malassezia* yeasts. Finally, all studies resulted in the same conclusion, namely that *Malassezia* colonization occurs more frequently in the hairy areas that are rich with sebaceous glands [93].

Disturbance of the mutual coexistence between the animal and *Malassezia* mycobiota can lead to *Malassezia* overgrowth and development of skin disorders, such as dermatitis, paronychia, facial or pedal pruritis, and OE in dogs [142]. In order to detect an infection caused by *Malassezia*, it is important to gain insight into the population sizes of *Malassezia* on healthy dog skin and mucosa and their ecology [93], because there is no uniform threshold value of the number of *Malassezia* yeasts that will result in exacerbation of skin disease.

The presence of viable *Malassezia* yeasts in the environment is considered to be low. The environment has, at least, an anecdotic role in our understanding of the epidemiology of *Malassezia*-related diseases. *Malassezia* yeasts have been detected in sand [131] and water (G. Midgley, personal communication). Renker et al. [175] detected DNA of lipid-dependent *Malassezia* yeasts (viz. *M. globosa* and *M. restricta*) from nematodes living in soils from forests in Central Europe. The biological significance of the possible presence of *Malassezia* yeasts in the soil remains to be investigated.

3.3.3 Factors Associated with Increased *Malassezia* Populations in Animals

3.3.3.1 Factors Related to the Animal Hosts

Many factors may be associated with the proliferation of *Malassezia* yeasts in animals, including breed, gender, and age. Moreover, alteration of microenvironmental skin conditions connected to a chemical mechanism (pH modifications), long-term treatments (antibiotics and glucocorticoids), or presence of other bacteria or fungi and coexisting diseases may contribute to *Malassezia* proliferation. Numerous studies recognized breed predisposition in dogs as an important factor. In 1982, Lukman [127] noted the highest frequency of *Malassezia* recovery in Dalmatian dog, German shepherd, and German shorthaired pointer. Hajsig et al. [108] reported that *M. pachydermatis* was found in 75% of examined German shepherd dog breed. Cytological examination of 39 different breeds of dogs resulted in a significantly higher prevalence of *Malassezia* in basset hounds and dachshunds when compared with all other breeds of dogs examined in the study. A similar, but statistically not significant tendency was observed in cocker spaniels, springer spaniels, and German shepherds [165]. Bond et al. [24] reported that basset hounds, cocker spaniels, and West Highland white terriers are breeds predisposed to develop *Malassezia* dermatitis (Fig. 3.11). Another study on healthy basset hounds showed a significantly higher number of *M. pachydermatis* recovered from the nose, mouth, and vulva in comparison with healthy mixed-breed dogs [18] (Fig. 3.12). In the

Fig. 3.11 West Highland white terriers are predisposed to *Malassezia* dermatitis (Service de Parasitologie, ENVA)

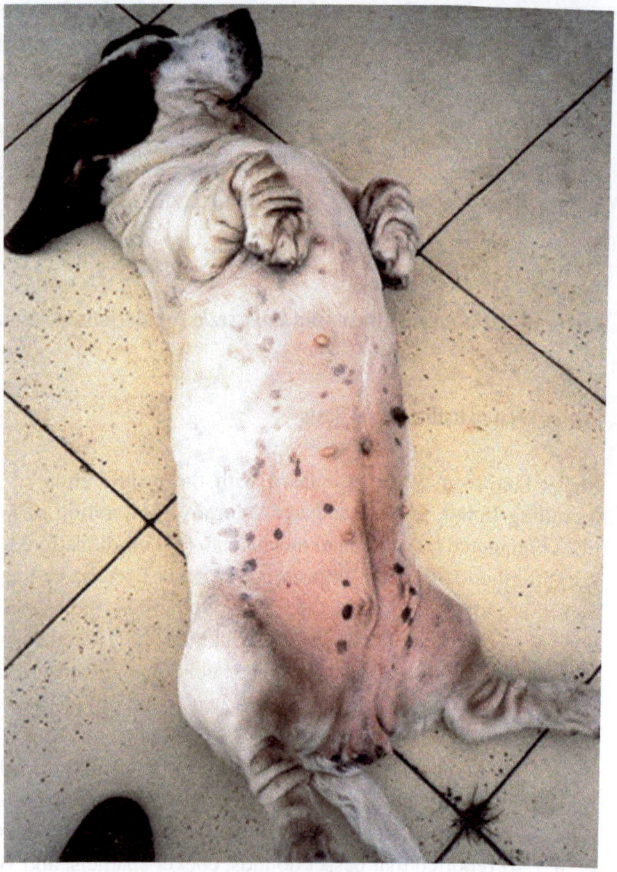

Fig. 3.12 The skin of basset hounds represents a very favorable biotope for the development of *Malassezia* yeasts (Service de Parasitologie, ENVA)

study exploring immunological responses of healthy dogs and dogs with dermatitis, a significantly higher number of *M. pachydermatis* was found in axilla of healthy basset hounds in comparison with beagles and setters [26]. Kumar et al. [121] reported the highest presence of *Malassezia* otitis in German shepherds and Labrador retrievers. Girão et al. [81] showed the highest prevalence of otitis connected to *M. pachydermatis* in the poodle. A large number of healthy German shepherds were found positive for *M. pachydermatis*. An epidemiological study on 1,370 dogs reported a higher occurrence of OE in breeds with pendulous ears when compared with breeds with erect ears [133]. Nobre et al. [152] reported a higher occurrence of *Malassezia* otitis in cocker spaniels, German shepherds, and Brazilian fila, while Mauldin et al. [136] found that West Highland white terriers, English setters, Shih Tzus, basset hounds, and American cocker spaniels were predisposed to *Malassezia* dermatitis. Some breed predispositions are related to physical characteristics of the skin, such as the presence of skin folds in basset hounds [18, 97] or pendulous ears [33, 81, 133, 168].

Several studies, performed to determine a possible age predisposition in dogs, reported contradictory results. A few investigations did not find a significant correlation between the number of *Malassezia* and the age of the animal [152, 165]. Other studies demonstrated that the age of the dogs could be considered as a predisposing factor, but the conclusions of these investigations were not the same. In a study by Mauldin et al., [135] more than 70% of dogs with *Malassezia* dermatitis were 4 years and older. Girão et al. [81] reported the highest occurrence of *Malassezia* otitis in 1–3-year-old dogs, Kumar et al. [121] in 2–4-year-old dogs, and Cafarchia et al. [33] detected higher *Malassezia* populations in dogs younger than 1 year.

Gender predisposition has been excluded by numerous studies [81, 121, 152, 165]. In most cases, equal numbers of *Malassezia* yeasts occurred in females and males (46.43 vs. 53.57%) as reported by Girão et al. [81]. However, studies conducted in Italy [33] observed a higher number of *Malassezia* recovered from males than from females with otitis, and a study from Brazil reported 60.7% of positive males in comparison with 33% of the females [169], thus indicating possible gender predisposition. In a retrospective study of *Malassezia* dermatitis in dogs, it has been reported that sterilized females and males were more susceptible to develop this skin disorder [135].

3.3.3.2 Factors Related to the Skin Microenvironment

The optimal temperature for *Malassezia* growth under laboratory conditions is 32–34°C. Some species can grow below these values (viz., *M. pachydermatis*); others develop at 37°C or above (viz. *M. pachydermatis*, *M. sympodialis*, *M. furfur*, *M. yamatoensis*, *M. slooffiae*, and *M. dermatis*). For optimal growth of *Malassezia* yeasts, an adequate pH value is required. It was reported that optimal pH values for successful *Malassezia* development range between 4.0 and 8.0 [134]. The highest occurrence of canine *Malassezia* otitis in India was reported in June with 15.2% of otitis cases [121]. On the contrary, Cafarchia et al. [33] reported higher recovery of *Malassezia* yeasts from dogs with otitis during the winter. In general, there are indications that warm and humid weather supports the exacerbation of *Malassezia* infection in animals.

Malassezia pachydermatis demonstrates a preference toward the external ear canal causing OE in many cases. Certain conditions in the host need to be satisfied to stimulate the growth of this opportunistic pathogen. The ear canal is a dark, humid place, rich with cerumen, and experiencing moderate air circulation depending on the ear shape. Such an environment represents an optimal place for the exacerbation of infection. The concentration of fatty acids in the earwax influences the susceptibility to OE [133]. Most dog breeds with pendulous ears showed significant higher amounts of fatty acids in the ear (with the exception of the pug) than breeds with erect ears, when Siberian husky was excluded from this group [133].

Besides the mentioned physical and chemical factors that enable the attachment of *Malassezia* and the consequent infection, other conditions that may contribute to *Malassezia* overgrowth are: hormonal disbalance, keratinization defects, excessive production of sebum, bacterial infections, and hypersensitivity processes [181].

Atopic dermatitis is a common skin disease from which *Malassezia* yeasts can be recovered in high numbers. However, a significant association between atopic lesions and *Malassezia* colonization has not been found, suggesting that yeast overgrowth occurs secondarily as a result of changes in the skin microenvironment [148]. Regarding the location, the highest recovery of *Malassezia* yeasts was observed in the interdigital areas of the skin with 70.7% of isolations, followed by the ears with 63.4%. Nail folds, mouth, and groin were places with lower frequency of *Malassezia* isolation, approximately 33% positive isolations on average [148].

Different immunodeficiency status may be connected to malfunction of the immune system and inadequate immunologic responses to *Malassezia* yeasts. Chen and Hill [43] described immunological responses that occurred after the interaction of *M. pachydermatis* antigen and the skin immune system in dogs. Induction of Th1 lymphocytes or cell-mediated response leads to the production of IL-2 and IFN-γ cytokines, thus promoting the release of immunoglobulin IgG, while the induction of Th2 or humoral response leads to production of IL-4 and IL-13, followed by immunoglobulin IgE [43]. In the last decade, different surveys have been performed in order to explore cellular and humoral immune responses to *M. pachydermatis* in dogs. In an *in vitro* study, the possibility of proliferation of peripheral blood mononuclear cells as a result to *M. pachydermatis* antigen was explored. The results showed that dogs with and without incidence of disease were able to develop cell-mediated response to *M. pachydermatis* antigen. Additionally, this study indicated that there was no connection between low levels of total IgA in sera of dogs and the presence of high population densities of *M. pachydermatis*. Finally, high levels of specific IgG were found in the serum of dogs affected with *Malassezia* dermatitis [26].

3.3.3.3
Factors Related to *Malassezia* Species and Isolates

The evaluation of enzymatic activities of 33 different *Malassezia* strains belonging to various species showed that strains of *M. pachydermatis* had the highest enzymatic activity as they were capable to secrete 15 different enzymes. *M. furfur* produced eight, *M. sympodialis* and *M. slooffiae* five, *M. obtusa* four, and two *M. globosa* strains three and four enzymes, respectively. *M. restricta* secreted only naphtol-AS-BI-phosphohydrolase, thus demonstrating a narrow range of enzyme production [130]. Isolates of *M. pachydermatis* obtained from dogs with single or generalized skin lesions demonstrated significantly higher phospholipase activities than those obtained from healthy skin [38], thus suggesting the importance of enzymatic activity in the development of cutaneous *Malassezia* overgrowth.

In epidemiological studies, genotyping of *Malassezia* yeasts may be required in order to identify the source of infection and to discover a possible connection between genotypes and disease. Different studies targeted different rRNA or rDNA regions in order to distinguish different molecular patterns among *M. pachydermatis* resulting in a genotypic classification. Sequence analysis of the D1/D2 domains of the LSU rRNA of 100 isolates of *M. pachydermatis* resulted in the differentiation of seven different genotypes (or sequence types), namely, Ia, Ib, Ic, Id, Ie, If, and Ig [92]. Sequence types Ic, Id, and I g seemed to be specific for rhinoceros, dogs, and ferrets, respectively. Sequence type Ia was found in isolates from both animals and humans, while sequence type Id involved *M. pachydermatis* showing small colonies and poor growth on Sabouraud glucose agar. In general, all *M. pachydermatis* isolated from dogs belonged to sequence types Ia, Id, and Ie. Moreover, it was found that one animal could be the carrier of two or more *Malassezia* sequence types [92]. Sequence analysis of *M. pachydermatis* isolates from the external ear and cutaneous lesions of sea lion using the variable domains of the LSU rRNA showed that all strains belonged to sequence type Id [94].

Aizawa et al. [2] used random amplification of polymorphic DNA (RAPD) and chitin-synthase gene (*CHS2*) sequencing in order to discriminate 16 isolates of *M. pachydermatis*. Both methods yielded similar results, namely A, B, and C sequence types, and 13 isolates matched with the type strain of *M. pachydermatis*. Subsequently, the same research group used the same molecular method to differentiate 98 *M. pachydermatis* isolates from dogs and 12 from cats. RAPD analysis yielded four different genotypic groups of *M. pachydermatis* listed A, B, C, and D. RAPD type A was identified as the most common genotype and occurred in 81 isolates from dog skin and ear and was found to be identical to *M. pachydermatis*, while RAPD type B pattern was observed in 13 isolates and was close to *M. furfur*. RAPD type C of *M. pachydermatis* showed a different sequence pattern and was isolated only from one dog suffering from OE. RAPD type D was found in two isolates of dogs with OE and one with AD. The 12 isolates from ear and skin form cats belonged to RAPD type A [3].

Sugita et al. [192] explored the variable intergenic spacer region (IGS1) as a possible genotypic marker for *M. pachydermatis*. Forty-three strains isolated from carnivores (dogs and cats) were analyzed and three different groups with ten subtypes were identified. The authors stated that their IGS groups 1, 2, and 3 matched with the sequence types Ie, Id, and Ia as identified previously by Guillot et al. [92, 192].

RAPD analysis was performed by Castellá et al. [40] in order to genotype isolates of *M. pachydermatis* originating from dogs, cats, horses, goat, and pig. Thirty five from 55 isolates belonged to RAPD type I showing 12 bands. RAPD types II, III, and IV were found only among dog isolates. RAPD types II and IV displayed nine bands and were recovered from two animals that suffered from otitis, and RAPD type III had 11 bands and was isolated from a healthy dog. The authors suggested that their RAPD type I corresponded to sequence type Ia as determined by Guillot et al. [92] or RAPD type A as described by Aizawa et al. [3]. In addition, they reported that one animal can host different genotypes of *M. pachydermatis* [40].

Cafarchia et al. [36] analyzed 104 *Malassezia* isolated from dogs using sequence analysis of three different loci, namely *CHS2*, the LSU rRNA gene, and the ITS1 region of the rDNA. The authors distinguished three different sequence types: A, B, and C, but analysis of the ITS-1 region yielded an additional sequence type. In 185 samples of *M. pachydermatis* isolated from 30 healthy and affected dogs, three *CHS2* (A_c, B_c and C_c) and eight ITS1 (A_{I1}–A_{I4}, B_{I1}, C_{I1}–C_{I3}) genotypes and subgenotypes were identified [38]. All these observations indicate the presence of a considerable genetic diversity among *M. pachydermatis*, which, at least in part, may be relevant for our understanding of the interplay between the yeast and the patient.

3.4 Conclusions

In the last decade, a broad variety of *Malassezia* identification and (bio)typing tools has emerged. Owing to the different growth requirements by *Malassezia* spp. culture-independent molecular methods seem most accurate to assess the diversity of *Malassezia* yeasts in culture, on human, or animal skin. A wide array of PCR-based methods, such as

RAPD, PCR-typing, *t*FLP, RFLP, and PCR-derived methods comprising DGGE, SSCH, and the application of Luminex technology, has emerged to answer different research questions. In addition, the usefulness of the analysis of clone libraries and real-time PCR with clinical specimens is demonstrated in numerous studies. Recent advances in *Malassezia* genomics should pave the way for the design of a Multi Locus Sequence Typing (MLST) scheme for the various *Malassezia* species. On another track, further automation in the molecular analysis of *Malassezia* mycobiota can be expected due to the development of diagnostic (micro-)array platforms.

Detection and identification of *Malassezia* species from clinical specimens by real-time PCR has indicated that the species are present in related dermatoses. *M. globosa* and *M. restricta* are the most common species in pityriasis versicolor (PV), seborrheic dermatitis (SD), atopic dermatitis (AD), and psoriasis (PS). However, the ratio of these two microorganisms differs depending on the disease psoriasis. Recent qualitative and quantitative studies suggest that *M. restricta* plays a significant role in SD and PS, *M. globosa* plays a significant role in PV, while both species may be responsible for exacerbating AD.

Malassezia yeasts may be isolated from the skin and mucosa of a very wide range of warm-blooded vertebrates (all the mammals and probably many birds). However, the prevalence of colonization and the density of *Malassezia* populations vary according to the animal species and the anatomical sites. The nonlipid-dependent species *M. pachydermatis* is a common inhabitant of skin and mucosa in dogs. The mean carriage of lipophilic yeasts varies from 10 to 100% and depends on the dog breed, the anatomical location, and the identification methods used in the studies. The presence of lipophilic yeasts is more rarely reported in other animals and the *Malassezia* species involved may be different. Although *M. pachydermatis* is found in the external auditory of cats, this does not occur with the same high frequency as in dogs. Lipid-dependent species seem to be more frequent in cats than in dogs. Lipid-dependent species are the major component of the cutaneous microbiota in ruminants and horses. In swine, both lipid-dependent and nonlipid-dependent species have been isolated. Such a complex situation is probably related to the specific composition of cutaneous lipids and the competition with different types of microbiota (e.g., bacteria and yeasts) within the skin ecosystem. The recent analysis of the genome and secretory proteome of *M. globosa* and *M. restricta* showed that their lipid dependence can be attributed to the absence of a fatty acid synthase gene, while harvest of host lipids can be aided by several secreted lipases. Extending genomic and proteomic research to all known *Malassezia* species, and comparing the lipid composition on the epidermis of the corresponding hosts should enhance our understanding of *Malassezia* biology and explain the role of the various species in the disease of different animals.

References

1. Ahman S, Perrins N, Bond R (2007) Carriage of *Malassezia* spp. yeasts in healthy and seborrhoeic Devon Rex cats. Med Mycol 45:449–455
2. Aizawa T, Kano R, Nakamura Y et al (1999) Molecular heterogeneity in clinical isolates of *Malassezia pachydermatis* from dogs. Vet Microbiol 70:67–75
3. Aizawa T, Kano R, Nakamura Y et al (2001) The genetic diversity of clinical isolates of *Malassezia pachydermatis* from dogs and cats. Med Mycol 39:329–334

4. Akaza N, Akamatsu H, Sasaki Y et al (2008) *Malassezia* folliculitis is caused by cutaneous resident *Malassezia* species. Med Mycol 23:1–7
5. Amaya M, Tajima M, Okubo Y et al (2007) Molecular analysis of *Malassezia* microflora in the lesional skin of psoriasis patients. J Dermatol 34:619–624
6. Ashbee HR, Evans EG (2002) Immunology of diseases associated with *Malassezia* species. Clin Microbiol Rev 15:21–57
7. Aspiroz C, Moreno LA, Rezusta A et al (1999) Differentiation of three biotypes of *Malassezia* species on normal human skin. Correspondence with *M. globosa, M. sympodialis* and *M. restricta.* Mycopathol 145:69-74
8. Batra R, Boekhout T, Guého E et al (2005) *Malassezia* (Baillon), emerging clinical yeasts. FEMS Yeast Res 5:1101–1113
9. Bauwens L, De Vroey C, De Meurichy W (1996) A case of exfoliative dermatitis in a captive Southern white rhinoceros (*Ceratotherium simum simum*). J Zoo Wildlife Med 27:271–274
10. Baxter M, Lawler DC (1972) Causes of otitis externa in dogs and cats. Mod Vet Pract 20:52–53
11. Bensignor E, Jankowski F, Seewald W et al (2002) Comparison of two sampling techniques to assess quantity and distribution of *Malassezia* yeasts on the skin of basset hounds. Vet Dermatol 13:237–241
12. Bliss EL (1984) Tinea versicolor dermatomycosis in the goat. J Am Vet Med Assoc 184:1512–1513
13. Boekhout T, Bosboom R (1994) Karyotyping of *Malassezia* yeasts: taxonomic and epidemiological implications. Syst Appl Microbiol 17:146–153
14. Boekhout T, Renting M, Scheffers WA et al (1993) The use of karyotyping in the systematics of yeasts. Antonie van Leeuwenhoek 63:157–163
15. Boekhout T, Kamp M, Guého E (1998) Molecular typing of *Malassezia* species with PFGE and RAPD. Med Mycol 36:365–372
16. Bond R, Anthony RM (1995) Characterization of markedly lipid-dependent *Malassezia pachydermatis* isolates from healthy dogs. J Appl Bacteriol 78:537–542
17. Bond R, Lloyd DH (1996) Comparison of media and conditions of incubation for the quantitative culture of *Malassezia pachydermatis* from canine skin. Res Vet Sci 61:273–274
18. Bond R, Lloyd DH (1997) Skin and mucosal populations of *Malassezia pachydermatis* in healthy and seborrhoeic basset hounds. Vet Dermatol 8:101–106
19. Bond R, Lloyd DH (1998) Effect of topical therapy of *Malassezia pachydermatis*-associated seborrhoeic dermatitis on oral carriage of *M. pachydermatis*. Vet Rec 142:725–726
20. Bond R, Collin NS, Lloyd DH (1994) Use of contact plates for the quantitative culture of *Malassezia pachydermatis* from canine skin. J Small Anim Pract 35:68–72
21. Bond R, Lloyd DH, Plummer JM (1995) Evaluation of a detergent scrub technique for the quantitative culture of *Malassezia pachydermatis* from canine skin. Res Vet Sci 58:133–137
22. Bond R, Saijonmaa-Koulumies LEM, Lloyd DH (1995) Population sizes and frequency of *Malassezia pachydermatis* at skin and mucosal sites on healthy dogs. J Small Anim Pract 36:147–150
23. Bond R, Anthony RM, Dodd M et al (1996) Isolation of *Malassezia sympodialis* from feline skin. J Med Vet Mycol 34:145–147
24. Bond R, Ferguson EA, Curtis CF et al (1996) Factors associated with elevated cutaneous *Malassezia pachydermatis* populations in dogs with pruritic skin disease. J Small Anim Pract 37:103–107
25. Bond R, Howell SA, Haywood PJ et al (1997) Isolation of *Malassezia sympodialis* and *Malassezia globosa* from healthy pet cats. Vet Rec 141:200–201
26. Bond R, Elwood CM, Littler RM et al (1998) Humoral and cell-mediated responses to *Malassezia pachydermatis* in healthy dogs and dogs with *Malassezia* dermatitis. Vet Rec 143:381–384
27. Bond R, Lamport AI, Lloyd DH (2000) Colonisation status of *Malassezia pachydermatis* on the hair and in the hair follicle of healthy beagle dogs. Res Vet Sci 68:291–293

28. Bond R, Patterson-Kane JC, Perrins N et al (2006) Patch test responses to *Malassezia pachydermatis* in healthy basset hounds and in basset hounds with *Malassezia* dermatitis. Med Mycol 44:419–427
29. Breuer-Strosberg R, Hochleithner M, Kuttin ES (1990) *Malassezia pachydermatis* isolation from a scarlet macaw. Mycoses 33:247–250
30. Cabañes FJ, Hernández JJ, Castellá G (2005) Molecular analysis of *Malassezia sympodialis*-related strains from domestic animals. J Clin Microbiol 43:277–283
31. Cabañes FJ, Theelen B, Castellá G et al (2007) Two new lipid-dependent *Malassezia* species from domestic animals. FEMS Yeast Res 7:1064–1076
32. Cafarchia C, Otranto D (2004) Association between phospholipase production by *Malassezia pachydermatis*. J Clin Microbiol 42:4868–4869
33. Cafarchia C, Gallo S, Capelli G et al (2005) Occurrence and population size of *Malassezia* spp. in the external ear canal of dogs and cats both healthy and with otitis. Mycopathol 160:143–149
34. Cafarchia C, Gallo S, Romito D et al (2005) Frequency, body distribution, and population size of *Malassezia* species in healthy dogs and in dogs with localized cutaneous lesions. J Vet Diagn Invest 17:316–322
35. Cafarchia C, Otranto D, Campbell BE et al (2007) Multilocus mutation scanning for the analysis of genetic variation within *Malassezia* (Basidiomycota: Malasseziales). Electrophoresis 28:1176–1180
36. Cafarchia C, Stefania Latrofa M, Testini G et al (2007) Molecular characterization of *Malassezia* isolates from dogs using three distinct genetic markers in nuclear DNA. Mol Cell Probes 21:229–238
37. Cafarchia C, Gallo S, Danesi P et al (2008) Assessing the relationship between *Malassezia* and leishmaniasis in dogs with or without lesions. Acta Trop 107:25–29
38. Cafarchia C, Gasser RB, Latrofa MS et al (2008) Genetic variants of *Malassezia pachydermatis* from canine skin: body distribution and phospholipase activity. FEMS Yeast Res 8:451–459
39. Canteros CE, Rivas MC, Lee W et al (2007) Concordance between phenotypical features and PCR-REA for the identification of *Malassezia* spp. Rev Iberoam Micol 24:278–282
40. Castellá G, Hernández JJ, Cabañes FL (2005) Genetic typing of *Malassezia pachydermatis* from different domestic animals. Vet Microbiol 108:291–296
41. Celis AM, Cepero de García MC (2005) Genetic polymorphisms of *Malassezia* spp. yeast isolates from individuals with and without dermatological lesions. Biomedica 25: 481–487
42. Chang HJ, Miller HL, Watkins N, et al (1998).) An epidemic of *Malassezia pachydermatis* in an intensive care nursery associated with colonization of health care workers' pet dogs. New Engl J Med 338:706-711
43. Chen TA, Hill PB (2005) The biology of *Malassezia* organisms and their ability to induce immune responses and skin disease. Vet Dermatol 16:4–26
44. Chryssanthou E, Broberger U, Petrini B (2001) *Malassezia pachydermatis* fungaemia in a neonatal intensive care unit. Acta Paediatr 90:323–327
45. Colombo S, Nardoni S, Cornegliani L et al (2007) Prevalence of *Malassezia* spp. yeasts in feline nail folds: a cytological and mycological study. Vet Dermatol 18:278–283
46. Coutinho SD, Paula CR (2000) Proteinase, phospholipase, hyaluronidase and chondroitin-sulphatase production by *Malassezia pachydermatis*. Med Mycol 38:73–76
47. Coutinho SD, Fedullo JD, Correa SH (2006) Isolation of *Malassezia* spp. from cerumen of wild felids. Med Mycol 44:383–387
48. Crespo MJ, Abarca ML, Cabañes FJ (1999) Isolation of *Malassezia furfur* from a cat. J Clin Microbiol 37:1573–1574
49. Crespo MJ, Abarca ML, Cabañes FJ (2000) Otitis externa associated with *Malassezia sympodialis* in two cats. J Clin Microbiol 38:1263–1266
50. Crespo MJ, Abarca ML, Cabañes FJ (2000) Atypical lipid-dependent *Malassezia* species isolated from dogs with otitis externa. J Clin Microbiol 38:2383–2385

51. Crespo MJ, Abarca ML, Cabañes FJ (2002) Occurrence of *Malassezia* spp. in horses and domestic ruminants. Mycoses 45:333–337
52. Crespo MJ, Abarca ML, Cabañes FJ (2002) Occurrence of *Malassezia* spp. in the external ear canals of dogs and cats with and without otitis externa. Med Mycol 40:115–121
53. Crespo-Erchiga V, Ojeda Martos A, Vera Casaño A et al (2000) *Malassezia globosa* as the causative agent of pityriasis versicolor. Br J Dermatol 143:799–803
54. Damoser J, Flatscher J, Mikula M et al (1992) Mykologische Untersuchungen desäuβeren Gehörgangs von Füchsen. Tierärztl Umsch 47:193–195
55. Darabi K, Hostetler SG, Bechtel MA et al (2009) The role of *Malassezia* in atopic dermatitis affecting the head and neck of adults. J Am Acad Dermatol 60:125–136
56. De Baere T, Claeys G, Swinne D et al (2002) Identification of cultured isolates of clinically important yeast species using fluorescent fragment length analysis of the amplified internally transcribed rRNA spacer 2 region (ITS2). BMC Microbiol 2:21
57. Diaz MR, Boekhout T, Theelen B et al (2006) Barcoding and flow cytometry as a high-throughput identification system for *Malassezia* species. J Med Microbiol 55:1197–1209
58. Dinsdale JR, Rest JR (1995) Yeast infection in ferrets. Vet Rec 137:647–648
59. Dizotti CE, Coutinho SD (2007) Isolation of *Malassezia pachydermatis* and *M. sympodialis* from the external ear canal of cats with and without otitis externa. Acta Vet Hung 55:471–477
60. Dodge CW (1935) Fungous diseases of men and other mammals. Medical mycology. CV Mosby Co, Saint Louis
61. Duarte ER, Melo MM, Hahn RC et al (1999) Prevalence of *Malassezia* spp. in the ears of asymptomatic cattle and cattle with otitis in Brazil. Med Mycol 37:159–162
62. Duarte ER, Resende JC, Rosa CA et al (2001) Prevalence of yeasts and mycelial fungi in bovine parasitic otitis in the State of Minas Gerais, Brazil. J Vet Med B Infect Dis Vet Public Health 48:631–635
63. Duarte ER, Lachance MA, Hamdan JS (2002) Identification of atypical strains of *Malassezia* spp. from cattle and dog. Can J Microbiol 48:749–752
64. Duarte ER, Batista RD, Hahn RC et al (2003) Factors associated with the prevalence of *Malassezia* species in the external ears of cattle from the state of Minas Gerais, Brazil. Med Mycol 41:137–142
65. Duarte ER, Hahn RC, Hamdan JS (2004) Detection of *Malassezia* spp. from healthy cattle and cattle with otitis by direct examination and culture. Acta Sci Vet 32:83–87
66. Dufait R (1985) Présence de *Malassezia pachydermatis* (syn. *Pityrosporum canis*) sur les poils et les plumes d'animaux domestiques. Bull Soc Fr Mycol Méd 14:19–22
67. Farr PM, Krause LB, Marks JM et al (1985) Response of scalp psoriasis to oral ketoconazole. Lancet 2:921–922
68. Fell JW, Scorzetti G, Connell L et al (2006) Biodiversity of micro-eukaryotes in Antarctic Dry Valley soils with <5% soil moisture. Soil Biol Biochem 38:3107–3115
69. Fraser G (1961) *Pityrosporum pachydermatis* Weidman of canine origin. Trans Brit Mycol Soc 44:441–448
70. Fry L, Baker BS (2007) Triggering psoriasis: the role of infections and medications. Clin dermatol 25:606–615
71. Gaitanis G, Velegraki A, Frangoulis E et al (2002) Identification of *Malassezia* species from patient skin scales by PCR-RFLP. Clin Microbiol Infect 8:162–173
72. Gaitanis G, Robert V, Velegraki A (2006) Verifiable single nucleotide polymorphisms of the internal transcribed spacer 2 region for the identification of 11 *Malassezia* species. J Dermatol Sci 43:214–217
73. Gaitanis G, Velegraki A, Alexopoulos EC et al (2006) Distribution of *Malassezia* species in pityriasis versicolor and seborrhoeic dermatitis in Greece. Typing of the major pityriasis versicolor isolate *M. globosa*. Br J Dermatol 154:854–859
74. Gaitanis G, Velegraki A, Alexopoulos EC et al (2009) *Malassezia furfur* fingerprints as possible markers for human phylogeography. ISME J 3:498–502

75. Galuppi R, Tampieri MP (2008) Epidemiology and variability of *Malassezia* spp. Parasitologia 50:73–76
76. Gandra RF, Simão RC, Matsumoto FE et al (2006) Genotyping by RAPD-PCR analyses of *Malassezia furfur* strains from pityriasis versicolor and seborrhoeic dermatitis patients. Mycopathologia 162:273–280
77. Gandra RF, Gambale W, de Cassia Garcia Simao R et al (2008) *Malassezia* spp. in acoustic meatus of bats (*Molossus molossus*) of the Amazon Region, Brazil. Mycopathologia 165:21–26
78. Garau M, del Palacio A, García J (2005) Prevalence of *Malassezia* spp. in healthy pigs. Mycoses 48:17–20
79. Gasser RB, Hu M, Chilton NB et al (2006) Single-strand conformation polymorphism (SSCP) for the analysis of genetic variation. Nat Protoc 1:3121–3128
80. Gemmer CM, DeAngelis YM, Theelen B et al (2002) Fast, noninvasive method for molecular detection and differentiation of *Malassezia* yeast species on human skin and application of the method to dandruff microbiology. J Clin Microbiol 40:3350–3357
81. Girão MD, Prado MR, Brilhante RS et al (2006) *Malassezia pachydermatis* isolated from normal and diseased external ear canals in dogs: a comparative analysis. Vet J 172:544–548
82. Giusiano G, Bustillo S, Mangiaterra M et al (2003) Identification of *Malassezia* species by PCR-REA. Rev Argent Microbiol 35:162–166
83. Gordon MA (1979) *Malassezia (Pityrosporum) pachydermatis* (Weidman) Dodge 1935. Sabouraudia 17:305–309
84. Gründer S, Mayser P, Redmann T et al (2005) Mycological examinations on the fungal flora of the chicken comb. Mycoses 48:114–119
85. Guého E, Guillot J (1999) Comments on *Malassezia* species from dogs and cats. Mycoses 42:673–674
86. Guého E. Meyer SA (1989) A reevaluation of the genus *Malassezia* by means of genome comparison. Antonie van Leeuwenhoek 55:245–253
87. Guého E, Midgley G, Guillot J (1996) The genus *Malassezia* with description of four new species. Antonie van Leeuwenhoek 69:337–355
88. Guillot J (1993) Importance du genre *Malassezia* chez les carnivores domestiques. Faculté Méd Créteil, Univ Paris XII, Alfort, ENV Alfort
89. Guillot J (1995) Taxonomie et phylogénie des levures du genre *Malassezia*. Thèse de Doctorat d'Université, Univ Paris XII, Sciences de la Vie et de la Santé, option Parasitologie, Paris
90. Guillot J, Guého E (1995) The diversity of *Malassezia* yeasts confirmed by rRNA sequence and nuclear DNA comparisons. Antonie van Leeuwenhoek 67:297–314
91. Guillot J, Chermette R, Guého E (1994) Prévalence du genre *Malassezia* chez les mammifères. J Mycol Méd 4:72–79
92. Guillot J, Guého E, Chévrier G et al (1997) Epidemiological analysis of *Malassezia pachydermatis* isolates by partial sequencing of the large subunit ribosomal RNA. Res Vet Sci 62:22–25
93. Guillot J, Guého E, Mialot M et al (1998) Importance des levures du genre *Malassezia* en dermatologie vétérinaire. Le Point Vétérinaire 29:21–31
94. Guillot J, Petit T, Degorce-Rubiales F et al (1998) Dermatitis caused by *Malassezia pachydermatis* in a California sea lion (*Zalophus californianus*). Vet Rec 142:311–312
95. Guillot J, Breugnot C, de Barros M et al (1998) Usefulness of modified Dixon's medium for quantitative culture of *Malassezia* species from canine skin. J Vet Diagn Invest 10:384–386
96. Guillot J, Deville M, Berthelemy M et al (2000) A single PCR-restriction endonuclease analysis for rapid identification of *Malassezia* species. Lett Appl Microbiol 31:400–403
97. Guillot J, Bensignor E, Jankowski F et al (2003) Comparative efficacies of oral ketoconazole and terbinafine for reducing *Malassezia* population sizes on the skin of basset hounds. Vet Dermatol 14:153–157
98. Gupta AK, Bluhm R (2004) Seborrheic dermatitis. J Eur Acad Dermatol Venereol 18:13–26
99. Gupta AK, Kohli Y (2004) Prevalence of *Malassezia* species on various body sites in clinically healthy subjects representing different age groups. Med Mycol 42:35–42
100. Gupta AK, Kohli Y, Summerbell RC (2000) Molecular differentiation of seven *Malassezia* species. J Clin Microbiol 38:1869–1875
101. Gupta AK, Kohli Y, Summerbell RC et al (2001) Quantitative culture of *Malassezia* species from different body sites of individuals with or without dermatoses. Med Mycol 39:243–251

102. Gupta AK, Batra R, Bluhm R et al (2004) Skin diseases associated with *Malassezia* species. J Am Acad Dermatol 51:785–798
103. Gupta AK, Boekhout T, Theelen B et al (2004) Identification and typing of *Malassezia* species by amplified fragment length polymorphism (AFLP) and sequence analysis of the internal transcribed spacer (ITS) and large subunit (LSU) regions of ribosomal DNA. J Clin Microbiol 42:4253–4260
104. Gustafson BA (1955) Otitis externa in the dog. A bacteriological and experimental study, Royal Veterinary College of Sweden, Stockholm
105. Gustafson BA (1959) Lipophilic yeasts belonging to genus *Pityrosporum*, found in swine. Acta Pathol Microbiol Scand 45:275–280
106. Gustafson BA (1960) The occurrence of yeasts belonging to genus *Pityrosporum* in different kinds of animals. Acta Pathol Microbiol Scand 48:51–55
107. Hajsig M, Topolko S, Hajsig D et al (1982) *Pityrosporum canis* in some wild mammals in Croatia. Vet Arhiv 52:183–187
108. Hajsig M, Tadić V, Lukman P (1985) *Malassezia pachydermatis* in dogs: significance of its location. Vet Arhiv 55:259–266
109. Hajsig D, Hajsig M, Svoboda-Vuković D (1990) *Malassezia pachydermatis* in healthy cats. Vet Arhiv 60:69–73
110. Hall L, Wohlfiel S, Roberts GD (2003) Experience with the MicroSeq D2 large-subunit ribosomal DNA sequencing kit for identification of commonly encountered, clinically important yeast species. J Clin Microbiol 41:5099–5102
111. Heng MC, Henderson CL, Baker DC et al (1990) Correlation of *Pityrosporum ovale* density with clinical severity of seborrheic dermatitis as assessed by simplified technique. J Am Acad Dermatol 23:82–86
112. Hirai A, Kano R, Makimura K et al (2002) A unique isolate of *Malassezia* from a cat. J Vet Med Sci 64:957–959
113. Hirai A, Kano R, Makimura K et al (2004) *Malassezia nana* sp. nov., a novel lipid dependent yeast species isolated from animals. Int J Syst Evol Microbiol 54:623–627
114. Hossain H. Landgraf V, Weiss R, et al (2007) Genetic and biochemical characterization of *Malassezia pachydermatis* with particular attention to pigment-producing subgroups. Med Mycol 45:41–49
115. Howell SA, Quin C, Midgley G (1993) Karyotype of oval forms of *Malassezia furfur*. Mycoses 36:263–266
116. Kalinowska-Pujdak A, Schmalreck A et al (2006) Species differentiation of yeasts of the genus *Malassezia* with Fourier transform infrared spectroscopy. Hautarzt 57:127–136
117. Kano R, Aizawa T, Nakamura Y et al (1999) Chitin synthase 2 gene sequence of *Malassezia* species. Microbiol Immunol 43:813–815
118. Kato H, Sugita T, Ishibashi Y et al (2006) Detection and quantification of specific IgE antibodies against eight *Malassezia* species in sera of patients with atopic dermatitis by using an enzyme-linked immunosorbent assay. Microbiol Immunol 50:851–856
119. Kennis RA, Rosser EJ Jr, Olivier NB et al (1996) Quantity and distribution of *Malassezia* organisms on the skin of clinically normal dogs. J Am Vet Med Assoc 208:1048–1051
120. Kocková-Kratochvílová A, Ladzianská K, Bučko S (1987) *Malassezia pachydermatis* in small animals. Mykosen 30:541–543
121. Kumar A, Singh K, Sharma A (2002) Prevalence of *Malassezia pachydermatis* and other organisms in healthy and infected dogs ears. Israel J Vet Med 57:145–148
122. Kuttin ES, Glas I (1985) Mycotic otitis externa in animals. Mykosen 28:61–68
123. Kuttin ES, Müller J (1994) The fungal flora of zoo animals' ears. Mycoses 37:59–60
124. Lai X, Cao L, Tan H et al (2007) Fungal communities from methane hydrate-bearing deep-sea marine sediments in South China sea. ISME J 1:756–762
125. Lee YW, Kim SM, Oh BH et al (2008) Isolation of 19 strains of *Malassezia dermatis* from healthy human skin in Korea. J Dermatol 35:772–777
126. Leśniak M, Dworecka-Kaszak B (2006) RAPD differentiation of yeast like fungi *Malassezia pachydermatis*. Wiad Parazytol 52:299–304
127. Lukman P (1982) Findings of *Pityrosporum canis* in the organism of healthy and sick dogs. Vet Arhiv 52:37–44

128. Lyskova P, Vydrzalova M, Mazurova J (2007) Identification and antimicrobial susceptibility of bacteria and yeasts isolated from healthy dogs and dogs with otitis externa. J Vet Med A 54:559–563
129. Makimura K, Tamura Y, Kudo M et al (2000) Species identification and strain typing of *Malassezia* species stock strains and clinical isolates based on the DNA sequences of nuclear ribosomal internal transcribed spacer 1 regions. J Med Microbiol 49:29–35
130. Mancianti F, Rum A, Nardoni S et al (2001) Extracellular enzymatic activity of *Malassezia* spp. isolates. Mycopathol 149:131–135
131. Marcelou-Kinti O, Hadzivassiliou M, Kapetanakis I et al (1973) Isolation of yeast-like organisms of the genus *Pityrosporum* from the sand. Acta Microbiol Hellenica 18:293–297
132. Marouteix L (1994) La flore fongique cutanée du chien séborrhéique. Rôle des levures du genre *Malassezia* et essai de traitement par l'énilconazole, Thesis, Faculté de Médecine, Nantes
133. Masuda A, Sukegawa T, Mizumoto N et al (2000) Study of lipid in the ear canal in canine otitis externa with *Malassezia pachydermatis*. J Vet Med Sci 62:1177–1182
134. Matousek JL, Campbell KL, Kakoma I et al (2003) Evaluation of the effect of pH on *in vitro* growth of *Malassezia pachydermatis*. Can J Vet Res 67:56–59
135. Mauldin EA, Scott DW, Miller WH et al (1997) *Malassezia* dermatitis in the dog: a retrospective histopathological and immunopathological study of 86 cases (1990–95). Vet Dermatol 8:191–202
136. Midgley G (1989) The diversity of *Pityrosporum* (*Malassezia*) yeasts *in vivo* and *in vitro*. Mycopathol 106:143–153
137. Midgley G, Clayton YM (1969) The yeast flora of birds and mammals in captivity. Antonie van Leeuwenhoek 35:23–24
138. Midreuil F, Guillot J, Guého E et al (1999) Genetic diversity in the yeast species *Malassezia pachydermatis* analysed by multilocus enzyme electrophoresis. Int J Syst Bacteriol 49:1287–1294
139. Mirhendi H, Makimura K, Zomorodian K et al (2005) A simple PCR-RFLP method for identification and differentiation of 11 *Malassezia* species. J Microbiol Methods 61:281–284
140. Montero CI, Shea YR, Jones PA et al (2008) Evaluation of pyrosequencing technology for the identification of clinically relevant non-dematiaceous yeasts and related species. Eur J Clin Microbiol Infect Dis 27:821–830
141. Morishita N, Sei Y, Sugita T (2006) Molecular analysis of *Malassezia* microflora from patients with pityriasis versicolor. Mycopathol 161:61–65
142. Morris DV (1999) *Malassezia* dermatitis and otitis. Vet Clin North Am Small Anim Pract 29:1303–1310
143. Muyzer G, Smalla K (1998) Application of denaturing gradient gel electrophoresis (DGGE) and temperature gradient gel electrophoresis (TGGE) in microbial ecology. Antonie van Leeuwenhoek 73:127–141
144. Muyzer G, de Waal EC, Uitterlinden AG (1993) Profiling of complex microbial populations by denaturing gradient gel electrophoresis analysis of polymerase chain reaction-amplified genes coding for 16 S rRNA. Appl Environ Microbiol 59:695–700
145. Nakabayashi A, Sei Y, Guillot J (2000) Identification of *Malassezia* species isolated from patients with seborrhoeic dermatitis, atopic dermatitis, pityriasis versicolor and normal subjects. Med Mycol 38:337–341
146. Nakagaki K, Hata K, Iwata E et al (2000) *Malassezia pachydermatis* isolated from a South American sea lion (*Otaria byronia*) with dermatitis. J Vet Med Sci 62:901–903
147. Nardoni S, Mancianti F, Rum A et al (2005) Isolation of *Malassezia* species from healthy cats and cats with otitis. J Feline Med Surg 7:141–145
148. Nardoni S, Dini M, Taccini F et al (2007) Occurrence, distribution and population size of *Malassezia pachydermatis* on skin and mucosae of atopic dogs. Vet Microbiol 122:172–177
149. Nell A, James SA, Bond CJ et al (2002) Identification and distribution of a novel *Malassezia* species yeast on normal equine skin. Vet Rec 150:395–398
150. Nicklas W (1979) Untersuchungen zur Keimflora in äusseren Gehörgang von Katzen. Kleintier Praxis 24:107–111
151. Nikkels AF, Piérard GE (2003) Framing the future of antifungals in atopic dermatitis. Dermatol 206:398–400

152. Nobre MO, Castro AP, Nascente PS et al (2001) Occurrency of *Malassezia pachydermatis* and other infectious agents as cause of external otitis in dogs from Rio Grande do Sul State, Brazil (1996/1997). Braz J Microbiol 32:245–249
153. O'Donnell K, *Fusarium* and its near relatives, In: Reynolds DR and Taylor JW (1993), The fungal holomorph: mitotic, meiotic and pleomorphic speciation in fungal systematic. CAB International, Wallingford, UK, pp 225–233
154. Omodo-Eluk AJ, Baker KP, Fuller H (2003) Comparison of two sampling techniques for the detection of *Malassezia pachydermatis* on the skin of dogs with chronic dermatitis. Vet J 165:119–124
155. Ordeix L, Galeotti F, Scarampella F et al (2007) *Malassezia* spp. overgrowth in allergic cats. Vet Dermatol 18:316–323
156. Pappalardo E, Martino PA, Noli C (2002) Macroscopic, cytological and bacteriological evaluation of anal sac content in normal dogs and in dogs with selected dermatological diseases. Vet Dermatol 13:315–322
157. Paulino LC, Tseng CH, Strober BE et al (2006) Molecular analysis of fungal microbiota in samples from healthy human skin and psoriatic lesions. J Clin Microbiol 44: 2933–2941
158. Paulino LC, Tseng CH, Blaser MJ (2008) Analysis of *Malassezia* microbiota in healthy superficial human skin and in psoriatic lesions by multiplex real-time PCR. FEMS Yeast Res 8:460–471
159. Perrins N, Gaudiano F, Bond R (2007) Carriage of *Malassezia* spp. yeasts in cats with diabetes mellitus, hyperthyroidism and neoplasia. Med Mycol 45:541–546
160. Peterson SW, Kurtzman CP (1991) Ribosomal RNA sequence divergence among sibling species of yeasts. System Appl Microbiol 14:124–129
161. Pin D (2004) Seborrhoeic dermatitis in a goat due to *Malassezia pachydermatis*. Vet Dermatol 15:53–56
162. Pinter L, Noble WC (1998) Stomatitis, pharyngitis and tonsillitis caused by *Malassezia pachydermatis* in a dog. Vet Dermatol 9:257–261
163. Pinter L, Anthony RM, Glumac N et al (2002) Apparent cross-infection with a single strain of *Malassezia pachydermatis* on a pig farm. Acta Vet Hung 50:151–156
164. Pitkäranta M, Meklin T, Hyvärinen A et al (2008) Analysis of fungal flora in indoor dust by ribosomal DNA sequence analysis, quantitative PCR, and culture. Appl Environ Microbiol 74:233–244
165. Plant JD, Rosenkrantz WS, Griffin CE (1992) Factors associated with and prevalence of high *Malassezia pachydermatis* numbers on dog skin. J Am Vet Med Assoc 201:879–882
166. Pochi PE, Strauss JS, Downing DT (1979) Age-related changes in sebaceous gland activity. J Invest Dermatol 73:108–111
167. Prado MR, Brito EHS, Girão MD et al (2004) Higher incidence of *Malassezia pachydermatis* in the eyes of dogs with corneal ulcer than in healthy dogs. Vet Microbiol 100:115–120
168. Prado MR, Brilhante RSN, Cordeiro RA et al (2008) Frequency of yeasts and dermatophytes from healthy and diseased dogs. J Vet Diagn Invest 20:197–202
169. Preziosi DE, Morris DO, Johnston MS et al (2006) Distribution of *Malassezia* organisms on the skin of unaffected psittacine birds and psittacine birds with feather-destructive behavior. J Am Vet Med Assoc 228:216–221
170. Prohić A (2003) Identification of *Malassezia* species isolated from scalp skin of patients with psoriasis and healthy subjects. Acta Dermatovenerol Croat 11:10-16
171. Prohić A, Ozegovic L (2007) *Malassezia* species isolated from lesional and non-lesional skin in patients with pityriasis versicolor. Mycoses 50: 58–63
172. Raabe P, Mayser P, Weiss R (1998) Demonstration of *Malassezia furfur* and *M. sympodialis* together with *M. pachydermatis* in veterinary specimens. Mycoses 41:493–500
173. Raabe P, Mayser P, Weiss R (2000) Further comments on *Malassezia* species from dogs and cats. Mycoses 43:437
174. Randjandiche M (1979) Le genre *Pityrosporum* Sabouraud 1904. Université Liège, Faculté de Médecine Vétérinaire, Liège, Thesis

175. Renker C, Alphei J, Buscot F (2003) Soil nematodes associated with the mammal pathogenic fungal genus *Malassezia* (Basidiomycota: Ustilaginomycetes) in Central European forests. Biol Fertil Soils 37:70–72
176. Roé E, Garcia Muret MP, Marcuello E et al (2006) Description and management of cutaneous side effects during cetuximab or erlotinib treatments: a prospective study of 30 patients. J Am Acad Dermatol 55:429–437
177. Saëz H (1982) *Pityrosporum pachydermatis*, caractéristiques morpho-biochimiques; fréquence comparée chez l'animal et chez l'homme. Ann Méd Vét 126:645–650
178. Salkin IF, Gordon MA, Stone WB (1978) *Pityrosporum pachydermatis* in a black bear (*Ursus americanus*). Sabouraudia 16:35–38
179. Salkin IF, Stone WB, Gordon MA (1980) Association of *Malassezia (Pityrosporum) pachydermatis* with sarcoptic mange in New York State. J Wildlife Dis 16:509–514
180. Sandström Falk MH, Tengvall Linder M, Johansson C et al (2005) The prevalence of *Malassezia* yeasts in patients with atopic dermatitis, seborrhoeic dermatitis and healthy controls. Acta Derm Venereol 85:17–23
181. Scott DW, Miller WH, Griffin CE, Fungal skin diseases, In: Scott DW, Miller WH, Griffin CE (2001) Muller and Kirk's small animal dermatology. WB Saunders, Philadelphia, pp 363-422
182. Senczek D, Siesenop U, Böhm KH (1999) Characterization of *Malassezia* species by means of phenotypic characteristics and detection of electrophoretic karyotypes by pulsed-filed gel electrophoresis (PFGE). Mycoses 42:409–414
183. Spoor HJ, Traub EF, Bell M (1954) *Pityrosporum ovale* types cultured from normal and seborrhoeic subjects. Arch Derm Syphilol 63:323–330
184. Sugita T (2007) The 50th Anniversary Educational Symposium of the Japanese Society for Medical Mycology: Mycological study on *Malassezia* Nippon Ishinkin Gakkai Zasshi 48:179–182
185. Sugita T, Nishikawa A (2003) Molecular and quantitative analyses of *Malassezia* microflora on the skin of atopic dermatitis patients and genotyping of *M. globosa* DNA. Nippon Ishinkin Gakkai Zasshi 44:61–64
186. Sugita T, Nishikawa A, Ikeda R et al (1999) Identification of medically relevant *Trichosporon* species based on sequences of internal transcribed spacer regions and construction of a database for *Trichosporon* identification. J Clin Microbiol 37:1985–1993
187. Sugita T, Takashima M, Ikeda R et al (2000) Phylogenetic and taxonomic heterogeneity of *Cryptococcus humicolus* by analysis of the sequences of the internal transcribed spacer regions and 18 S rDNA, and the phylogenetic relationships of *C. humicolus*, *C. curvatus*, and the genus *Trichosporon*. Microbiol Immunol 44:455–461
188. Sugita T, Suto H, Unno T et al (2001) Molecular analysis of *Malassezia* microflora on the skin of atopic dermatitis patients and healthy subjects. J Clin Microbiol 39:3486–3490
189. Sugita T, Nakajima M, Ikeda R et al (2002) Sequence analysis of the ribosomal DNA intergenic spacer 1 regions of *Trichosporon* species. J Clin Microbiol 40:1826–1830
190. Sugita T, Kodama M, Saito M et al (2003) Sequence diversity of the intergenic spacer region of the rRNA gene of *Malassezia globosa* colonizing the skin of patients with atopic dermatitis and healthy individuals. J Clin Microbiol 41:3022–3027
191. Sugita T, Tajima M, Amaya M et al (2004) Genotype analysis of *Malassezia restricta* as the major cutaneous flora in patients with atopic dermatitis and healthy subjects. Microbiol Immunol 48:755–759
192. Sugita T, Takeo K, Hama K et al (2005) DNA sequence diversity of intergenic spacer I region in the non-lipid-dependent species *Malassezia pachydermatis* isolated from animals. Med Mycol 43:21–26
193. Sugita T, Tajima M, Tsubuku H et al (2006) Quantitative analysis of cutaneous *Malassezia* in atopic dermatitis patients using real-time PCR. Microbiol Immunol 50:549–552
194. Sugita T, Suzuki M, Goto S et al (2009) Quantitative analysis of the cutaneous *Malassezia* microbiota in 770 healthy Japanese by age and gender using a real-time PCR assay. Med Mycol PMID: 19462267

195. Tajima M, Sugita T, Nishikawa A, et al (2008) Molecular analysis of *Malassezia* microflora in seborrheic dermatitis patients: comparison with other diseases and healthy subjects.J Invest Dermatol 128:345-351
196. Takahata Y, Sugita T, Hiruma M (2007) Quantitative analysis of *Malassezia* in the scale of patients with psoriasis using a real-time polymerase chain reaction assay. Br J Dermatol 157:670–673
197. Takahata Y, Sugita T, Kato H et al (2007) Cutaneous *Malassezia* flora in atopic dermatitis differs between adults and children. Br J Dermatol 157:1178–1182
198. Tarazooie B, Kordbacheh P, Zaini F et al (2004) Study of the distribution of *Malassezia* species in patients with pityriasis versicolor and healthy individuals in Teheran. Iran. BMC Dermatol 4:5
199. Theelen B, Silvestri M, Guého E et al (2001) Identification and typing of *Malassezia* yeasts using amplified fragment length polymorphism (AFLP^{Tm}) and denaturing gradient gel electrophoresis (DGGE). FEMS Yeast Res 1:79–86
200. Uzal FA, Paulson D, Eigenheer AL et al (2007) *Malassezia slooffiae*-associated dermatitis in a goat. Vet Dermatol 18:348–352
201. Van Belkum A, Boekhout T, Bosboom R (1994) Monitoring spread of *Malassezia* infections in a neonatal intensive care unit by PCR-mediated genetic typing. J Clin Microbiol 32:2528–2532
202. Vos P, Hogers R, Bleeker M et al (1995) AFLP: a new technique for DNA fingerprinting. Nucl Acids Res 23:4407–4414
203. Wagner R, Schadler S (2000) Qualitative study of *Malassezia* species colonisation in young puppies. Vet Rec 147:192–194
204. Weidman F (1925) Exfoliative dermatitis in the Indian rhinoceros (*Rhinoceros unicornis*) with description of a new species, *Pityrosporum pachydermatis*. Rep Lab Mus Comp Pathol Zoo Soc 53:36–43
205. Wesche P, Bond R (2003) Isolation of *Malassezia pachydermatis* from the skin of captive rhinoceroses. Vet Rec 153:404–405
206. White TJ, Burns T, Lee S et al, Amplification and direct sequencing of fungal ribosomal RNA genes for phylogenetics, In: Innis MA, Gelfand DH, Sninsky JJ et al (1990), PCR protocols: a guide to methods and applications. Academic Press, San Diego, California, pp 315–322
207. White SD, Vandenabeele SI, Drazenovich NL et al (2006) *Malassezia* species isolated from the intermammary and preputial fossa areas of horses. J Vet Intern Med 20:395–398
208. Xu J, Saunders CW, Hu P et al (2007) Comparative genomics of dandruff-associated *Malassezia* species reveals convergent and divergent virulence traits with plant and human fungal pathogens. Proc Nat Acad Sc USA 104:18730–18735
209. Yamada Y, Makimura K, Ueda K (2003) DNA base alignment and taxonomic study of genus *Malassezia* based upon partial sequences of mitochondrial large subunit ribosomal RNA gene. Microbiol Immunol 47:475–478
210. Yamamoto A, Serizawa S, Ito M, Sato Y (1987) Effect of aging on sebaceous gland activity and on the fatty acid composition of wax esters. J Invest Dermat 89:507–512
211. Zhang N, Suh SO, Blackwell M (2003) Microorganisms in the gut of beetles: evidence from molecular cloning. J Invertebr Pathol 84:226–233
212. Zomorodian K, Mirhendi H, Tarazooie B et al (2008) Molecular analysis of *Malassezia* species isolated from hospitalized neonates. Pediatr Dermatol 25:312–316
213. Zomorodian K, Mirhendi H, Tarazooie B et al (2008) Distribution of *Malassezia* species in patients with psoriasis and healthy individuals in Tehran. J Cutan Pathol 35:1027–1031

Physiology and Biochemistry 4

Peter Mayser and George Gaitanis

Core Messages

- Due to the ecologic niche of *Malassezia* yeasts, host-specific adaptations are an important issue in their biology. The commensal status of *Malassezia* yeasts is not clearly distinguished from the pathogenic stage, as transition from the one to the other is probably a continuum and not an on/off condition. Part of this pathogenic potential is determined by the activities of the enzymatic systems of *Malassezia* yeasts. The efficiency with which *Malassezia* yeasts utilize nutrients on the skin surface and in the sebaceous gland determines the size of their population, and also the quality and quantity of the produced metabolic by-products. This organic material ranges from free fatty acids that apply their action through an irritating or toxic effect to highly bioactive indole derivatives that bind to specific cellular receptors and regulate the expression of downstream metabolic pathways. The recently reported genome and secretory proteome of *M. globosa* and, in part, of *M. restricta* provide a molecular basis to understand the adaptions of *Malassezia* yeasts to their environment and to identify factors of pathogenicity.
- As many experiments on *Malassezia* biochemistry and physiology had been performed before the current taxonomy of *Malassezia* species was developed, and the corresponding strains were not deposited in official fungal collections, the genus name *Malassezia* will be used to include strains that were categorized as *Malassezia furfur* (*sensu lato*), *Pityrosporum ovale* and *Pityrosporum orbiculare* (see Chap. 2).

P. Mayser (✉)
Department of Dermatology and Andrology, Justus Liebig University Giessen, Gaffkystrasse 14, 35385 Giessen, Germany
e-mail: peter.mayser@derma.med.uni-giessen.de

T. Boekhout et al. (eds.), *Malassezia and the Skin*
DOI: 10.1007/978-3-642-03616-3_4,

4.1 Nutritional Requirements

With the exception of *M. pachydermatis*, all *Malassezia* species require an external lipid source for growth, i.e., they are obligatorily lipid-dependent. On the other hand, these lipids are thought to be an important virulence factor of the fungus as the lipid-rich cell wall seems to protect *Malassezia* from phagocytosis [1] and to downregulate the inflammatory immune response [2]. In addition, adhesion to host cells may be mediated by the hydrophobicity of this highly lipophilic cell wall.

4.1.1 Lipid Metabolism

Lipid dependence is known to be based on a defect in the synthesis of myristic acid, which serves as the precursor of long-chained fatty acids [3–5]. At the molecular level, this metabolic deficiency can be explained by the absence of a gene encoding fatty acid synthase in the genomes of *M. globosa* and *M. restricta* [6], which is compensated by an abundance of genes encoding hydrolases that could provide fatty acids. Longer chain fatty acids can be synthesized from mid-chain fatty acids. Transformation from saturated fatty acids into unsaturated ones is possible [3]. Nonetheless, the lipid composition of these yeasts is not constant, but reflects to a great extent the nutritive lipid offer [4, 7, 8]. In most cases, triacylglycerides, free fatty acids or polyoxysorbitan fatty acid esters (i.e., Tween®) are used as substrates. Apart from hydrolysis of neutral fat, which has been known for a long time, assimilation of phospholipids has also been described [9]. Assimilation of steroids has been demonstrated for cholesterol and cholesterol esters, leading to the induction of hyphae [8].

4.1.2 General Metabolism

Except the lipid metabolism, little is known about the growth conditions and metabolic activity of *Malassezia* species. As standard assimilation tests cannot be performed because of the lipid dependence, only few studies are available. Carbohydrate assimilation has been determined solely for *M. pachydermatitis,* which can assimilate mannitol, glycerol, and sorbitol as the sole source of carbohydrate [10, 11]. The capability to ferment sugars has not been found in any of the *Malassezia* species, which confirms the basidiomycetous nature of *Malassezia* yeasts [12]. *Malassezia* yeast cells were able to grow in vitamin-free basal media that contained lipids as the sole source of carbon [3]. Enhanced growth was observed after addition of asparagine, thiamine and pyridoxine, but these are not essential growth agents [3, 13]. Anorganic sulfate and sulfite were not utilized [14]. *Malassezia* yeasts utilize both organic and anorganic nitrogen sources [11], but are unable to assimilate potassium nitrate.

4.2 Cellular Wall

The cell wall of *Malassezia* spp. shows a unique structure [15–17]. It is approximately 0.12 µm thick, multilayered, consisting of the outer lamella, the multilayered wall, and the plasma membrane [16, 18–21]. A characteristic feature is the corrugate pattern of the plasma membrane, which may be visible by light microscopy and is surrounded by an electron lucent band [11] (see also Chap. 2.3). Functions of single compounds of the cell wall have not yet been defined. With a lipid percentage of 15%, the cell wall of *Malassezia* spp. shows a significantly higher proportion of lipids than that of *Saccharomyces* spp. (viz., 1–2%) [22]. The construction of the whole cell sheath with the high lipid content is presumably responsible for the great resistance of *Malassezia* cells to external influences, e.g., they show a high mechanical rigidity and osmoresistance [14].

4.3 Production of Filaments

The hyphal stage, observed by Eichstedt in 1846 [23], is difficult to induce by culture, although some isolates of *M. furfur* can spontaneously produce hyphae. Dorn and Roehnert [24] achieved this by adding glycine to the medium, and in the same year, Porro et al. [8] successfully used a medium that contained cholesterol and cholesterol esters. Faergemann and Bernander [25] induced hyphal growth under increased carbon dioxide tension and observed filamentous growth of *P. ovale* on human stratum corneum *in vitro*. Most hyphae were produced by the *P. ovale* strains ATCC 44341 and ATCC 44031, both of which are now reclassified as *M. sympodialis*. This is in line with a later finding by Saadatzadeh et al. [26] who tested several *Malassezia* strains and found that only strains of serovar A = *M. sympodialis* produced hyphae under their conditions. Although *M. globosa* is thought to be the main causative agent for PV [27], mycelial growth has not yet been induced *in vitro* for this species. However, initial hyphal formation is occasionally observed in primary cultures [11, 25, 28–33] of this species isolated from PV. Only little is known about the hyphal stage. These hyphae, which do not present any septal pores, are now considered to be pseudohyphae and are discussed in more detail in Chap. 2.3. The hyphal stage is thought to play a role in the pathogenesis of PV, as hyphae are abundantly present in scales taken from lesions [30]. The frequency of hyphae in patients with PV is 100% in lesions, 42% in nonlesional skin of the trunk and 50% on the head. However, in 6–7%, they are also observed to occur on healthy skin [34].

4.4 Enzymatic Activities

The need for *Malassezia* species to assimilate fatty acids from external sources is reflected in the presence of multiple genes for secreted lipases and phospholipases. RT-PCR and proteomics experiments confirmed their expression on human scalp [6].

4.4.1 Lipase

In *Malassezia* yeasts, hydrolysis of triacylglycerides was demonstrated during release of glycerol and free fatty acids [12, 29]. A membrane-bound extracellullar lipase was shown *in vitro* by Catterall et al. [35]. The enzyme is extracellularly located on the membrane surface after generation within the cell and extruded through the membrane. However, in scales of PV, identical histochemistry and electron microscopy techniques failed to demonstrate this enzyme *in vivo*.

Using different fatty acid (mono-)esters, Mayser et al. [36] found that yeast-dependent hydrolysis was critically dependent on alcohol moiety of these synthetic fatty acid (mono-) esters, while growth was best stimulated by unsaturated fatty acids. The authors observed greater extracellular lipase activity and more pronounced production of fatty acid ethyl esters in *M. sympodialis* than in *M. furfur* [37]. Differences occur in the lipid metabolism of *Malassezia* species. In particular, *M. furfur* was observed to assimilate ricinoleic acid (12-hydroxy oleic acid) and PEG-35 castor oil (Cremophor EL), and these features are now used for species differentiation [38, see also Chap. 2.1]. Ran et al. [39] did not detect lipases in the supernatant, but localized the main lipolytic activity in the insoluble fraction of cell extracts. In contrast, Plotkin et al. [40] demonstrated lipolytic activity in the supernatant, and in intracellular soluble and insoluble extracts of *M. furfur,* and further characterized three different lipolytic activities in the soluble fraction.

Cloning and characterization of the first gene encoding a secreted lipase of *M. furfur* was reported by Brunke and Hube [41]. The gene, *MfLIP1*, shows high sequence similarities to other known extracellular lipases, but is not a member of the lipase gene family in *M. furfur. MfLIP1* consists of 1,464 base pairs coding for the protein with a molecular weight of 54.3 kDa, which has a conserved lipase motif and an N-terminal signal peptide of 26 amino acids. The cDNA of *MfLIP1* was expressed in *Pichia pastoris* and the biochemical properties of the recombinant lipase were analyzed. MfLip1 is most active at 40°C and the pH optimum was found to be 5.8. The lipase hydrolyzed lipids, such as Tweens, which are frequently used as the source of carbon in growth media for *Malassezia*, and had minor phospholipase activities. Two more lipases were characterized shortly after the characterization of MfLip1. The first one was a secreted lipase from *M. pachydermatis* with molecular weight of 48.1 kDa and an optimum pH of 7.5 [42]. The lipolytic activity was detected with tributyrin as substrate and not with triolein, suggesting that the enzyme is an esterase. Another lipase gene was identified and characterized in *M. globosa* (termed *M. globosa LIP1*gene), which codes for a 32-kDa enzyme that is active at pH 5.5 [43]. The purified and recombinant *M. globosa* lipases were only able to hydrolyze monoglycerides and diglycerides, but not triglycerides. Interestingly, *M. globosa* whole cells showed lipase activity including the hydrolysis of triglycerides, which indicates the existence of additional lipases in *M. globosa*. Moreover, by RT-PCR the authors detected *M. globosa LIP1*gene sequences on human scalp, suggesting *in vivo* expression of this gene, and they linked lipase activity to the pathogenesis of dandruff and seborrheic dermatitis. Xu et al. [6] found that the *M. globosa* genomes encode 14 lipases, 13 of which are predicted to be secreted. Twelve of the secretory lipases can be divided into two families represented by the PFAM categories LIP and lipase 3; the latter contains the previously described *M. globosa* enzyme *LIP1*. According to Ro and Dawson [44], lipase activity of *Malassezia* species is fundamental in the development of dandruff and seborrheic dermatitis. *Malassezia* yeasts hydrolyze triglycerides of human sebum and the released free fatty acids cause irritation in susceptible individuals.

4.4.2 Lipoxygenase

Increased levels of lipid peroxides and their derivatives were demonstrated to occur in lesional, but not in nonlesional skin of patients with PV [45]. Therefore, they were thought to play a role in the pathogenesis of the disease. The enzyme, however, has not yet been synthesized. De Luca et al. [46] observed increased formation of lipid peroxides when the medium had been enriched with polyunsaturated fatty acids. The authors explained depigmentation accompanying PV by the formation of lipid peroxides, which might have toxic effects on melanocytes.

4.4.3 Azelaic Acid

In a medium containing olive oil, *Malassezia* yeasts (likely *M. furfur*) produced C_9-C_{11} dicarboxylic acids [47]. Especially azelaic acid, a C_9 dicarboxylic acid (HOOC-$(CH_2)_7$-COOH), competitively inhibits tyrosinase, an enzyme that is involved in the production of melanin. However, subsequent studies showed that these fatty acids did not influence melanocytes *in vivo* and *in vitro* [48, 49]. Furthermore, their therapeutic use in the treatment of hyperpigmentation and melanoma, based on tyrosinase inhibition, yielded disappointing results [48]. Because of its antimicrobial activity, azelaic acid is now predominantly used in the therapy of acne and inflammatory rosacea [50].

4.4.4 Gamma-Lactone

The characteristic odor of *Malassezia* cultures was first described by van Abbe [51]. Later, these volatile substances were identified as γ-decalactone [52], and they are thought to be responsible for the fruity and strong odor of the cultures.

4.4.5 Phospholipase

Phospholipase production by pathogenic isolates of *M. pachydermatis* from animals has been investigated, and higher *in vitro* production of phospholipase was observed in strains isolated from lesional skin when compared with strains isolated from healthy skin [53]. In another study, phospholipase activity was correlated with *M. pachydermatis* subgenotypes, identified by sequencing two regions of the chitin synthase-2 gene and the first internal transcribed spacer (ITS1) of the nuclear ribosomal DNA. Subgenotype B appears to be present mainly on healthy skin without producing phospholipase, whereas other genotypes and subgenotypes seem to be predominantly associated with skin lesions and show a high phospholipase activity [54]. Moreover, upregulation of phosholipase production by β-endorphin in *M. pachydermatis* strains isolated from healthy dogs and nonlesional skin of dogs with dermatitis has recently been reported [55]. Among the anthropophilic *Malassezia* species, phospholipase production has been observed in *M. furfur* [9], but the involvement of this enzyme in the development of human skin disease is not well understood. Proteomics experiments have shown that three phospholipases are secreted in *M. globosa*, thus confirming the extracellular location of these enzymes [6]. Yet, the production of opioid peptides, e.g., β-endorphin, in AD and its association with pruritus [56] and the effect these compounds may have on the altered phospholipase production by lipophilic *Malassezia* yeasts colonizing lesional skin, is a prospective area of research that could shed light on pathogen–host interactions.

4.5 Production of Pigments

The feature of pigment formation in higher fungi, particularly in macromycetes, has long been used for taxonomic purposes. However, elucidation of some of the complex and instable compounds, their synthesis pathways and industrial usage was only achieved during the last 30 years. The investigation of pigment-free mutants, especially of *Aspergillus fumigatus*, showed that pigment formation can be correlated with pathogenicity (excellently reviewed by Langfelder et al. [57]). Two metabolic pathways leading to pigment synthesis have hitherto been demonstrated to occur in *Malassezia* yeasts.

4.5.1 Melanin

Production of melanin is well-studied in relation to pathogenicity in fungi [57]. Many properties related to evasion of the host immune system and antifungal drug resistance have been attributed to melanin or melanin-like pigments. Gaitanis et al. [58] tested all *Malassezia* strains semiquantitatively for their abilities to produce pigment when grown in lipid-supplemented and lipid-depleted L-DOPA and tyrosine agars. No oxidation of tyrosine was detected when *Malassezia* yeasts had been grown on tyrosine agar, indicating that melanogenesis may occur via a tyrosinase-independent pathway. In contrast, *Malassezia* strains tested on L-DOPA agar produced a pigment with various melanization intensities. *Malassezia dermatis* strains and *M. furfur* demonstrated highest and lowest levels of pigment production, respectively. However, in *Malassezia* spp. L-DOPA was oxidized only after disruption of membranes, suggesting that phenoloxidase, the enzyme that mediates melanin production, is not secreted, but either attached to the cell wall or bound to the plasma membrane as in *Cryptococcus neoformans* [58]. Skin scales originating from hyperpigmented PV lesions, even those from patients displaying both hypo- and hyperpigmented lesions, showed Fontana-Masson staining-positive *Malassezia* cells and hyphae, which led the authors to conclude that melanization takes place *in vivo,* and may be responsible for hyperpigmentation observed within the skin lesions. Thus, *M. sympodialis* was isolated from hyperpigmented lesions, and *M. furfur* from hypopigmented lesions of the same patient. No melanin-like pigment was detected in yeast cells and hyphae in skin scales from hypopigmented PV lesions.

4.5.2 Tryptophan-Derived Indole Pigments

Investigations on the metabolic requirements of *Malassezia* species showed that *M. furfur* was able to produce a brownish pigment diffusing into the agar when cultured on a selective agar medium that consists of the amino acid tryptophan (Trp) and a lipid source [59]. Under UV light, the produced pigment showed green-yellowish fluorescence [60]. Pigment production is specifically dependent on the presence of Trp, but independent of its optical activity (i.e., S or R). The lipid source can be variable. Thus, there are similarities to the pigment synthesis of another basidiomycetous yeast, *C. neoformans*, in which, depending on the nitrogen source (i.e., diphenyls, in particular L-dopa), a secondary metabolism can be induced, which results in the synthesis of melanin [61]. By induction of this pathway in *C. neoformans,* it is possible to demonstrate increased pathogenicity and in particular, inhibition of phagocytosis [62].

The capability to produce Trp-derived pigment is typical of *M. furfur* and occurs in reference strains as well as clinical strains isolated from foci. Some *M. pachydermatis* strains were also found to produce pigment, though after markedly longer incubation periods, with a lower yield and a limited "color spectrum" [10, 63]. Thin layer chromatography (TLC) showed a complex pigment composition with a variety of differently colored single bands and numerous fluorochromes.

The chemistry and pharmacology of these compounds (Table 4.1) were found to be especially interesting in that they might explain phenomena of PV. Because of the complex composition of the pigment, priority was given to the isolation of biologically active metabolites detected by means of bioassays that are based on clinical phenomena of PV.

Pityriasis versicolor is one of the commonest human skin diseases, which is characterized by lesions of varying color, a phenomenon that is still unclarified [64]. Considering the fact that *Malassezia* belongs to the normal cutaneous mycobiota and regularly produces azelaic acid, it is difficult to understand why PV (alba) does not occur more commonly but only under special conditions (e.g., high air humidity and excessive sweating). Thus, the role of azelaic acid in the pathogenesis of PV is not yet fully elucidated [65].

The following tryptophan-derived indole pigments have been characterized, which may help clarify the pathogenesis of PV and allow further insights into the pathogenicity factors, especially of *M. furfur* (Table 4.1).

Pityriacitrin, a yellow compound eluting from the column (LiChrospher-RP-8 column; 30×250 mm; Merck, Darmstadt, FRG) with 64% acetonitrile, was found to be a

Table 4.1 Bioactive indole derivatives synthesized by *M. furfur* when grown on agar containing L-tryptophan as the single nitrogen source

Pityriacitrin $C_{20}H_{13}N_3O$ MW 311.34	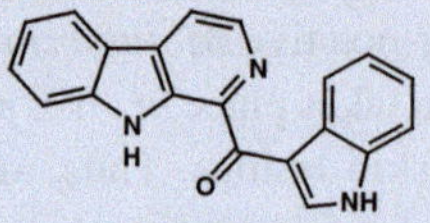	Malassezin $C_{18}H_{14}N_2O$ MW 274.32	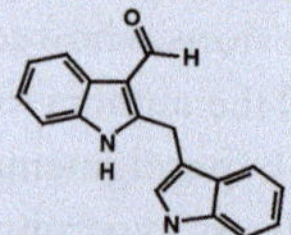
Pityrialactone $C_{20}H_{12}N_2O_3$ MW 328.32	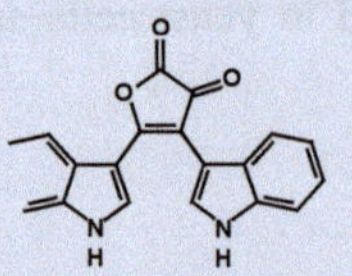	*Malassezia* Indole A $C_{21}H_{17}N_3O_3$ MW 358.38	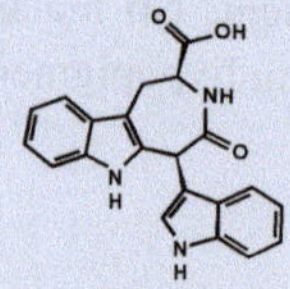
Pityriaanhydride	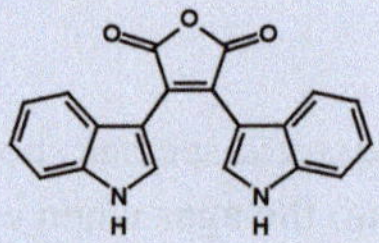	Keto-Malassezin	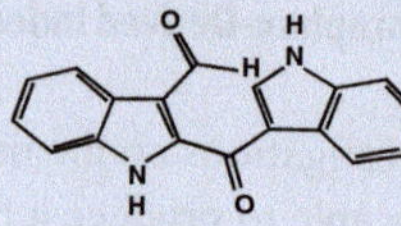
Pityriarubin A $C_{32}H_{22}N_4O_4$ MW 526.55	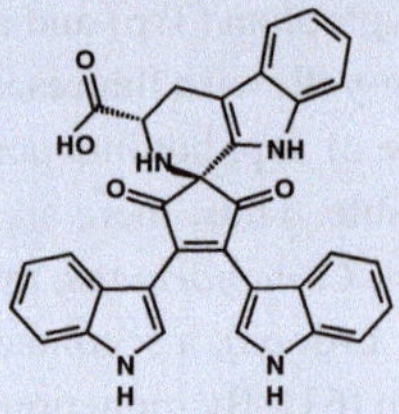	Malasseziacitrin	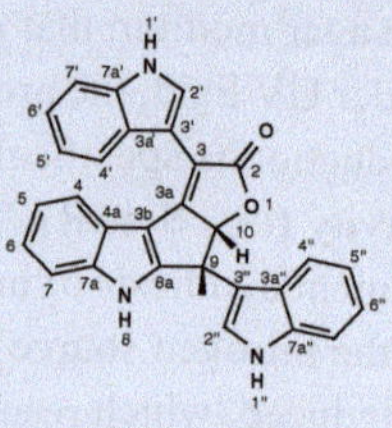
Pityriarubin B $C_{32}H_{20}N_4O_4$ MW 524.53	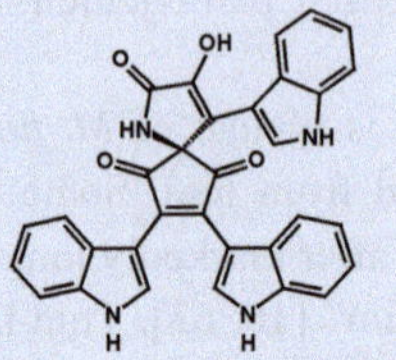	Indolo[3,2 b]carbazole MW 282.296	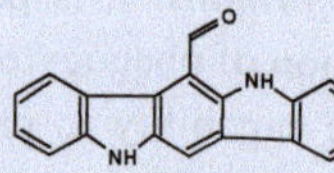
Pityriarubin C $C_{32}H_{19}N_3O_5$ MW525.51	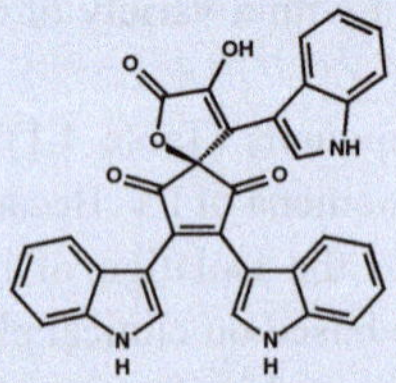	Indirubin $C_{16}H_{10}N_2O_2$ MW 262.263	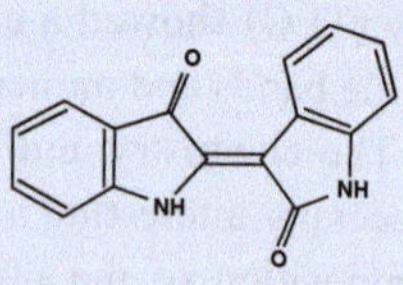
O52	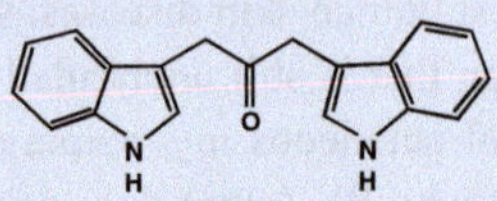		
Malassezia-carbazole A–D	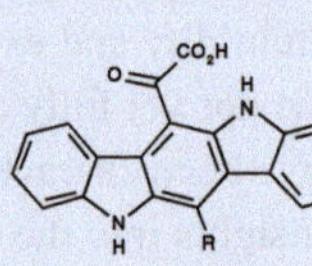 18, R = H; 19, R = CHO 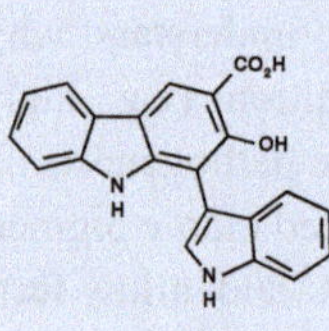 20	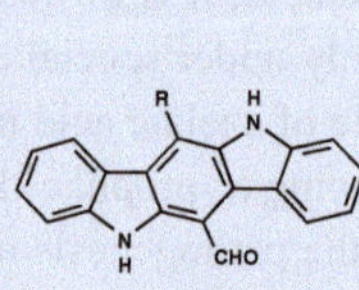 21, R = CO_2H	

potent UV filter due to its broad absorption (λ_{max} 389, 315, 289, 212 nm). The compound is an indole derivative (9*H*-Pyrido[3,4-b]indole-1-yl)(1'*H*-indole-3'-yl)methanon; $C_{20}H_{13}N_3O$) named pityriacitrin, which was also demonstrated to function in bacteria as a potent UV filter. Its UV-protective properties were confirmed by means of a yeast model and in humans [66–68]. Pityriacitrin produced by *M. furfur* might explain UV protection in depigmented foci of PV alba [65, 69].

Pityrialactone is a fluorescent bisindole compound that might serve as a free scavenger and also as a light protector due to its mesomeric structure [70]. Remarkably, the substance shows blue fluorescence in a lipophilic environment and yellow fluorescence in watery milieus. This might explain the multicolor fluorescence of PV lesions described in the literature (viz., varying from yellow–orange to green–yellow), depending on whether the substance is dissolved in sweat or epidermal lipids.

Pityriaanhydride, With the isomeric structure, the brick-red pityriaanhydride, another bisindole component of Trp-derived pigments of *M. furfur,* was demonstrated for the first time to occur in nature.

Pityriarubins. The likewise brick-red are the first compounds of the new substance class of bisindolyl cyclo-pentendiones [71]. In their basic structure, they are related to pityriaanhydride, and also to bisindolylmaleinimides and indolocarbazoles, which have gained much interest as protein kinase inhibitors [72]. The isolated pityriarubins A, B and C can suppress the release of reactive oxygen species (ROS, oxidative "burst") from activated granulocytes [73]. Inhibition was observed on activation with calcium ionophore A23187, formyl-met-leu-phe (FMLP) and interleukin 3. No inhibition occurred on activation with the classical protein kinase C (PKC) activators phorbol ester and dioctanoyl glycerol, and also after unspecific activation of G-proteins with sodium fluoride, via the alternative complement way using zymosan [73]. In all these cases, apart from zymosan, the unspecific arcyriarubin A, a representative of the class of bisindolylmaleinimides, showed inhibition of oxidative burst. Pityriarubins had dose-effectiveness curves similar to those of the highly potent arcyriarubin A, so that a comparable potent, but very specific effect on the same target(s) can be assumed (IC_{50} between 2 and 5 μM). Despite high fungal load, these effects of the described substances might explain the missing granulocyte infiltration in lesions of PV [74], helping the yeast to escape the immune response. Based on their specific mechanism of action, they might therefore present a new anti-inflammatory principle.

Malassezin is a colorless substance that shows the feature of an aryl hydrocarbon receptor (AHR) agonist ($EC\ 50_{Erod}$ 1.6×10^{-6}) and induces cytochrome P450 ($EC_{50} = 1.57$ μM) in cultures of rat hepatocytes [75]. Because of its acyclic structure, malassezin does not meet the criteria of a potent (highly affinic) AHR agonist such as indolo[3,2-*b*]carbazole. Remarkably, the yellow to orange–red *Malassezia* carbazoles A–D [*Malassezia* carbazol A (18), B (20), C (21) and D (19)], compounds that fulfill these criteria, have meanwhile been isolated from the raw extract of *M. furfur* [71]. Of these, 18, 19 and 21 show high similarity to synthetic indolo[3,2-*b*]carbazol, which is thought to be the best ligand for the AHR [75]. They have not yet been investigated for their ability to act as AHR agonists, because of insufficient yields of the substance. Interestingly, malassezin causes dose-dependent induction of apoptosis in cultivated human melanocytes, which may explain the damage to melanocytes in lesions of PV alba [76]. Cytoskeletal changes have also been observed. By activity determination of caspases 8 and 9, the apoptosis pathway was characterized as being "intrinsic". It is assumed that changes in the cytoskeletal structure lead to impaired melanosome transport. Both apoptosis induction and cytoskeletal changes may be connected with depigmentation as observed in PV alba. Furthermore, through interaction with the AHR receptor, malassezin might induce changes in the differentiation and growth behavior of keratinocytes, which would be responsible for the characteristic desquamation observed with PV.

With its unusual azepine structure, *Malassezia* indole A dose-dependently inhibits tyrosinase, which is a key enzyme of melanin synthesis, so that both effects may complement each other in their mechanism of depigmention [77, 78].

A remarkable feature of the two substances, O52, a hitherto unknown 1,3-Bis(indole-3-yl)acetone, and Keto-Malassezin (Table 4.1), which were isolated and described for the first time, is their capability to inhibit the dopa reaction on human epidermal melanocytes *in situ*. They are probably also tyrosinase inhibitors [77]. Both mechanisms can contribute to the depigmentation observed in lesions of PV alba. Keto-Malassezin represents a remarkable substance: As a malassezin derivative, it combines the features of an AHR agonist with those of a tyrosinase inhibitor, and thus appears most interesting with regard to its effects on melanocytes (i.e. apoptosis induction inhibition of melanin synthesis).

Further metabolites, which have not yet been assigned to specific effects, have been summarized by Irlinger et al. [79]. The most complicated and structurally most interesting indole alkaloid, from cultures of *M. furfur* appears to be Malasseziacitrin. It is probably biosynthetically produced by three tryptophan molecules.

Recently, the production of two potent AHR ligands, indolo[2,3 *b*] carbazole (EC50 = 2.6×10^{-7} M) (ICZ) and indirubin (EC50 = 0.2×10^{-9} M), by *M. furfur* strains was described [58, 80]. It should be noted that the binding capacity of indirubin to human AHR expressed in *Saccharomyces cerevisiae* is up to 50-fold higher than that of dioxin (EC50 = 9×10^{-9} M) [81]. Moreover, indirubin has been identified in human urine and is considered as a candidate endogenous ligand of AHR. Its identification as a cyclin-dependent kinase inhibitor and the possibility of employing indirubin as an anticancer agent further highlight the significance of this metabolic pathway in *M. furfur*. Higher quantities of ICZ were produced by *M. furfur* strains isolated from SD and PV, as compared to strains isolated from healthy skin [82, 83], and the production was higher in SD strains than in those from PV. Interestingly, the UV-protective pityriacitrin was produced by all *M. furfur* strains, suggesting a broader significance of this substance for the survival of this species under UV exposure on human skin. The pathway leading to the formation of Trp-derived pigments in the plant pathogenic basidiomycete *Ustilago maydis*, which is phylogenetically close to *Malassezia,* was recently described [84]. A tryptophan aminotransferase termed Tam1 was recognized as the single enzyme necessary to catalyze the first step of indole pigment production, i.e., tryptophan deamination and formation of indolepyruvate. Production of the biologically active indole derivatives (viz., pityrianhydride, pityriarubins A, B and C, pityriacitrin, malasseziazole A) was a result of subsequent spontaneous production of pigments. Yet, as stated by the authors [84], these observations could not explain the differences in pigment production between *M. furfur* and *U. maydis*, and additional factors of the ecological niche (i.e. different amino acids, different strains) could be responsible for this discrepancy. The variability in indole compound synthesis is further supported by the association of pigment production with virulence in *M. furfur* strains

[82]. Understanding of the possible *in vivo* significance of this pathway would be achieved by investigating *M. furfur* strains from different disease backgrounds, geographical origins and/or by knockout studies.

4.6 New Findings on the Biosynthesis of *Malassezia* Compounds

By labelling with [1-^{13}C]-DL-tryptophan, it was investigated whether the spiro carbon atom of the pityriarubins originates from the carboxyl group of tryptophan (Fig. 4.1) [71]. The labeled precursor [1-^{13}C]-DL-tryptophan was demonstrated according to Erlenmeyer's

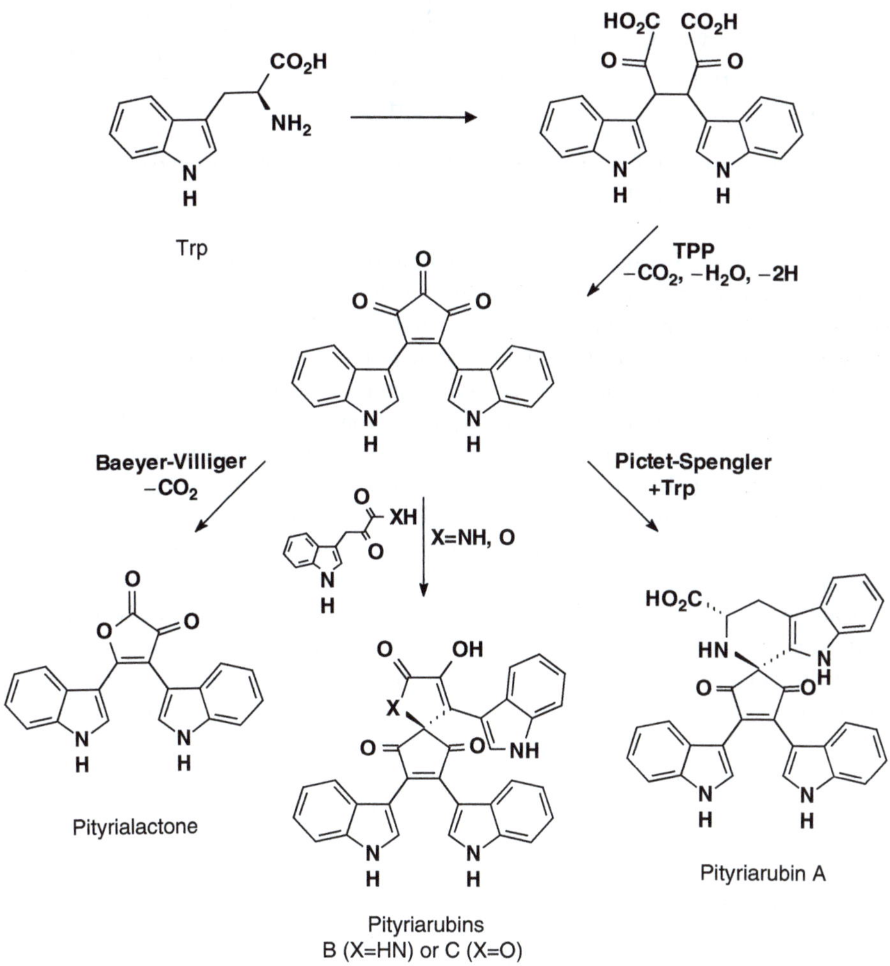

Fig. 4.1 Proposal on the biosynthesis of pityriarubins and pityrialactone

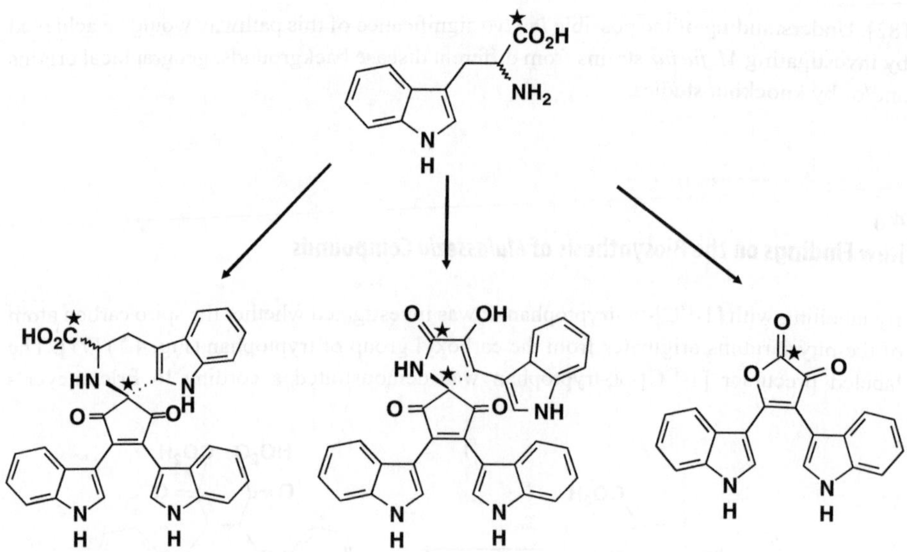

Fig. 4.2 Confirmation of the proposed biosynthesis of pityrialactone and the pityriarubins A and B by feeding with racemic [1-13C] Trp

azlactone synthesis. In both pityrialacton and pityriarubins, incorporation of [1-^{13}C]-DL-tryptophan was confirmed by means of the ^{13}C-NMR spectra. All three pityriarubins showed ^{13}C enrichment of the spiro carbon atom, and also labeling of the functions of carboxyl, lactam and lactone (Fig. 4.2). Pityriacitrin and pityriaanhydride failed to show ^{13}C enrichment, which is in accordance with the proposed synthesis pathway. Thus it was demonstrated that the spiro carbon atom originates from the carboxyl group. Lack of optical activity in pityriarubin A and *Malassezia* indole A indicates that the configuration of tryptophan does not play a role in the biosynthesis of these compounds. Furthermore, biosynthesis of these highly complex indole compounds is in fact feasible only with Trp.

Altogether, the Trp-derived pathway found in *M. furfur* and in some strains of *M. pachydermatis* comprises a complex and hitherto unknown metabolic pathway, which might explain some phenomena of PV and may help understand pathogenicity factors of *Malassezia* yeasts. *M. furfur* produces a variety of substances, and most of the secondary metabolites that have been found are unknown structures which may act as pathogenicity factors. On the other hand, *M. furfur* is rarely isolated from lesions of PV, so that the impact of the indolic pathway on the disease cannot yet be clearly assessed. As the pigment itself has not been demonstrated previously in lesions of PV because of the use of poor and contaminated material, an interesting approach is to identify genes associated with pigment production and demonstrate the expression of these genes in an *in vitro* skin model and subsequently, in lesions of PV. By means of cDNA subtraction technology, according to Diatchenko et al. [85], several genes identified in the Trp culture have been sequenced and compared with already annotated genes in gene databanks [86]. First results indicate that tryptophan apparently induces an almost toxin-like stress metabolism in *M. furfur,* as several genes, involved in detoxifying singlet oxygen and nitrosative stress were identified. It is possible that the new, hitherto scarcely investigated pigment pathway represents an energy consuming attempt of *M. furfur* to transform tryptophan into "less toxic" metabolites. This would be in line with the observation that production of pigments is associated with slow growth [87]. The relevance of these genes in the disease has to be investigated by demonstration of the corresponding transcripts in skin samples of patients with PV and an *in vitro* infection model. Moreover, a surprisingly high number of fungi are able to synthesize the mentioned compounds. Among pigment forming Ascomycetes, *Candida glabrata* [88] is suitable for the identification of pigment-associated genes. Identification of pigment-associated genes in these fungi will help to get an insight into the evolution of the pathways and, subsequent generation of knockout mutants, to determine whether these are factors of pathogenicity in human, animal or plant disease.

The recent observation by Fritsche et al. [80] that the potent AHR ligand 6-formylindolo [3,2-b]carbazole is formed spontaneously from tryptophan in cells radiated by UVB, and is partly responsible for UVB damage in association with the production of diverse bioactive indoles by *M. furfur*, highlights the importance of understanding the complex interactions of this yeast with the skin of the human or mammalian host.

References

1. Kesavan S, Holland KT, Ingham E (2000) The effects of lipid extraction on the immunomodulatory activity of *Malassezia* species *in vitro*. Med Mycol 38:239–247
2. Kesavan S, Walters CE, Holland KT et al (1998) The effects of *Malassezia* on pro-inflammatory cytokine production by human peripheral blood mononuclear cells *in vitro*. Med Mycol 36:97–106
3. Porro MN, Passi S, Caprill F et al (1976) Growth requirement and lipid metabolism of *Pityrosporum orbiculare*. J Invest Dermatol 66:178–182
4. Shifrine M, Marr AG (1963) The requirement of fatty acids by *Pityrosporum ovale*. J Gen Microbiol 32:263–270
5. Wilde PF, Stewart PS (1968) A study of the fatty acid metabolism of the yeast *Pityrosporum ovale*. Biochem J 108:225–231
6. Xu J, Saunders CW, Hu P et al (2007) Dandruff-associated *Malassezia* genomes reveal convergent and divergent virulence traits shared with plant and human fungal pathogens. Proc Natl Acad Sci USA 104:18730–18735
7. Huang HP, Little CJ, Fixter LM (1993) Effects of fatty acids on the growth and composition of *Malassezia pachydermatis* and their relevance to canine otitis externa. Res Vet Sci 55:119–123
8. Porro M, Passi S, Caprilli F et al (1977) Induction of hyphae in cultures of *Pityrosporum* by cholesterol and cholesterol esters. J Invest Dermatol 69:531–534
9. Riciputo RM, Oliveri S, Micali G et al (1996) Phospholipase activity in *Malassezia furfur* pathogeneic strains. Mycoses 39:233–235
10. Hossain HM, Landgraf V, Weiß R et al (2007) The genetic and biochemical characterization of the species *Malassezia pachydermatis* with particular attention on pigment-producing subgroups. Med Mycol 45:41–49
11. Slooff WC (1970) Genus 6 *Pityrosporum* Sabouraud. In: Lodder J (ed) The yeasts, a taxonomic study, 2nd edn. North-Holland, Amsterdam, pp 1167–1186
12. Weary PE (1970) Comedogenic potential of the lipid extract of *Pityrosporum ovale*. Arch Dermatol 102:84–91

13. Benham RW (1945) *Pityrosporum ovale*. A lipophilic fungus. Thiamin and oxalo-acetic acid as growth factors. Proc Soc Exp Biol Med 58:199–201
14. Brotherton J (1967) Lack of swelling and shrinking of *Pityrosporum ovale* in media of different osmotic pressures and its relationship with survival in the relatively dry condition of the scalp. J Gen Microbiol 48:305–308
15. Mittag H (1994) Fine structural investigation of *Malassezia furfur*. I. Size and shape of the yeast cells and a consideration of their ploidy. Mycoses 37:393–399
16. Mittag H (1995) Fine structural investigation of *Malassezia furfur*. II. The envelope of the yeast cells. Mycoses 38:13–21
17. Simmons RB, Ahearn DG (1987) Cell wall ultrastructure and diazonium blue B reaction of *Sporopachydermia quercum*, *Bullera tsugae*, and *Malassezia* spp. Mycologia 79:38–43
18. Barfatani M, Munn RJ, Schjeide OA (1964) An ultrastructure study of *Pityrosporum orbiculare*. J Invest Dermatol 43:231–233
19. Breathnach AS, Gross M, Martin B (1976) Freeze-fracture replications of cultured *Pityrosporum orbiculare*. Sabouraudia 14:105–113
20. Keddie FM (1966) Electron microscopy of *Malassezia furfur* in tinea versicolor. Sabouraudia 5:134–137
21. Swift JA, Dunbar SF (1965) The ultrastructure of *Pityrosporum ovale* and *Pityrosporum canis*. Nature 206:1174–1175
22. Thompson E, Colvin JR (1970) Composition of the cell wall of *Pityrosporum ovale* (Bizzozero) Castellani and Chalmers. Can J Microbiol 16:263–265
23. Eichstedt CF (1846) Pilzbildung in der Pityriasis versicolor. Froriep's Neue Notizen aus dem Gebiete der Natur- und Heilkunde 853:270–271
24. Dorn M, Roehnert K (1977) Dimorphism of *Pityrosporum orbiculare* in a defined culture medium. J Invest Dermatol 69:224–248
25. Faergemann J, Bernander S (1981) Micro-aerophilic and anaerobic growth of *Pityrosporum orbiculare*. Sabouraudia 17:171–179
26. Saadatzadeh MR, Ashbee HR, Cunliffe WJ et al (2001) Cell-mediated immunity to the mycelial phase of *Malassezia* spp. in patients with pityriasis versicolor and controls. Br J Dermatol 144:77–84
27. Crespo Erchiga V, Ojeda Mertos A, Vra Casano A et al (2000) *Malassezia globosa* as the causative agent of pityriasis versicolor. Br J Dermatol 143:799–803
28. Burke RC (1961) Tinea versicolor: susceptibility factors and experimental infections in human beings. J Invest Derm 36:389–401
29. Caprilli F, Mercantini R, Nazzaro-Porro M et al (1973) Studies of the genus *Pityrosporum* in submerged culture. Mycopathol Mycol Appl 51:171–189
30. Gordon MA (1951) The lipophilic mycoflora of the skin. I. *In vitro* culture of *Pityrosporum orbiculare* n.sp. Mycologia 43:524–535
31. Guého E, Midgley G, Guillot J (1996) The genus of *Malassezia* with description of four new species. Antonie von Leeuwenhock 69:337–355
32. Guého E, Boekhout T, Ashbee HR et al (1998) The role of *Malassezia* species in the ecology of human skin and as pathogens. Med Mycol 36:220–229
33. Roberts SOB (1969) Pityriasis versicolor: a clinical and mycological investigation. Br J Derm 81:315–326
34. McGinley KJ, Leyden LJ, Marples RR et al (1975) Quantitative microbiology of the scalp in non-dandruff, dandruff and seborrheic dermatitis. J Invest Dermatol 64:401–405
35. Catterall MD, Ward MW, Jacobs P (1978) A reappraisal of the role of *Pityrosporum orbiculare* in pityriasis versicolor and the significance of extracellular lipase. J Invest Dermatol 71:398–401
36. Mayser P, Führer D, Schmidt R et al (1995) Hydrolysis of fatty acid esters by *Malassezia furfur*: different utilization depending on alcohol moiety. Acta Derm Venereol 75:105–109

37. Mayser P, Pickel M, Haze P et al (1998) Different utilization of neutral lipids by *Malassezia furfur* and *Malassezia sympodialis*. Med Mycol 36:7–14
38. Mayser P, Haze P, Papavassilis C et al (1997) Differentiation of *Malassezia* spp. selectivity of cremophor EL, castor oil and ricinoleic acid for *Malassezia furfur*. Br J Dermatol 137:208–213
39. Ran Y, Yoshike T, Ogawa H (1993) Lipase of *Malassezia furfur*: some properties and their relationship to cell growth. J Med Vet Mycol 31:77–85
40. Plotkin LI, Squiquera L, Mathov I et al (1996) Characterization of the lipase activity of *Malassezia furfur*. J Med Vet Mycol 34:43–48
41. Brunke S, Hube B (2006) MfLIP1, a gene encoding an extracellular lipase of the lipid-dependent fungus *Malassezia furfur*. Microbiology 152:547–454
42. Shibate N, Okanuma N, Hirai K et al (2006) Isolation, characterization and molecular cloning of a lipolytic enzyme secreted from *Malassezia pachydermatitis*. FEMS Microbiol Lett 256:137–144
43. DeAngelis YM, Saunders CW, Johnstone KR et al (2007) Isolation and expression of a *Malassezia globosa* lipase gene, LIP1. J Invest Dermatol 127:2138–2146
44. Ro BI, Dawson TL (2005) The role of sebaceous gland activity and scalp microfloral metabolism in the etiology of seborrheic dermatitis and dandruff. J Invest Dermatol Symp Proc 10:194–197
45. Nazzaro-Porro M, Passi S, Picardo M et al (1986) Lipoxygenase activity of *Pityrosporum in vitro* and *vivo*. J Invest Dermatol 87:108–112
46. De Luca C, Picardo M, Breathnach A et al (1996) Lipoperoxidase activity of *Pityrosporum*: characterisation of by-products and possible role in pityriasis versicolor. Exp Dermatol 5:49–56
47. Nazzaro-Porro M, Passi S (1978) Identification of tyrosinase inhibitors in cultures of *Pityrosporum*. J Invest Dermatol 71:205–208
48. Breathnach AS, Nazzaro-Porro M, Passi S (1984) Azelaic acid. Br J Dermatol 111:115–120
49. Robins EJ, Breathnach AS, Bennet D et al (1985) Ultrastructural observations on the effect of azelaic acid on normal human melanocytes and human melanoma cell line in tissue culture. Br J Dermatol 113:687–697
50. Elewski B, Thiboutot D (2006) A clinical overview of azelaic acid. Cutis 77(suppl 2):12–16
51. van Abbe NJ (1964) The investigation of dandruff. J Soc Cosmet Chem 15:609–630
52. Labows JN, McGinley KJ, Leyden JJ et al (1979) Characteristic gamma-lactone odor production of the genus *Pityrosporum*. Appl Environ Microbiol 38:412–415
53. Cafarchia C, Otranto D (2004) Association between phospholipase production by M. pachydermatitis and skin lesions. J Clin Microbiol 42:4868–4869
54. Cafarchia C, Gasser RB, Latrofa MS et al (2008) Genetic variants of *Malassezia pachydermatis* from canine skin: body distribution and phospholipase activity. FEMS Yeast Res 8: 451–459
55. Cafarchia C, Dell'Aquila ME, Capelli G et al (2007) Role of beta-endorphin on phospholipase production in *Malassezia pachydermatis* in dogs: new insights into the pathogenesis of this yeast. Med Mycol 45:11–15
56. Bigliardi-Qi M, Lipp B, Sumanovski LT et al (2005) Changes of epidermal mu-opiate receptor expression and nerve endings in chronic atopic dermatitis. Dermatology 210:91–99
57. Langfelder K, Streibel M, Jahn B et al (2003) Biosynthesis of fungal melanins and their importance for human pathogenic fungi. Fungal Genet Biol 38:143–58
58. Gaitanis G, Chasapi V, Velegraki A (2005) Novel application of the masson-fontana stain for demonstrating *Malassezia* species melanin-like pigment production *in vitro* and in clinical specimens. J Clin Microbiol 43:4147–4151
59. Mayser P, Imkampe A, Winkeler M et al (1998) Growth requirements and nitrogen metabolism of *Malassezia furfur*. Arch Dermatol Res 290:277–282

60. Mayser P, Wille G, Imkampe A et al (1998) Synthesis of fluorochromes and pigments in *Malassezia furfur* by use of tryptophan as single nitrogen source. Mycoses 41:265–271
61. Polacheck I, Kwon-Chung KJ (1988) Melanogenesis in *Cryptococcus neoformans*. J Gen Microbiol 134:1037–1041
62. Polak A (1990) Melanin as a virulence factor in pathogenic fungi. Mycoses 33:215–224
63. Mayser P, Töws A, Krämer HJ et al (2004) Further characterization of pigment-producing *Malassezia* strains. Mycoses 47:34–39
64. Gupta AK, Batra R, Bluhm R et al (2003) Pityriasis versicolor. Dermatol Clin 21:413–429
65. Thoma W, Krämer HJ, Mayser P (2005) Pityriasis versicolor alba. JEADV 19:147–152
66. Machowinski A, Krämer HJ, Hort W et al (2006) Pityriacitrin – a potent UV filter produced by *Malassezia furfur* and its effect on human skin microflora. Mycoses 49:388–392
67. Mayser P, Pape B (1998) Decreased susceptibility of *Malassezia furfur* to UV light by synthesis of tryptophan derivatives. Antonie van Leeuwenhoek 73:315–319
68. Mayser P, Schäfer U, Krämer HJ et al (2002) Pityriacitrin – an ultraviolet-absorbing indole alkaloid from the yeast *Malassezia furfur*. Arch Dermatol Res 294:131–134
69. Hiram De Almeida Jr, Mayser P (2006) Absence of sunburn in lesions of pityriasis versicolor alba. Mycoses 49:516
70. Mayser P, Stapelkamp H, Krämer HJ et al (2003) Pityrialacton – a new fluorochrome from the tryptophan metabolism of *Malassezia furfur*. Antonie van Leeuwenhoek 84:185–191
71. Irlinger B, Krämer HJ, Mayser P et al (2004) Pityriarubins, biologically active Bis(indolyl) spirans from cultures of the lipophilic yeast *Malassezia furfur*. Angew Chem Int Ed Engl 43:1098–1100
72. Davis PD, Hill CH, Lawton G et al (1992) Inhibitors of protein kinase C. 1. 2, 3-Bisindolyarylmaleimides. J Med Chem 35:177–184
73. Krämer HJ, Kessler D, Hipler UC et al (2005) Pityriarubins, novel highly selective inhibitors of respiratory burst from cultures of the yeast *Malassezia furfur*: comparison with the bisindolylmaleimide arcyriarubin A. ChemBioChem 6:2290–2297
74. Wroblewski N, Bär S, Mayser P (2005) Missing granulocytic infiltrate in pityriasis versicolor – indication of specific anti-inflammatory activity of the pathogen? Mycoses 48(suppl 1): 66–71
75. Wille G, Mayser P, Thoma W et al (2001) Malassezin - A novel agonist of the arylhydrocarbon receptor from the yeast *Malassezia furfur*. Bioorg Med Chem 9:955–960
76. Krämer HJ, Podobinska M, Bartsch A et al (2005) Malassezin, a novel agonist of the arylhydrocarbon receptor from the yeast *Malassezia furfur*, induces apoptosis in primary human melanocytes. ChemBioChem 6:860–865
77. Dahms K, Krämer HJ, Thoma W et al (2002) Tyrosinaseinhibition durch Tryptophanmetabolite von *Malassezia furfur* in humaner epidermis. Mycoses 45:230
78. Thoma W, Dahms K, Krämer HJ et al (2001) Tyrosinase-Inhibition durch KO27- einem Stoffwechselmetaboliten von *Malassezia furfur*. Mycoses 44:238
79. Irlinger B, Bartsch A, Krämer HJ et al (2005) New tryptophan metabolites from cultures of the lipophilic yeast *Malassezia furfur*. Helv Chim Acta 88:1472–1485
80. Fritsche E, Schäfer C, Calles C et al (2007) Lightening up the UV response by identification of the arylhydrocarbon receptor as a cytoplasmatic target for ultraviolet B radiation. Proc Natl Acad Sci 104:8851–8856
81. Adachi J, Mori Y, Matsui S et al (2001) Indirubin and indigo are potent aryl hydrocarbon receptor ligands present in human urine. J Biol Chem 276:31475–31478
82. Gaitanis G, Magiatis P, Stathopoulou K et al (2008) AhR Ligands, malassezin, and indolo[3, 2-b]carbazole are selectively produced by *Malassezia furfur* strains isolated from seborrheic dermatitis. J Invest Dermatol 128:1620–1625
83. Giakoumaki D, Stathopoulou K, Melliou E et al (2008) Identification of indirubin as a metabolite of *Malassezia furfur* strains isolated from diseased skin. 7th Joint Meeting of AFERP,

ASP, GA, PSE, SIF. Natural products with pharmaceutical, neutraceutical, cosmetic and agrochemical interest. Athens, Greece
84. Zuther K, Mayser P, Hettwer U et al (2008) The tryptophan aminotransferase Tam1 catalyses the single biosynthetic step for tryptophan-dependent pigment synthesis in *Ustilago maydis*. Mol Microbiol 68:152–172
85. Diatchenko L, Lau YF, Campbell AP et al (1996) Suppression subtractive hybridization: a method for generating differentially regulated or tissue-specific cDNA probes and libraries. Proc Natl Acad Sci USA 93:6025–6030
86. Hort W, Lang S, Brunke S et al (2008) Analysis of differentially expressed genes associated with tryptophan-dependent pigment synthesis in *M. furfur* by cDNA subtraction technology. Med Mycol 25:1–11
87. Barchmann T, Hort W, Mayser P (2005) Untersuchungen zur Regulation des Tryptophan-abhängigen Sekundärmetabolismus von *Malassezia furfur*. Mycoses 48:307
88. Mayser P, Wenzel M, Krämer HJ et al (2007) Production of indole pigments by *Candida glabrata*. Med Mycol 45:519–524

Malassezia Species and Immunity: Host–Pathogen Interactions

5

H. Ruth Ashbee and Ross Bond

Core Messages

- *Malassezia* species stimulate both the cellular and humoral immune responses in healthy humans and also in patients with *Malassezia*-associated diseases. The cellular immune responses to *Malassezia* yeasts often appear to be heightened in patients with disease.
- It has also become clear that *Malassezia* has two phenotypes – one, which is immuno-stimulatory and may occur in disease states, and the second, which is immunosuppressive and may occur during commensal carriage.
- Although there are fewer studies related to dogs when compared with humans, dogs also develop a spectrum of humoral and cellular immune responses to their commensal yeast, *Malassezia pachydermatis*, both in health and in disease.

5.1 Introduction

Malassezia species are found on normal human skin as members of the cutaneous commensal flora and can be isolated from the scalp and trunk of most individuals. In contrast to their commensal roles, they are also known to be the aetiological agents of pityriasis versicolor (PV) and are implicated in a range of other cutaneous diseases, including seborrhoeic dermatitis (SD) and dandruff, folliculitis, atopic/eczema dermatitis syndrome (AEDS) and psoriasis [1]. This apparently paradoxical behaviour, and their existence at the edge of the commensal – pathogen divide means that understanding their interaction with the immune system is very important. Although many groups have studied this relationship, it is only within the last 10 years that the detailed molecular interactions have been

H. Ruth Ashbee (✉)
Mycology Reference Centre, Department of Microbiology, Leeds General Infirmary, Leeds LS1 3EX, UK
e-mail: h.r.ashbee@leeds.ac.uk

T. Boekhout et al. (eds.), *Malassezia and the Skin*
DOI: 10.1007/978-3-642-03616-3_5, © Springer Verlag Berlin Heidelberg 2010

elucidated. We are now beginning to unravel how the organism sometimes stimulates and sometimes downregulates the immune response against it, so that in most people it remains a commensal, but in some it causes disease. (Note that the discussion of immunological responses in patients with AEDS and the role of *Malassezia* in this disease will be discussed in Chap. 7 and is not included here).

Studying host–pathogen interactions necessarily means that certain antigens must be selected from the pathogen against which to measure the host's response. Thus, antigen selection may become pivotal in determining what responses are found. Examining the literature in this area, the diversity of antigenic preparations from *Malassezia* that have been used by different groups quickly becomes apparent. Whole yeast cells [2–6], physically disrupted yeast cells [4, 5, 7–9], cytoplasmic antigens [10–13], cell surface proteins [14] and other crude preparations have all been used in assays of cellular and humoral immunity.

An important study by Zargari et al. [15] showed that the duration of culture of *Malassezia* during antigen production affects which antigens are present; protein antigens were gradually lost over time, but carbohydrate antigens were maintained over the 21-day culture period. The stability of antigens also varies on storage, with degradation of many protein antigens at room temperature or higher temperatures after 1 month of storage [16]. Additionally, different extraction methods also release differing antigens, with more protein antigens released by cell disruption with micro-glass beads than with sonication only [17].

The final problem compounding this variability is that many of the studies used strains classified according to different classification systems. Until 1996, when a unified classification system was defined [18], the strains used for immunological studies were classified in different systems, and hence, the relationship between strains remains obscure. Although strains from some studies were deposited in culture collections and are still available, many have been lost and so, cannot be classified according to the new scheme. For example, ATCC strain 42132, which was initially considered to be *Malassezia furfur*, is now classified as *Malassezia sympodialis*; so, many studies using antigens from '*M. furfur*' may, in fact, have been using other species that were not recognised at that time. This is always a potential problem, but the genus *Malassezia* has undergone a particularly intense period of taxonomic review, with only 2 species recognised in 1989 (*M. furfur* and *M. pachydermatis*), compared to the 13 that are currently recognised. Using crude antigenic extracts in immunological assays has, therefore, led to a bewildering array of results and findings, with little consensus between different research groups.

In 1997, Schmidt et al. [19] reported the expression and characterisation of the first defined antigen from *M. furfur*, Mal f1. Since then, a further 11 antigens have been described and characterised from *Malassezia* species; most of them are antigens recognised by patients with AEDS. Further details on these antigens and their characterisation can be found in Chap. 7 in the Sect. 7.2.1.

5.2 The Skin Immune System

Before examining how *Malassezia* interacts with the immune system, this section will review what is currently known about the skin and the immunologically active role that it plays. *Malassezia* is a commensal of the skin, and most of the conditions with which it has been associated affect the skin.

Cutaneous cells are known to be an integral part of the immune system and the skin immune system (SIS) is a well-recognised entity [20]. The SIS contributes to both innate immune responses and those of the adaptive immune response [21]. Because the skin is the outermost part of the body, it is subjected to a constant barrage of microbial, physical and chemical insults, and its physical integrity and barrier function is an important part of the innate skin immune defence [22]. In addition, the relatively acidic pH of sweat and sebum on human skin and the presence of various fatty acids all help defend against cutaneous microbial assault, and the stable commensal flora help to prevent transient organisms becoming established on the skin.

There are several chemicals present in the skin that have activity against micro-organisms: most notably, anti-microbial peptides, which are present in abundance in the skin [21]. Anti-microbial peptides are polypeptides of low-molecular weight that have activity against bacteria, viruses and fungi at physiological concentrations [23]. They are produced by keratinocytes, both constitutively and in response to micro-organisms and cytokines. To date, at least 10 anti-microbial peptides have been found in skin, including the human beta defensins (HBD), psoriasin, the cathelicidin LL37 and dermicidin [24]. Anti-microbial peptides are not only able to kill a range of organisms, but they also act as mediators of inflammation and cytokine production, thus influencing many aspects of the skin milieu.

The complement system is another component of the innate immune system and is made up of over 30 proteins that are present in human serum as well as in the skin [25]. There are three complement pathways, the classical pathway, activated by immune complexes; the lectin pathway, activated by terminal mannose residues found on bacteria; and the alternative pathway, activated by lipopolysaccharide from Gram negative bacteria and the cell walls of some yeasts. Activation of the complement cascade results in the lysis of susceptible cells and also (a) the production of opsonic fragments, which enhance the ability of neutrophils and macrophages to phagocytose material; and (b) production of peptides that induce local and systemic inflammatory responses. Several organisms that occur on the skin can activate the complement cascade, including *Candida* species [26], propionibacteria [27] and *Malassezia* yeasts [28], and so may initiate inflammation and the recruitment of neutrophils into the skin.

The importance of neutrophils in defence against fungi is highlighted by the predisposition of neutropenic patients to fungal infections [29]. Within the skin, neutrophils are not normally present, although they may occur in conditions where significant inflammation occurs [30].

Although the innate immune response is characterised by its lack of specificity, it utilises pattern recognition receptors to recognise microbial compounds. Toll-like receptors (TLR) present on neutrophils, macrophages and many other cells can recognise microbial structures, known as 'pathogen-associated molecular patterns' (PAMPs) on bacteria, viruses and fungi [31]. Within the normal cutaneous cell population, keratinocytes and Langerhans cells express TLR's, although other cells that can infiltrate the skin, e.g. neutrophils also express them. PAMPs present on fungi include β-glucans on *Candida albicans* and *Aspergillus fumigatus*, mannan and phospholipomannan on *C. albicans* and galactomannan on *A. fumigatus*, which act as ligands for TLR2, TLR4 and other receptors [32]. Interaction of the fungal PAMP and corresponding TLR or receptor triggers signalling cascades within the receptor-bearing cell, leading to production of inflammatory cytokines and, in phagocytes, triggering of the killing mechanisms via both the oxidative and non-oxidative pathways [21]. The inflammatory cytokines released may activate or recruit further cells, leading to an escalation of the response.

In addition to the innate immune mechanisms in the skin, adaptive immune responses also occur. Specific, adaptive immune responses are only elicited if the innate response fails to remove the 'foreign' material. Antigens from the skin surface, including those of commensal or pathogenic organisms are taken up by the resident dendritic cell (DC) populations in the skin – Langerhans' cells and dermal dendritic cells. DCs express several TLRs, as well as serving as antigen-presenting cells and so function as the link between innate and adaptive immune responses. Several fungi, including *C. albicans, Cryptococus* (*Cr.*) *neoformans, A. fumigatus, Histoplasma capsulatum* and *Malassezia* spp. can be internalised by DCs and their antigens presented [33]. After uptake of antigens, DCs migrate to the draining lymph nodes and interact with naïve CD4 T-cells, inducing differentiation of the T-cells into either Th1 or Th2 subsets [21]. The T-cells specific to the antigen are stimulated and undergo clonal expansion, after which they leave the lymph node and enter the skin. Once in the skin, they can interact with their antigen and produce cytokines according to their Th profile. Th1 cells are associated with protective responses, synthesising TNFα, IFNγ, IL1, IL6 and IL12, while Th2 cells are not protective, synthesising IL10, IL4 and IL5 [33].

Although many immunological cells are either present in the skin or can migrate there under appropriate cytokine or chemotactic influences, B lymphocytes are absent. Despite this, immunoglobulins specific to the commensal flora are produced in normal individuals [3, 34, 35] and are able to bind to the eliciting organism on the skin [36]. Although it was previously thought that B-cells and humoral immunity were of limited importance in defence against fungal infections, recent data have demonstrated that, at least for some fungi, particularly *C. albicans* and *Cr. neoformans*, humoral immunity may influence the course of the infection [37].

5.3 Innate Immune Responses to *Malassezia*

5.3.1 Complement System

Malassezia was first suggested as having the ability to activate the complement system by Martin-Scott [38] in 1952, but it was not until the 1980s when it was demonstrated that it could activate the alternative [28] and classical pathways [39]. The activation of both pathways is time and cell concentration dependent, but the molecule responsible for the activation has not been elucidated, although β-glucan in the cell wall may be involved [40].

Complement is involved in the inflammation associated with several dermatoses, including acne vulgaris [41] and psoriasis [9]. The role of complement in *Malassezia*-associated diseases has not been extensively investigated, but C3 has been found deposited in the lesions of seborrhoeic dermatitis [42, 43] and hence, at least for this disease, inflammation may be partly complement-mediated.

5.3.2 Anti-microbial Peptides

Anti-microbial peptides are important components of the innate immune system and LL-37, a cathelicidin, is present in the skin. It has been shown that this anti-microbial peptide is active against *Malassezia in vitro*, with an MIC of 20–30 μM [44] and it was found to be present in higher concentrations in lesions of PV than in healthy skin. The LL-37 was processed from an inactive precursor in the lesions and co-localised with the *Malassezia*, suggesting that it had a role in innate immune defence against *Malassezia* in PV.

Most of the anti-microbial peptides in the skin are produced by keratinocytes either constitutively or as a result of infection or inflammation [21]. Human β defensin 2 (HBD), another anti-microbial peptide found in skin, was initially described as being induced in psoriatic cells and is highly active against Gram negative bacteria and *C. albicans*. Donnarumma et al. [45] demonstrated that when *M. furfur* is co-cultured with keratinocytes, the keratinocytes internalised the yeast, which resulted in the upregulation of HBD2 via a protein kinase pathway. HBD2 can act as a chemoattractant for neutrophils and macrophages and activate these cells to release the inflammatory cytokines, IL6 and IL8. Thus, the interaction of *Malassezia* with keratinocytes can lead to the initiation of inflammatory changes in the skin.

5.3.3 Transferrin

Transferrin is an iron-binding β-globulin plasma protein synthesised by the liver that transports iron to mammalian cells; in human plasma, transferrin is normally present at 2–3.2 mg/mL and is typically one-third saturated with iron [46]. Transferrin may contribute to innate host defence against bacterial and fungal pathogens by limiting microbial access to iron [47]. Transferrin may also inhibit bacterial adhesion [48] and have a direct, iron-independent antifungal effect [49]. Among pathogenic fungi, transferrin has been shown to inhibit the growth of dermatophytes [49–51], *Candida* spp. [51, 52], *Cr. neoformans* [53, 54], *Rhizopus oryzae* [49] and *H. capsulatum* [55].

Transferrin has been demonstrated in canine serum by high-resolution protein electrophoresis [56] and in bovine serum by radial immunodiffusion [57] and immunoelectrophoresis [48]. Jenkinson et al. [58] showed that transferrin was one of the soluble proteins present within the stratum corneum of cattle and sheep and suggested that it was mainly derived from sweat. It has also been demonstrated in human sweat [59]. Since the microbiota of healthy mammalian skin is located in the superficial layers of the stratum corneum and in the pilosebaceous units [60–63], transferrin is present at sites where it may influence cutaneous microbial populations in humans and animals. In addition, it has been postulated that the diffusion of serum into the inflamed epidermis enables unsaturated transferrin to promote recovery from dermatophytosis and to prevent the systemic growth of dermatophytes [50, 64].

Canine and bovine unsaturated transferrin was shown to have a dose-dependent, reproducible, inhibitory effect on the growth of *M. pachydermatis in vitro* [65] at concentrations that were comparable to plasma concentrations reported in humans [46]. The spongiotic dermatitis reaction that occurs in cases of *Malassezia* dermatitis in dogs [66] might promote the accumulation of transferrin from serum and, in association with other components of the SIS, serve to restrict yeast proliferation at the sites of infection in the skin. Saturated transferrin was equally effective at inhibiting the growth of *M. pachydermatis* in that study, indicating that the inhibitory effect on this yeast is not dependent upon iron depletion. Further studies of the mechanism of inhibition of growth and of the potential role of transferrin in innate immunity to *Malassezia* spp. yeasts are warranted.

5.3.4 Cellular Responses and Cytokine Production

Phagocytes form an important part of the innate immune defence against fungi and the interaction of neutrophils and *Malassezia* has been examined in several studies. The uptake and killing of *Malassezia* by neutrophils *in vitro* is a complement-dependent process [67]. Uptake is maximal at 40 min, after which it plateaus, but killing of the ingested yeast is very inefficient, with only 5% of the cells killed after 2 h. This lack of killing is in contrast to other yeast and fungal genera, where killing of up to 80% of cells occurs [68, 69]. In this study, if *Malassezia* cells were pre-treated with ketoconazole, killing increased to 23%, with the amount of killing proportional to increasing drug concentrations. Several possible explanations have been proposed for this inability of neutrophils to kill *Malassezia*. When grown in the presence of oleic acid, *Malassezia* produces azelaic acid, a C9 dicarboxylic acid [70]. Azelaic acid decreases production of O_2^- and OH· by neutrophils in a dose-dependent manner and also reduces H_2O_2 production, due to inhibition of cellular metabolism [71]. It is also an oxygen radical scavenger [72], and if azelaic acid is produced by *Malassezia in vivo*, this may help to protect it from the oxygen-dependent killing mechanisms used by neutrophils. Another factor that has been suggested to help protect *Malassezia* from phagocytic killing is the presence of a lipid-rich capsular-like layer around the yeast cells [73]. Preliminary data suggest that the capsular-like layer reduced *Malassezia* uptake and activation of neutrophils and that this may contribute to their protection during phagocytosis [74].

More recently, the role of some of the tryptophan metabolites of *Malassezia* yeasts in this lack of neutrophil activity against *Malassezia* have been demonstrated [75]. *M. furfur* can convert tryptophan into a range of indole alkaloids, including the pityriarubins (A, B and C). Neutrophils were obtained from healthy donors and stimulated with a calcium ionophore to stimulate superoxide release. When the pityriarubins were added, they caused a decrease in the production of reactive oxygen species (ROS) in a concentration dependent manner, with pityriarubin C being most active in the assay. The authors concluded that this ability to produce such compounds would enable *Malassezia* to downregulate the immune response directed against it and contribute to the lack of inflammation seen in PV.

Binding between *Malassezia* and phagocytic cells has been studied in the human monocytic cell line THP-1, and is mediated by the mannose receptor, β-glucan receptor and complement receptor 3 [76]. When the THP1 cells were stimulated with live or heat-killed *Malassezia* cells, the production of IL8 increased in a time and concentration-dependent pattern [77]. Stimulation of the granulocyte cell line, HL60, by *Malassezia* increased the levels of both IL8 and IL1α, again in a time and concentration dependent pattern. This stimulation of monocytic cells by *Malassezia* and the resultant production of IL8 and IL1α will lead to the chemotaxis of neutrophils and T-cells and hence augment the inflammatory response.

Chemoattraction of neutrophils from psoriatic, atopic and healthy controls in response to *Malassezia* was studied by Bunse et al. [78]. Using a Boyden chamber, cells of *M. furfur* were incubated in the lower part of the chamber, with the test neutrophils in the upper part and after three hours, the neutrophils that had crossed the membrane separating the two parts of the chamber were counted. Neutrophils from patients with psoriasis showed significantly more chemoattraction than those from atopic or control subjects. Both the supernatant and washed cells of *M. furfur* produced the same effects; the factor causing chemoattraction was destroyed by acid hydrolysis and was hydrophilic, indicating that it was a protein. This chemoattraction of neutrophils by colonising *Malassezia* was suggested as contributing towards the Koebner phenomenon seen in patients with psoriasis.

One interesting facet of *Malassezia* is its ability to act as an adjuvant [79]. Mice injected intraperitoneally with different amounts of live or heat-killed *Malassezia* cells at various time points were subsequently challenged intraperitoneally with *Salmonella enterica*. Pretreatment with even small numbers of *Malassezia* conferred resistance to infection and optimal protection occurred when the *Malassezia* was injected 4 days before the *Salmonella*. Protection was due to the activation of intra-peritoneal macrophages, which showed increased bactericidal activity as well as increases in number. Protection by *Malassezia* was equivalent to that of *Propionibacterium acnes*, a known activator of the reticuloendothelial system. *Malassezia* pre-treatment was also able to protect mice against subsequent challenge with tumour cell lines, and this protection was due to stimulation of macrophages to produce oxygen intermediates [80]. Thus, *Malassezia* is able to upregulate the immune system and give protection against subsequent bacterial or tumour challenge.

The interaction of *Malassezia* and peripheral blood mono-nuclear cells (PBMNC) has been studied in some detail. One of the first studies was by Neuber et al. [8], looking at the effects of a strain of '*P. ovale*' on the PBMNC from patients with seborrhoeic dermatitis. They collected the PBMNC from 10 patients with SD and10 healthy controls and measured the IL2, IL10 and IFNγ released when the cells were stimulated with *P. ovale* compared with the levels of the cytokines that were produced constitutively. Stimulation of PBMNC from normal donors enhanced both IL2 and IFNγ production, but this was not seen when PBMNCs from patients with SD were stimulated. In contrast, IL10 synthesis was significantly higher by PBMNC from patients with SD compared with normals. The level of constitutive IL10 production was also higher from patients with SD. The finding of this increased production of IL10, which can inhibit T-cell cytokine synthesis, was suggested as a possible cause of the decreased production of IL2 and IFNγ and suggested by the authors as an explanation for the 'disturbed T-cell function in SD'.

Later studies examined cytokine profiles in the PBMNC of normal, healthy donors when co-cultured with *Malassezia*. An initial observation made by Walters et al. [81]

reported that *M. globosa* significantly reduced the production of IL1β by PBMNC. However, in a later study, using formalised whole *Malassezia* yeast cells, harvested in mid-exponential phase in co-culture experiments, it was found that when they were used at a yeast:PBMNC ratio of 1:1; *M. sympodialis* and *M. globosa* were able to significantly increase production of IL1β, IL6 and TNFα above base line; while *M. restricta* caused significantly increased production of IL6 and TNFα [82]. In contrast, when the ratio of yeast cells/PBMNC was 20:1, all the species caused suppression of IL1β, IL6 and TNFα. When viable cells were used, also at a ratio of 20:1, suppression of all three cytokines was also seen. This ability of *Malassezia* to suppress the pro-inflammatory cytokines is important and seems to correlate with the situation seen *in vivo*. *Malassezia* is a member of the normal cutaneous flora and elicits minimal inflammation in this situation. In PV too, there is little inflammatory infiltrate, suggesting that despite a large fungal burden, there is comparatively little stimulation of the immune system [11]. The role of the capsular-like layer in the downregulation of cytokine production by PBMNC has been studied [83]. Yeast cells of *M. sympodialis, M. globosa* and *M. restricta* were treated with chloroform : methanol to extract the lipids around and within the cell wall; these cells were then co-cultured with PBMNC at a yeast:PBMNC ratio of 20:1. Untreated yeast cells from the same species were also tested in parallel. The lipid-extracted cells of all the species were found to induce either significantly higher ($P<0.05$) or constitutive levels of IL1β, IL6 and TNFα in PBMNC after 24 h of co-culture. Therefore, by removing the lipid layer around the yeast cells, the immunosuppressive phenotype had been reversed and resulted in an immunostimulatory phenotype. The lipid layer appears to have some parallels with the capsule of *Cr. neoformans*, giving *Malassezia* some protection from phagocytic uptake and killing and also limiting its ability to stimulate cytokine production by PBMNC. This ability to limit immune stimulation may be the way in which *Malassezia* can exist as a commensal on the skin of most people.

5.3.5 Interaction of *Malassezia* spp. with Cutaneous Cells

Over the last 10 years, the details of the interaction of *Malassezia* with many cutaneous cell types have been investigated and these are detailed below (Note that details of the interaction of *Malassezia* with DCs will be covered in the Sect. 7.2.1 in Chap. 7).

5.3.5.1 Infiltrates Within *Malassezia*-Associated Diseases

Cutaneous conditions with which *Malassezia* spp. have been associated are numerous, and some of the evidence that has implicated *Malassezia* in these conditions has come from histological studies. Histological studies have also provided information about how *Malassezia* interacts *in situ* with cutaneous cells within the lesions.

The inflammatory cells seen in lesions of PV consist mainly of mature T cells [84], although the phenotype has been reported to be predominantly helper/inducer [84] or suppressor [85]. Langerhans cells are also commonly found, often adjacent to the fungi within the lesions [85].

A recent study compared the inflammatory infiltrate in PV with that in tinea corporis and noted that the infiltrate in PV was significantly less than that in the dermatophyte infection, despite a heavy fungal burden [86]. The authors concluded that this might be due to the lipids associated with the *Malassezia* yeast cells, and also possibly the production of pityriarubins, which can suppress ROS production by neutrophils. This correlates with the work of other authors who found that the lipid layer around the cells was correlated with the ability to downregulate cytokine responses in PBMNC [82] and keratinocytes [87] (see Sect. 5.3.5.3).

The infiltrate in lesions of SD consists of lymphocytes [88], most of which are CD4+ Th cells [89]. Numbers of Langerhans' cells are not increased in lesional skin. A study more recently found that in addition to lymphocytes and Langerhans' cells, macrophages, monocytes and granulocytes are also present [42]. Large numbers of NK1+ and CD16+ cells were present indicating an irritant-like reaction. ICAM1 (intercellular adhesion molecule 1) and C1q positive cells were found in larger numbers in the skin of patients than that of controls. ICAM 1 is present on endothelial cells and leucocytes and is involved in leucocyte adhesion and inflammation, being required for neutrophil migration into inflamed tissue, which may explain the presence of neutrophils in the infiltrate of SD.

The other condition associated with *Malassezia* where the lesional infiltrate has been extensively characterised is folliculitis. The extent of the infiltrate is dependent on whether the follicle has ruptured. In the absence of follicular rupture, lymphocytes are the main cell type in the lesions [42, 90–93], but once the follicle ruptures, the infiltrate is more extensive and varied, including neutrophils [94], macrophages, eosinophils and plasma cells [95].

In dogs with dermatitis associated with *M. pachydermatis*, dermal inflammation is characterised by a superficial perivascular or interstitial infiltrate of mono-nuclear cells, with focal accumulations of neutrophils, eosinophils and mast cells [66, 96–98]. Exocytosis of lymphocytes is frequent whereas eosinophilic or neutrophilic micro-abscesses are less often seen. Mast cells may be aligned in a linear pattern at the dermo-epidermal junction in some cases [66]. A variety of different pathological patterns have been reported in cats with abundant cutaneous *Malassezia* spp., depending primarily on the nature of the concurrent diseases; dermal infiltrates ranged from sparse in cats with an atrophic follicular pattern associated with pancreatic paraneoplastic alopecia, to a more prominent mixed dermal infiltrate at the dermo-epidermal junction in cats with interface dermatitis of unknown aetiology [99].

5.3.5.2 Interaction with Melanocytes

The interaction between *Malassezia* and melanocytes is of particular interest in PV as the alteration of pigmentation seen in this condition is usually the presenting sign for most patients. Several groups have attempted to unravel the mechanisms by which either hypo- or hyper-pigmentation can occur. An early study found that in patients with the hypopigmented form of disease, although there were similar numbers of melanocytes in lesions and normal skin, the melanocytes in affected skin had a decreased deposition of melanin on the melanosomes and that the melanosome granules were not evenly distributed throughout the keratinocytes in the epidermis [100], a finding confirmed by subsequent studies [101, 102]. This alteration of melanosome distribution has been reported in hyper-pigmented lesions too [102–104]; however, one group found no differences in melanosome distribution and suggested that the hyper-pigmentation in PV was not due to melanin [105]. The mechanisms by which *Malassezia* may affect melanocytes are not fully understood. Early work indicated that azelaic acid was an inhibitor of tyrosinase, an enzyme involved in melanogenesis, and might have a cytotoxic effect on melanocytes, resulting in hypo-pigmentation [70]. Subsequent work by the same group, demonstrated that *Malassezia* has lipoxygenase activity, which could lead to the production of lipoperoxides, similar to those seen in PV, which damage melanocytes and result in hypopigmentation [106, 107]. More recently, a tryptophan metabolite of *Malassezia*, malassezin, has been shown to induce apoptosis in melanocytes, causing decreased melanin synthesis and resultant hypo-pigmentation [108]. Although these mechanisms may cause hypo-pigmentation, they are unlikely to explain the hyper-pigmentation seen in some PV lesions. One study found that the keratin was thicker in hyper-pigmented lesions, but the number of melanocytes and melanosomes was not different to normal skin [104], although they only studied five patients.

5.3.5.3 Interaction with Keratinocytes

Adherence, the specific attachment of micro-organisms to cells and tissues, is thought to play an important role in the colonisation and infection of mammals by pathogenic fungi [109–112]. Adherence allows fungi to resist physical forces, which may otherwise result in its removal from the host and may precede other pathogenic events such as germination and invasion [111]. The factors which influence the adherence of *Candida* spp. to epithelial cells of both mucosal and cutaneous origin have been extensively investigated [110, 113], and several adhesins are now well-characterised [114], whereas the adherence of *Malassezia* spp. has received much less attention.

The interaction of *Malassezia* with keratinocytes is of great importance as keratinocytes are the largest cell populations within the skin, having both a structural and immunological role. An early study demonstrated that in PV, *Malassezia* were taken up by keratinocytes and could be seen intra-cellularly, mainly as hyphal elements [115]. The keratinocytes had undergone degradation, demonstrated by the presence of 'lipidic' material within the keratinocytes, possibly due to the action of enzymes produced by *Malassezia.*

5.3.5.4 Adherence to Human Keratinocytes

Malassezia furfur was found to adhere to human stratum corneum cells *in vitro* in a dose, time and temperature-dependent manner [116]. Stratum corneum cells from various sites were co-incubated with various species of *Malassezia*, at a concentration of 10^5 epithelial cells and varying concentrations of *Malassezia* cells, varying time periods and incubation temperatures. Epithelial cells were then washed and Gram stained and the number of adherent cells counted under light microscopy. Significantly more adherent cells were found at 37°C than at 25°C; at 90 and 120 min than at 60 min, but there were no differences in adherence between the different strains tested. Adherence was only seen from an inoculum of 10^4 organisms and increased significantly with increasing inoculum up to 10^7 organisms. However, as most areas of the body only have a population of 10^2–10^4 CFU/cm^2 of *Malassezia*, even in the presence of disease, it is unclear if the findings reported in this study can be extrapolated to normal human skin.

Using the same assay system, the group later examined the adherence of *Malassezia* to stratum corneum cells from 28 patients with SD and 30 healthy controls and found no significant differences in the adherence between the two groups, leading them to suggest that adherence is not a critical process in SD [117, 118].

Schechtman et al. [118] investigated the adherence of various *Malassezia* spp. (corresponding to *M. globosa*, *M. sympodialis* and *M. obtusa* in the current taxonomy) [1] to stratum corneum cells from human patients with SD of varying HIV status. The *Malassezia* strains evaluated adhered to human stratum corneum cells in a time-dependent manner. A single *Malassezia* strain showed comparable adherence to skin cells derived from HIV-positive and healthy individuals without SD. Isolates obtained from patients with SD, HIV-associated SD, PV and *Malassezia* folliculitis showed variable adherence, but overall, there was no relationship between the adhesive properties of the strains and the disease status of the host from which they originated. In addition, there was no correlation between strain adherence to human stratum corneum cells *in vitro* and the clinical severity of SD.

5.3.5.5
Adherence of *Malassezia pachydermatis* to Canine Stratum Corneum Cells

Bond and Lloyd reported a series of studies [119–121] investigating the *in vitro* adherence of *M. pachydermatis* to canine stratum corneum cells and observed a number of similarities with related studies of other pathogenic fungi. Adherence of *M. pachydermatis* to canine corneocytes *in vitro* peaked at 2 h [119]; this was in accordance with previous *in vitro* studies of the adherence of *M. furfur* and *C. albicans* to human corneocytes [116, 122], and of *C. albicans* to human buccal epithelial cells [123]. Early exponential phase cells of one *M. pachydermatis* strain were less adherent than stationary phase cells, presumably reflecting variable synthesis or surface expression of adhesin(s) during cell growth and development. Similarly, growth phase has been shown to have significant effects of the adherence of *Cr. neoformans* [124] and *C. albicans* [125, 126] to various cell types. Formalin treatment did not adversely affect the adherence of *M. pachydermatis*, indicating that the yeast adhesin(s) resist this fixative and that yeast viability is not required for attachment to canine corneocytes, in accordance with studies on *Cr. neoformans* [124].

Yeast adhesins are generally mosaic proteins because they require several domains with discrete functions [114]. In *Candida* spp, several adhesins are now well characterised, including the hydrophobic and peptide-binding Als proteins of *C. albicans* and the sugar binding *EPA* galectins of *C. glabrata*. Trypsin treatment (0.1% for 30 min) of four out of five *M. pachydermatis* strains led to reduced adherence, implying that trypsin-sensitive cell surface proteins or glycoproteins are important in the adherence of *M. pachydermatis* to canine corneocytes. Trypsin treatment of the corneocyte had less consistent effects with only two strains showing reduced adherence, but also providing some evidence for a role for epithelial cell proteins or glycoprotein ligands. However, in neither case did trypsin treatment abrogate yeast adherence, indicating that either protein or glycoprotein adhesins or ligands remained functional despite enzymatic attack, or that other mechanisms are involved, perhaps involving adhesins with broad specificities [114].

Both *N*-acetyl D-glucosamine and chitin, a component of the cell wall of *C. albicans*, which contains *N*-acetyl D-glucosamine, have been shown to inhibit the adherence of this species to epithelial cells [127, 128]. A variety of sugars, including *N*-acetyl D-glucosamine, mannose, and sucrose, has been reported to inhibit the adherence of *Cr. neoformans* to rat glial cells [124]. By contrast, mannose, sucrose and *N*-acetyl D-glucosamine did not consistently inhibit the adherence of *M. pachydermatis* to canine corneocytes, although the adherence of one strain was inhibited by *N*-acetyl D-glucosamine only [119].

Lectins are proteins or glycoproteins that bind specifically to various saccharides; for example, concanavalin A (con A) binds α-D-mannose. Con A pre-treatment inhibited the adherence of one *M. pachydermatis* to canine stratum corneum cells, suggesting that mannosyl-bearing carbohydrate residues on the epithelial cells serve as ligands for adhesins expressed by this strain. This conclusion was supported by the dose-related abrogation of the inhibitory effects of con A by pre-incubation with a 6 %, but not a 3 %, solution of methyl α-D-mannopyranoside, a hapten inhibitor of con A. Similar results were found in studies of the adherence of *C. albicans* to human buccal epithelial cells [129]. Further studies are required to identify the molecular basis of the adherence of *M. pachydermatis* to canine corneocytes *in vitro*, to determine whether similar mechanisms occur *in vivo*, and to investigate whether drugs or inhibitors of the adherence process can be used in the ecological control skin diseases of dogs associated with *M. pachydermatis*.

5.3.5.6
Malassezia pachydermatis Adherence in Canine *Malassezia* dermatitis

Changes in the adhesive capacities of epithelial cells have been associated with a number of pathological conditions and infectious diseases [130–132]. An assay of *M. pachydermatis* adherence to canine stratum corneum cells was used to examine whether the adhesive capacities of the epithelial cells could be related to population sizes of the yeast in the skin of seborrhoeic and healthy basset hounds, a breed predisposed to *Malassezia* dermatitis associated with very high yeast counts [133], and in healthy Irish setter control dogs [121]. The *M. pachydermatis* population sizes in the healthy basset hounds ($n=5$) were significantly greater ($P<0.001$) than those in the Irish setters ($n=6$) but were significantly lower ($P<0.001$) than those in the seborrhoeic basset hounds ($n=6$). The numbers of adherent yeast cells in the two strains tested were significantly greater ($P<0.001$) on corneocytes derived from healthy basset hounds, compared with corneocytes taken from healthy Irish setters and seborrhoeic basset hounds. The inverse relationship between population densities *in vivo* and stratum corneum cell adherence *in vitro* suggests that increased stratum corneum cell receptivity for the yeast is not important in the pathogenesis of SD associated with *M. pachydermatis* in basset hounds. These findings are similar to those reported in a previous investigation of *M. furfur* adherence in humans with SD; the yeast was better able to adhere to stratum corneum cells from healthy individuals than to cells from patients with SD [117]. It is possible that the reduced adherence to cells from the seborrhoeic dogs reflects an adaptive host response to the putative infection with *M. pachydermatis*.

The histopathological features of acanthosis with a compact or parakeratotic stratum corneum [134] in basset hounds with *Malassezia* dermatitis suggests that the epidermis is hyperproliferative, and the rapid turnover of epidermal cells may preclude normal maturation and reduce the expression of ligands for *M. pachydermatis* adhesins. This hypothesis is supported by the observations of Bibel et al. [131] who showed a two-fold increase in the adherence of *Staphylococcus aureus* to fully keratinised human nasal mucosal cells, when compared to cells from the upper granular layer.

5.3.5.7
Cytokine Production by Keratinocytes in Response to *Malassezia*

Keratinocytes produce a range of cytokines and immunological mediators, some of which are produced constitutively and some as a result of infection or trauma [135]. Four studies have now been published describing the interactions of *Malassezia* species and keratinocytes, although the conditions used and results described differ widely between the studies (See Table 5.1). The first study to examine this interaction co-cultured keratinocytes with yeast cells of *M. furfur, M. pachydermatis, M. slooffiae* and *M. sympodialis* for 1–24 h and

Table 5.1 Conditions used and cytokines measured in keratinocyte-*Malassezia* co-culture studies

References	Watanabe et al. [136]	Baroni et al. [137]	Ishibashi et al. [138]	Thomas et al. [87]
Species	*M. furfur* *M. pachydermatis* *M. slooffiae* *M. sympodialis*	*M. furfur*	*M. globosa* *M. restricta* *M. furfur* *M. sympodialis*	*M. furfur* *M. globosa* *M. obtusa* *M. restricta* *M. slooffiae* *M. sympodialis*
Keratinocytes	Normal human	HaCat cell line	Normal human and cell line	Normal human
Ratio Y:K (multiplicity of infection)	1:1	30:1	20:1	27:1
Duration of co-culture	1–24 h	–	24 h	12–96 h
Cytokines measured	IL1β, IL6, IL8, MCP-1, TNFα	IL1α, TNFα, IL6, IL10, TGF β1	IL3, IL4, IL5, IL6, IL7, IL8, IL10, IL13, GM-CSF, BMP-4, BMP-6, FGF-6, MIP-3a, GCSF, TIMP-1, TIMP-2, Leptin	IL6, IL1 α, TNFα, IL8, IL10

assayed the cytokines produced [136]. *M. furfur* stimulated only low or undetectable levels of all cytokines (IL1β, IL6, IL8, TNFα and monocyte chemotactic protein 1 (MCP)), whereas the other species stimulated increasing levels over time, with the highest levels being seen with *M. pachydermatis*. Induction of cytokines only occurred in the presence of the yeast cells and could not be induced by culture supernatants, indicating the need for cell–cell contact.

However, conflicting results were reported by Baroni et al. [137], who found that *M. furfur* had significant immunomodulatory effects, downregulating IL1α and TNFα, leading to a lack of IL6; and upregulation of IL10 and TGFβ1. Two main differences between these studies, which may in part contribute to their divergent results, are the ratio of yeast : keratinocyte (multiplicity of infection) used and the source of keratinocytes. A further study concentrated on the interaction of *M. globosa* and *M. restricta* with keratinocytes and found that both species stimulated Th-2 type cytokine profiles, with *M. globosa* resulted in increased levels of IL5, IL10 and IL13, while *M. restricta* resulted in increased levels of IL4 [138]. These results were the same with keratinocytes from normal human skin and from a cell line. The authors suggested that this induction of a Th-2 type cytokine profile by *M. globosa* and *M. restricta*, both found on the skin of patients with atopic dermatitis (AD) [139], might be involved in the inflammation associated with the condition. The most recent study examining this interaction reported the effects of six species of *Malassezia* on normal human keratinocytes, using exponential and stationery phase yeast cells, some of which had been treated to remove the lipid-layer that surrounds the yeast [87]. *Malassezia* yeast cells, which retained their lipid layer, resulted in little or no increase in levels of IL1α, IL6, TNFα or IL8, but significantly increased the intracellular levels of IL10. If the cells had been treated and the lipid-layer removed, co-culture resulted in increased levels of IL6, IL8 and IL1α, but not TNFα, and decreased levels of IL10. The most noticeable effects were a 66-fold increase in levels of IL8 with viable, decapsulated stationary phase *M. globosa*, a 38-fold increase in levels of IL6 and a 12-fold decrease in intracellular IL10 levels with non-viable, decapsulated stationary phase *M. furfur*. Thomas et al. [37] concluded that this ability of *Malassezia* species to upregulate cytokine production could contribute significantly to the inflammation seen especially in SD, with IL6 and IL8 leading to increased proliferation of keratinocytes and migration of leucocytes into the skin.

The variations in the patterns of cytokine production by keratinocytes seen in the differing studies are hard to reconcile, but what is clear is that if these same effects are seen *in vivo*, *Malassezia* will have a significant effect on the cytokine milieu in the skin and so may influence the development and progress of skin inflammation and hence exacerbations or improvements in cutaneous conditions.

It has been shown that TLR2 is upregulated when keratinocytes are infected with *M. furfur* [140], leading to the induction of IL8 and human beta-defensin 2, an anti-microbial peptide, by the keratinocytes. IL8 is a pro-inflammatory cytokine and a neutrophil chemoattractant, and this mechanism may be involved in the inflammation seen in *Malassezia*-associated conditions.

Examining the interaction of *Malassezia* with keratinocytes from psoriatic patients, Baroni et al. [141] found that in patients colonised with *Malassezia*, levels of transforming growth factor β (TGF-β) and heat shock protein 70, were significantly higher than culture negative psoriatic patients or controls. TGF-β affects keratinocyte migration and proliferation and its upregulation by *Malassezia* was thought to be a contributory factor in the hyper-proliferation seen in exacerbations of psoriasis.

In dogs, epidermal hyperplasia is often prominent in skin biopsy specimens obtained from lesions of *Malassezia* dermatitis [66, 96–98]. Chen and others showed that extracts of and culture supernatants from *M. pachydermatis* were unable to induce the proliferation of cultured canine keratinocytes directly *in vitro* [142]. However, in a subsequent study, whole, viable *M. pachydermatis* cells did not induce proliferation of cultured canine keratinocytes either, contrary to an unpublished preliminary report [143], suggesting that the epidermal hyperplasia results from either a direct effect that does not operate in the absence of a blood supply or functioning skin immune system, or indirectly as a consequence of inflammation [144]. High concentrations of *M. pachydermatis* cells were associated with increased keratinocyte apoptosis.

5.3.5.8
Interaction with Dermal Fibroblasts

The interaction of *Malassezia* with dermal fibroblasts has also been studied, but only to a limited extent [145]. *Malassezia* adhered to the fibroblasts within 4 h and was taken into the cytoplasm within 24 h, a process that caused damage to the membrane of the fibroblasts. Cytochalasin D, an inhibitor of actin polymers, inhibited this uptake, suggesting that the engulfment was an active process, dependent on F-actin. Although these data are interesting, in normal skin, *Malassezia* is not likely to come into contact with dermal fibroblasts, so the relevance of these findings is unknown.

5.4
Adaptive Immune Responses to *Malassezia*

5.4.1
Humoral Responses in Humans

The humoral immune response to *Malassezia* species has been extensively studied since the early 1980s, both in healthy individuals with no history of disease and also in patients with *Malassezia*-associated conditions [146–153]. A wide range of methods have been used to detect the antibodies, including indirect-immunofluorescence (IIF), dot blots and enzyme-linked immunosorbent assays (ELISA), with IIF considered to be a less sensitive method than ELISA. The antigens used in the different studies included whole cells, soluble antigens, cytoplasmic extracts, protein antigens and carbohydrate antigens.

Antibodies specific to the yeast phase of *Malassezia* can be measured in the serum of healthy individuals who have no history of *Malassezia*-associated diseases. The first study to demonstrate the presence of antibodies in normal subjects was that of Alexander [154]. Using IIF and whole cells antigens, she found that 10 normal subjects had significant titres of specific immunoglobulins. DaMert et al. [5] also found 'high' titres in normal individuals.

Sohnle et al. [155] were the first group to study different classes of antibodies. They studied two groups of differing ages (23–44 years and 70–88 years), and determined antibodies to *Malassezia*, *C. albicans* and *Trichophyton rubrum* using a tube ELISA. Antigens from *Malassezia* were prepared by ether extraction and grinding and titres of IgG, IgA and IgM were recorded. Both groups of individuals had measurable *Malassezia*-specific antibodies of all three classes tested; levels of IgM were significantly lower in the older group.

In 1983, titres of IgG were measured in 21 adults and 36 children (range 6 months to 15 years) using IIF against whole yeast cells [6]; the titres in adults were found to be significantly higher than those in children. The authors suggested that the antibody levels were a reflection of colonisation levels, which are higher in adults than children. Similar findings were reported in levels of IgG in a group of healthy individuals, ranging in age from 29 to 81 years [156]. Levels of IgG were highest in the age group 29–31-years-old and lower in older subjects. As population levels of *Malassezia* on the skin are lower in older individuals, this study also tends to suggest that antibody levels may correlate with population densities of the organism.

The first study examining antibody levels to different species of *Malassezia* in the same individual used an ELISA to determine titres of IgM, IgA and IgM to *M. sympodialis, M. globosa* and *M. restricta* [157]. Five groups of non-atopic females, with no history of skin disease, grouped according to age (2–3 years, 7–10 years, 20–24 years, 33–40 years and 60–64 years) were included. Titres of IgG and IgM were determined in all individuals and titres of IgA in 36 subjects. IgG and IgM were found to all the species by the age of 2–3 years. Titres were similar for all the age groups, except those in the 60–64 year group, which were significantly lower. For IgG, titres were similar over the whole age range studied. Only 18 subjects had detectable levels of IgA and the titre was low in all of them.

One of the largest studies to measure antibodies to *Malassezia* obtained sera from 868 subjects, ranging in age from 0 to 80 years [158]. Subjects were included regardless of whether they had *Malassezia*-associated diseases and so the results are neither representative of healthy individuals nor patients. Culture filtrate from *M. furfur* was used in immunoprecipitation to test the sera. Antibodies were only found in 31% of the sera tested. They were not detected in subjects less than 11 years of age and highest levels were found in the 31–40-year old group. The findings of antibodies in such a low percentage of the study population is probably a reflection of the lack of sensitivity of the method used and the lack of antibodies in children contrasts with the findings of other groups.

The only study to examine antibody responses to the mycelial phase of *Malassezia* in healthy individuals found that specific IgM, IgG, IgG subclasses and IgA could all be detected [159] with the highest titres for IgG. Despite the relative lack of mycelia on normal skin, the antibody response may be due to cross -reacting or common antigens in the yeast and mycelial phases, eliciting the production of antibodies.

Overall, the presence of antibodies to *Malassezia* in healthy individuals appears to correlate to its presence on the skin as a member of the commensal flora. Stimulation of the immune system occurs at a young age and is of sufficient duration that IgM and IgG antibody classes are produced. IgA titres are low, indicating that mucosal stimulation only occurs to a limited extent in healthy individuals.

Against this background of antibodies in normal individuals, many studies have also examined the humoral immune response in patients with various *Malassezia*-associated diseases. The findings from the studies are summarised in Table 5.2. The results reported

Table 5.2 Studies examining humoral immunity in normal individuals and those with *Malassezia*-associated conditions

Findings of the study	No of subjects	Method used	Antigenic preparation	Reference
Studies which found higher antibody levels in patients than controls				
2 of 6 PV patients had higher titres, the other 4 and controls had low titres	6 PV, 10 controls	IIF	Tape strips from PV patients	[152]
Higher mean titres in patients	14 PV, 14 controls	IIF	Whole cells	[5]
Titres of IgG and IgM higher in patients	30 PV, 30 controls	IIF		[153]
	12 PV, 12 controls	IIF	Mycelial form	[159]
	20 PV, 10 controls	ELISA	Whole cells and exoantigens	[150]
Titre of IgG higher in patients	40 SD, 40 controls	ELISA	Cytoplasmic antigens	[12]
Titres of IgG higher in patients; titres of IgM higher in patients when measured against oval form	25 SD, 10 controls	ELISA	Whole cells and exoantigens	[150]
Titres of IgG higher in patients	32 folliculitis, 25 controls	IIF	Whole cells	[148]
IgG to 100 kDa and 120 kDa proteins found only in psoriasis patients	15 psoriasis, 10 SD, 9 PV, 10 controls	Western blot	Cytoplasmic and cell wall preparations	[151]
Studies which found no difference in antibody titres between patients and controls				
No difference in the titre of antibody titres between patients and controls	40 PV, 31 controls	Haem[n]	Soluble antigen	[7]
	30 PV, 22 controls	IIF	Whole cells	[6]
	6 PV, 6 Controls	Dot blot	Soluble antigens	[11]
	10 PV, 10 Controls	ELISA	Whole cells	[2, 3]
	10 SD, 10 controls	ELISA		[3]
	30 SD, 60 controls	IIF		[147]
	10 SD, 10 controls	IIF, FACS	Whole cells and carbohydrate antigen	[4]
	30 SD (historical controls used)	IIF	Whole cells	[89]
	19 SD, 19 controls	IIF		[149]
	19 SD, 19 controls	ELISA	Whole cells and cytoplasmic extract	[13]
	18 acne, 18 controls	IIF	Whole cells	-
Studies which found lower antibody titres in controls than patients				
	10 SD, 10 controls	ELISA	Protein antigen	[4]

PV pityriasis versicolor; *SD* seborrhoeic dermatitis; *IIF* indirect immunofluorescence; *ELISA* enzyme-linked immunosorbent assay are conflicting, with many studies finding increased antibody titres in patients, whilst others have found no differences between patients and controls. Even when the studies are examined in detail, there does not appear to be any pattern to these results. Antibody production to *Malassezia* is probably governed by a number of factors, not just the presence or absence of disease. Factors that could lead to variation in antibody production in a population include the density of *Malassezia* on the skin of individuals, the extent and frequency with which *Malassezia* antigens gain access to stimulate an immune response and also differences in the sensitivity and specificity of methods used to measure the antibody titres. The only conclusion which can, therefore, be drawn is that antibody levels do not appear to correlate with the presence of *Malassezia* associated disease and are not a useful immunological marker to study.

5.4.2 Humoral Responses in Dogs

Several studies have examined humoral immune responses to *M. pachydermatis* in dogs. Serum levels of *M. pachydermatis*-specific IgG and IgA measured by ELISA in 21 seborrhoeic basset hounds and 11 dogs of various breeds with *Malassezia* dermatitis exceeded ($P<0.01$ for IgG; $P<0.05$ for IgA) those of 14 healthy basset hounds and eight healthy beagles [160]. These results indicate that high serum IgG and IgA levels are not protective in *M. pachydermatis*-associated SD in basset hounds and other breeds. Increased IgA levels might be associated with antigen exposure due to mucosal carriage [133, 161]. Atopic dogs were also shown to have higher serum titres of *M. pachydermatis* – specific IgG when compared with healthy dogs, irrespective of whether there was cytological evidence of *Malassezia* overgrowth at the time of sampling [162].

Total serum IgA concentrations in healthy bassets, seborrhoeic bassets and affected dogs of mixed breeds were significantly greater ($P<0.01$) than those of beagles, and concentrations in affected basset hounds exceeded ($P<0.01$) those of the Irish setters [160]. These data indicate that serum IgA deficiency is not a factor in the pathogenesis of the disease.

Western immunoblotting was used to further investigate and compare humoral immune responses in healthy dogs and dogs with *Malassezia* dermatitis by blotting of *M. pachydermatis* antigens separated by SDS-PAGE [163]. A total of 14 bands of immunoreactivity were identified with approximate molecular weights of 219, 204, 132, 110, 76, 71, 66, 60, 50–54, 43–50, 42, 37–40, 35 and 32 kDa. The number of bands (mean±se) of immunoreactivity observed in the seborrhoeic basset hounds (10.7±0.4) was significantly greater ($P<0.01$) than those of beagles (3.0±1.0), Irish setters (5.5±1.1) and healthy basset hounds (5.6±0.7). Band numbers in the mixed-breed affected dogs (9.4±0.9) were significantly greater than those of beagles ($P<0.01$), healthy basset hounds ($P<0.01$) and Irish setters ($P<0.05$). There were no significant differences between beagles, Irish setters and healthy basset hounds. In general, staining intensity was greatest in specimens where multiple bands were evident.

There were clear differences in the individual protein bands recognised between groups. Most dogs displayed immunoreactivity towards the 132, 66 and 50–54 kDa proteins and few dogs recognised the 204 and 76 kDa proteins. The majority of the affected dogs showed reactivity towards the 219, 110, 71 and 42 kDa proteins, whereas reactivity to these proteins was seldom detected in healthy dogs. Less well-defined differences between groups were seen with the 40–37, 35 and 32 kDa proteins. There was a highly significant positive correlation (r_s=0.76, $P<0.001$) between the number of bands identified and the titre of *M. pachydermatis*-specific IgG antibodies in each sample as determined previously using ELISA.

In another study of healthy and atopic dogs, the relatively weak reactivity to a small number of bands in healthy dogs contrasted with the stronger reactivity to a larger number of bands seen in atopic dogs [164]. Many atopic dogs' sera recognised bands of 90, 82, 61, 45–51, 42, 29 and 25 kDa, whereas the 25 kDa band was not present with sera from healthy dogs. The identity and function of the 25 kDa protein and its significance in the pathogenesis of yeast infection in canine AD is not known. The differences in the reported molecular weights of the resolved proteins between these two studies [109, 130] might reflect variation in antigen expression that occurs during different growth phases of *M. pachydermatis* [165] in a manner analogous to that reported previously for *Pityrosporum orbiculare* (possibly *M. globosa* in the current taxonomy) [15], or other technical differences between the laboratories.

Significantly higher concentrations of *M. pachydermatis* specific IgE were detected by ELISA in atopic dogs when compared with healthy dogs, although there was no significant difference in concentrations between atopic dogs with and without evidence of yeast infection [162]. Sera from more than half of the atopic dogs recognised allergens of 45, 52, 56 and 63 kDa [166]. In a more recent study of serum concentrations of anti-*M. pachydermatis* IgE antibodies in atopic and healthy dogs, no significant differences were found between the affected and control groups, despite immediate intradermal test reactivity being much more frequent in the atopic dogs [167]. Serum from dogs with atopic dermatitis and immediate intradermal test reactivity to *M. pachydermatis* or high *M. pachydermatis* specific IgE concentrations was able to induce passive cutaneous anaphylaxis in healthy recipient beagle dogs that was abrogated by heat inactivation or IgE-absorption, indicating that anti-*Malassezia* IgE is functional in immediate hypersensitivity reactions in atopic dogs with *Malassezia* dermatitis.

5.4.3 Cellular Responses in Humans

Cellular immune responses are important in defence against fungi and the increased incidence of several *Malassezia*-associated diseases has been documented in patients with various cellular immune deficiencies. Patients who have received renal transplants and are treated with immunosuppressive agents have a higher than normal incidence of PV [168, 169]; bone marrow transplant [91] and heart transplant recipients [170] have higher than normal incidence of folliculitis; and SD is very common in HIV/AIDS patients [171–173].

Cellular immune responses can be measured using the lymphocyte transformation assay (LT) and the leucocyte migration inhibition assay (LMI). The LT measures the proliferation of lymphocytes in response to specific antigens, or mitogens, which cause non-specific stimulation of the cells and hence are used as measure of how much the lymphocytes are capable of proliferating. The LT response to antigens is much lower than to mitogens because only those lymphocytes specific to the stimulating antigen will proliferate. In the LMI, the leucocytes are stimulated by antigens and if they recognise the antigen, the migration of the leucocytes will be inhibited and the size of the inhibition can be measured [174].

5.4.3.1 Healthy Individuals

In contrast to the many studies which have examined humoral immunity to *Malassezia* in healthy individuals, there are very few studies that have examined cellular immunity in healthy subjects per se. A small study by Cunningham used the LT and LMI to study cellular responses in nine individuals, ranging in age from 8 to 57 years to *M. sympodialis*, *M. globosa* and *M. restricta* [175]. She found positive LT responses in all the age groups, but responses to *M. sympodialis* were lower than those to *M. globosa* and *M. restricta*. Responses in the LMI did not differ between age groups or species. Although this is a small study, it does demonstrate that cellular immunity is present in individuals in whom there is no history of *Malassezia*-associated disease.

Further data on cellular responses in healthy individuals can be obtained by studying control data from various studies on cellular responses in disease. The findings are summarised in Table 5.3. Cellular immunity is obviously stimulated in healthy subjects who have no history of skin diseases and from childhood until late middle age cellular immunity is measurable. No studies have included subjects older than 61 years, so we do not know if this immunity persists into old age, although it may be that as the densities of *Malassezia* decrease into old age, as the skin lipid content decreases, the stimulation they provide is lost and hence cellular immunity may also wane.

Table 5.3 Cellular immune responses specific to *Malassezia* in healthy individuals

No of subjects (age range)	Method used	Results	References
15 (22–40 years)	LMI	Mean response = 31.5% (significant response is >20%)	[176]
	LT	Mean response = 21–23 (significant response is >4)	[176]
32 (20–42 years)	LT	Mean response = 26.5 (significant response is >4)	[9]
20 (23–45 years)	LMI	8/19 positive (1 subject not tested)	[2]
	LT	3/20 positive	[2]
15 (23–61 years)	LT	Mean response of 31[a]	[179]
12 (19–49 years)	LT	Mean response of 27[a]	[179]
16 (mean = 56 years)	LT	Responses ranged from 0 to 50 with different antigen preparations[a]	[13]

[a]There was no discussion in these studies as to what the authors took as a significant response
LT Lymphocyte transformation assay; *LMI* Leucocyte migration inhibition assay

5.4.3.2
Pityriasis Versicolor

The first studies to examine the cellular immune response to *Malassezia* in patients with PV resulted in the widely quoted finding that these patients have a cellular immune deficiency to *Malassezia.* Studying 12 patients with PV and 15 controls, it was found that while responses in the LT were not different between patients and controls, the responses in patients in the LMI were significantly lower than controls [176]. In a subsequent study, involving 18 PV patients and 42 controls, the authors reported that the response of patients in the LT assay showed a 'modest, but statistically significant decrease in the magnitude of the lymphocyte transformation response' to *Malassezia.* However, this significant difference was only seen on day 6 of the assay, but the peak response was not reached until day 9, at which point there was no difference [177]. Because the rate of lymphocyte transformation may differ between individuals, it is important that comparisons are made at the peak response, and hence the conclusion that patients have a cell-mediated immune deficiency specific to *Malassezia* is not consistent with the data reported.

Subsequent studies did not support these initial reports either. A study of 31 PV patients and 30 healthy controls found that patients had higher responses in the LT assay than controls to *Malassezia* [178]. Using the LT and LMI with whole cells of *M. sympodialis, M.globosa* and *M. restricta* as antigens, Ashbee et al. [2] studied the cellular responses of 10 patients with PV and 10 age-and-sex-matched controls. There was no significant difference in the LT responses between patients and controls. Of note, none of the subjects responded to cells of *M. sympodialis* and it was suggested that if the early studies had used this species, the apparent lack of reactivity in the assay may be a reflection of lack of response to a certain species, rather than a general cell-mediated immune deficiency. In the LMI, a significantly greater proportion of patients with PV responded to *M. globosa* than controls, but their responses to *M. sympodialis* and *M. restricta* were not different. Thus, patients did not appear to have a cell-mediated deficiency and, in this study, appeared to be more responsive to *Malassezia.* This study also suggests that there are differences in responses between the different species of *Malassezia* and thus the organism selected as the antigen in studies will have a profound effect on the results obtained. It, therefore, seems desirable to select multiple species for immunological studies in an attempt to overcome these possible differences in responses between species.

One study which also reported a reduction in the LT response in PV patients, studied 12 patients and 15 controls [179]. Interestingly, this study only used one species, that was classified as *M. furfur* at that time (ATCC 42132), but which is now known to be *M. sympodialis* and hence maybe less reactive than other species. The mean response of all patients was compared to the mean response of all controls and found to be significantly different. However, one control had a very high response, which may have skewed the data, leading to this apparent difference.

The most recent study examining cellular responses in PV patients used *Malassezia* in the mycelial phase as antigens for the first time [180]. Mycelial transformation occurs in PV and is characteristic of the disease, and hence the mycelial antigens if they are different from the yeast antigens, will be almost specific to the disease. Using three strains that had been induced to form mycelia, the cellular responses of 12 patients with PV and 12 age-and-sex-matched controls were measured in the LT and LMI. The mean response of patients was significantly higher than that of controls in the LT assay for all three strains tested. In contrast, the responses of patients in the LMI assay were not different to those of controls. Again, this study does not support the contention that patients with PV have a cell-mediated deficiency to *Malassezia,* and adds weight to data suggesting that these patients' cells are more reactive in cellular assays than controls.

5.4.3.3 Seborrhoeic Dermatitis

The first study to determine the *Malassezia*-specific cellular response in patients with SD used whole cells of *M. sympodialis, M. globosa* and *M. restricta* in the LT and LMI assays [2]. In 10 patients with SD and 10 age-and-sex-matched controls, more patients than controls responded to *M. globosa* and *M. restricta,* but there was no difference in the responses to *M. sympodialis* in the LT assay. In the LMI, more patients than controls responded to *M. restricta,* but the responses to *M. sympodialis* and *M. globosa* were not different between the patients and controls.

Another study, examining 10 patients with SD and 10 healthy controls, used '*P. ovale*' as the antigen in proliferation assays [8]. *P. ovale* from older classifications is thought to correspond to *M. slooffiae, M. obtusa* or *M. sympodialis*, and hence the strain used here may have been an immunologically less-reactive *M. sympodialis*. Stimulation of lymphocytes was carried out with a protein extract of the strain at four concentrations and the response recorded after four days. The extract did not induce stimulation of lymphocytes from patients with SD at any concentration, but at two concentrations stimulated lymphocytes from healthy controls. At one concentration, the proliferation of lymphocytes from SD patients was significantly lower than healthy donors. It is of note that there appears to be considerable variation in the results presented, but as only means of patient and control data were presented, it is not possible to see what is the source of this variation.

A third study examined the responses of 16 patients with SD and 16 controls to *M. furfur* [13]. Five different antigenic preparations were used – whole cells, cytoplasmic antigens, cell wall antigens, a sonicated preparation and a commercially available preparation – also prepared by sonication. The results were diverse for the different antigen preparations, with the lowest response to the cell wall antigen preparation, but when analysed statistically there was no significant difference in responses between patients and controls.

The most recent study found that there was no difference in proliferation to *M. furfur* (an isolate now considered to be *M. sympodialis*) between 15 patients with SD and 15 controls [179].

Overall, despite some variation in results, most studies have either found no major difference in cellular responses specific to *Malassezia* in patients with SD when compared with healthy controls, or slightly increased responses in patients. Therefore, although SD is common in patients with gross deficiencies in cellular immunity, such as AIDS, the response specific to *Malassezia* does not appear to be deficient.

5.4.3.4 Psoriasis

The role that *Malassezia* plays in psoriasis has been a matter of debate and controversy. It is known that treatment of psoriatic scalp lesions with ketoconazole results in improvement of the lesions and concomitant decrease in *Malassezia* population densities [181, 182]. An attempt to understand their role was made by studying the response of T-cells from scalp psoriatic lesions and other sites, to see if *Malassezia*-specific T-cells were confined to the scalp [10]. Skin biopsies and blood were obtained from 13 patients with psoriasis and T-cell lines were derived from the skin by stimulation with '*P. orbiculare*' (synonymous with *M. globosa*). Blood T-cell lines were derived by stimulation with either '*P. ovale*' or '*P. orbiculare*.' Proliferation responses did not differ significantly between the two organisms. With the scalp T-cell lines and T-cell lines from other sites, more of them responded to '*P. orbiculare*' than '*P. ovale*.' T-cell lines generated from patients with alopecia areata yielded similar results, suggesting that these T-cell lines are not related to the pathogenesis of psoriasis and may simply be a reflection of sensitisation due to colonisation by *Malassezia* on the scalp and other sites.

5.4.4 Cellular Responses in Dogs

Cellular responses to *M. pachydermatis* have been studied in dogs. Peripheral blood mononuclear cell proliferative responses following exposure to *M. pachydermatis* antigen (500 μg/mL) *in vitro* in seven healthy basset hounds exceeded those of eight seborrhoeic basset hounds with high populations of *M. pachydermatis* and eight Irish setters with gluten-sensitive enteropathy ($P<0.05$) [160]. Stimulation indices in the latter two groups and eight healthy beagles were comparable. Stimulation indices following *M. pachydermatis* exposure at 50 μg/mL, and to the mitogen, phytohaemagglutinin in the four groups of dogs did not vary significantly. Further studies are required to determine whether the reduced peripheral blood mono-nuclear cell proliferative responses to *M. pachydermatis* in affected basset hounds, in comparison to healthy hounds, are a cause or an effect of this disease.

Morris and others [183] investigated peripheral blood mono-nuclear cell responses and intradermal test reactivity to a crude extract of *M. pachydermatis* amongst atopic and healthy dogs. Atopic dogs with cytological evidence of *Malassezia* dermatitis had an increased lymphocyte blastogenic response to the extract, compared with clinically normal dogs and dogs with *Malassezia* otitis. Responses in atopic dogs with and without *Malassezia* dermatitis or otitis did not vary significantly, and there was no correlation between PBMNC proliferation and intradermal test reactivity. The authors concluded that cell-mediated and/or humoral reactivity to *M. pachydermatis* might contribute to the pathogenesis of AD in dogs and modification of the dysregulated immune response to the yeast may be of clinical benefit in the management of canine atopic dermatitis. By contrast, although beagle dogs that had suspensions of *M. pachydermatis* applied to their skin once daily for 7 days developed transient mild skin lesions and delayed intradermal test reactivity to a crude *M. pachydermatis* extract, PBMC proliferation indices did not change during the study period

[184]. Cytokine production by reactive PBMNC after *M. pachydermatis* exposure in dogs does not appear to have been reported.

5.5 Models for *Malassezia*-Associated Diseases

5.5.1 Normal Skin

The factors that lead to the development of *Malassezia*-associated diseases are still not fully understood, despite decades of research. The ecosystem on human skin is complex and modelling the interactions between the different organisms is difficult. One of the first studies to look at the interaction of *Malassezia* with a skin model was that of Bhattacharyya et al. [185]. Skin equivalents were prepared and allowed to form stratified epidermis, after which they were inoculated with *Malassezia* yeasts cells (6×10^4 cells). Viable *Malassezia* could be recovered from the skin equivalents over the 4 day sampling period and an increase in cell count occurred. When examined under scanning electron microscope, the upper layers of the stratum corneum showed destruction; there were no hyphae formed with all the cells remaining as yeasts. The authors concluded that the model probably most closely simulated colonisation and commensal growth, rather than any disease process. A recent study using skin equivalents found that when *Malassezia* was inoculated at a population density similar to that found on normal skin (10^2 CFU), proliferation occurred reaching 10^4 CFU within 72 h [186]. This was despite the fact that *Malassezia* requires long chain fatty acids for growth, which were not provided, indicating that the skin equivalent provided these directly. Further development of such models may, with time, allow us to simulate the interactions between normal cutaneous commensals and so better understand the normal skin ecosystem and, in turn, changes which can give rise to disease.

5.5.2 Pityriasis Versicolor

Some of the earliest attempts to model PV were carried out by Moore [187], when he attempted to induce lesions on the skin of rabbits, guinea pigs, rats and mice. None of the animal experiments were successful, although he did succeed in inducing lesions on the normal skin of healthy volunteers.

Repeated application of lipid-dependent *Malassezia* yeasts onto the skin of guinea pigs induced cutaneous lesions characterised clinically by erythema, scaling and crusting [188, 189]. The lesions were characterised histologically by superficial inflammation with hyperkeratosis and parakeratosis. Swiss white mice also developed skin lesions following yeast application whereas the nude (nu/nu) mouse, the hairless mouse and the nude rat did not [188]. Later studies by Faergemann [190, 191] were successful in experimentally infecting rabbits. He inoculated the inner ear of the rabbits with 10^7 *Malassezia* yeasts, but found that the infection only developed if the inoculation site was occluded. The lesions fluoresced under Wood's light and when examined microscopically, showed the classical combination of yeasts and short hyphae seen in PV. Lesions generally resolved spontaneously.

5.5.3 Seborrhoeic Dermatitis

Recently, a mouse model of SD has been described [192]. D2C mice were found to spontaneously develop an inflammatory skin condition that resembles SD. The mice have severe CD4 and CD8 T-cell lymphopenia and a defect in their T-cell receptor, and developed the condition when they reached sexual maturity. When the skin was examined, there were small oval structures present, which were thought to be *Malassezia* yeasts and treatment with fluconazole resulted clinical improvement. At the age of 70–90 day the mice underwent spontaneous remission of the condition, when their lymphopenia resolved and their CD4+ population increased. The utility of this model has yet to be tested, particularly as mice do not normally carry *Malassezia* on their skin, so the oval structures seen may not be yeasts and, additionally, because the condition spontaneously resolves, it is not known how closely it actually resembles other aspects of SD.

5.5.4 Psoriasis

The role of *Malassezia* in psoriasis is still debated, but in 1980, a model using rabbits inoculated with *Malassezia* resulted in psoriasiform lesions [193]. The rabbits had the fur on their back clipped, and these areas were then inoculated twice daily from Monday to Friday and once daily on Saturday and Sunday with either live or heat-killed whole *Malassezia* cells. Three strains were used, one of which was *M. furfur*; the identity of the other two strains was '*M. ovalis.*' The inoculated areas were allowed to air dry but were not occluded. Within 48–72 h of the first application, redness developed at the site of inoculation, and by 5 days of continued application, the site was red and covered with a thick white scale. The lesions continued as long as the inoculation took place, but resolved within 72–96 h when the inoculation ceased. Both live and heat killed cells produced the same reaction and when examined under the microscope, there was a large collection of neutrophils, similar to that seen in psoriasis. In conclusion, the authors felt that this model not only showed that *Malassezia* had a role in psoriasis, but that it could be used to study the condition. As with mice, rabbits are not normally colonised by *Malassezia* spp. [194] and thus challenge experiments in those hosts might reflect events following novel exposure to *Malassezia*-derived antigens rather than the continuous low level exposure that can be anticipated in the skin of humans and dogs.

5.5.5 *Malassezia pachydermatis* Otitis and Dermatitis in Dogs

In a seminal study, Gustafson [195] showed that a single inoculation of a suspension of *M. pachydermatis* ('*Pityrosporum canis*') was able to induce a transient mild otitis externa

(OE) characterised by erythema and brown exudate in seven of seven 7 inoculated dogs, whereas a lipophilic strain ('*P. ovale*') did not induce OE in three dogs. Otitis associated with *M. pachydermatis* proliferation was also induced by the application of a mixture of malt agar and olive oil, which the author attributed to the 'activation' of the commensal *Malassezia* mycobiota. Mansfield and others [196] were also able to induce clinical signs of OE in dogs inoculated with either pure cultures of *M. pachydermatis* (two out of nine ears), a culture filtrate of *M. pachydermatis* in Sabouraud's broth (five of eleven ears), freshly prepared Sabouraud's broth (one of three ears), or sterile saline (one of eight ears). Histopathological examination of biopsies of affected ears showed epidermal hyperplasia and lymphocytic inflammation. The authors concluded that *M. pachydermatis* was an opportunistic aural pathogen and that moisture and other atraumatic manipulation of the ear canal promoted yeast proliferation. Similarly, Uchida and others [197] were able to induce an erythemato-ceruminous OE in eight out of eight healthy Beagle dogs within 3–4 days of the single application of a suspension of *M. pachydermatis*.

The effects of the daily application for 7 days of suspensions of *Malassezia pachydermatis* to normal canine skin were evaluated in 10 beagle dogs. Four out of six dogs challenged without occlusion developed transient lesions generally characterised clinically by mild erythema with papules and histologically by mild epidermal hyperplasia and a superficial perivascular dermatitis. The clinical features of erythema resembled those frequently seen in the early stages of the natural disease, but none of the dogs developed the matting of hair by greasy exudation that is often seen in intertriginous areas in naturally-occurring cases [198]. Saline-treated control sites showed no clinical signs. In four dogs challenged with occlusion, skin lesions occurred at both yeast and saline-treated sites; erythema and papules were more severe at the yeast-treated sites in three dogs. Occlusion induced more persistent lesions, which resolved within 24 days, again illustrating the importance of moisture in the proliferation of this yeast on skin [191, 196]. Population densities of the yeast were highest at day eight and declined rapidly following cessation of application. Peripheral blood mono-nuclear cell proliferation indices following *M. pachydermatis* exposure *in vitro* and serum concentrations of *M. pachydermatis*-specific IgG antibodies did not vary significantly during the study. Delayed (24 h) intradermal test reactivity to *M. pachydermatis* antigens developed in all eight dogs with clinical signs following yeast exposure. The authors concluded that the resistance of healthy canine skin to infection by *M. pachydermatis* is mediated by local delayed hypersensitivity responses and/or innate epidermal immune mechanisms. Further studies are required to determine how skin defences can be degraded to permit the development of persistent experimental skin infections in dogs.

5.6 Conclusions

The interaction of *Malassezia* species with the host immune system has been widely investigated and some key findings have emerged. In humans and dogs, it is known that *Malassezia* is able to stimulate humoral and cellular immunity both in healthy individuals and in patients with *Malassezia*-associated diseases. The magnitude of the humoral immune responses in humans does not appear to correlate with the presence or severity of disease, but may simply be related to commensal carriage and associated immune stimulation. Cellular immunity is important and in several diseases the magnitude of cellular responses is greater in patients than healthy people.

Malassezia appears to have two divergent phenotypes; one which stimulates the immune system resulting in significant activation of various immunological pathways, whilst the second phenotype allows it to limit immune stimulation to a large extent and perhaps contributes to its ability to exist in most people as a commensal. Understanding these two phenotypes and the factors that lead to changing between them are important if we are to understand how *Malassezia*-associated diseases arise and how they may be prevented.

References

1. Ashbee HR (2007) Update on the genus *Malassezia*. Med Mycol 45:287–303
2. Ashbee HR, Ingham E, Holland KT et al (1994) Cell-mediated immune responses to *Malassezia furfur* serovars A, B and C in patients with pityriasis versicolor, seborrheic dermatitis and controls. Exp Dermatol 3:106–112
3. Ashbee HR, Fruin A, Holland KT et al (1994) Humoral immunity to *Malassezia furfur* serovars A, B and C in patients with pityriasis versicolor, seborrheic dermatitis and controls. Exp Dermatol 3:227–233
4. Bergbrant IM, Johansson S, Robbins D et al (1991) The evaluation of various methods and antigens for the detection of antibodies against *Pityrosporum ovale* in patients with seborrhoeic dermatitis. Clin Exp Dermatol 16:339–343
5. DaMert GJ, Kirkpatrick CH, Sohnle PG (1980) Comparison of antibody responses in chronic mucocutaneous candidiasis and tinea versicolor. Inter Arch All Appl Immunol 63:97–104
6. Faergemann J (1983) Antibodies to *Pityrosporum orbiculare* in patients with tinea versicolor and controls of various ages. J Invest Dermatol 80:133–135
7. Furukawa F, Danno K, Imamura S et al (1981) Histological and serological studies of *Pityrosporum orbiculare* in cases of pityriasis versicolor. J Dermatol 8:27–30
8. Neuber K, Kroger S, Gruseck E et al (1996) Effects of *Pityrosporum ovale* on proliferation, immunoglobulin (IgA, G, M) synthesis and cytokine (IL-2, IL-10, IFN gamma) production of peripheral blood mononuclear cells from patients with seborrhoeic dermatitis. Arch Dermatol Res 288:532–536
9. Sohnle PG, Collins-Lech C (1980) Relative antigenicity of *P. orbiculare* and *C. albicans*. J Invest Dermatol 75:279–283
10. Baker BS, Powles A, Garioch JJ et al (1997) Differential T-cell reactivity to the round and oval forms of *Pityrosporum* in the skin of patients with psoriasis. Brit J Dermatol 136:319–325
11. Hashimoto K, Taniguchi Y, Simon MR et al (1989) Immunological aspects of superficial fungal infections. Jpn J Med Mycol 30:81–91
12. Midgley G, Hay RJ (1988) Serological responses to *Pityrosporum (Malassezia)* in seborrhoiec dermatitis demonstrated by ELISA and Western blotting. Bull Soc Fr Mycol Méd 17: 267–276
13. Parry ME, Sharpe GR (1998) Seborrhoeic dermatitis is not caused by an altered immune response to *Malassezia* yeast. Brit J Dermatol 139:254–263
14. Arzumanyan VG, Serdyuk OA, Kozlova NN et al (2003) IgE and IgG antibodies to *Malassezia* spp. yeast extract in patients with atopic dermatitis. Bull Exp Biol Med 135:460–463

15. Zargari A, Doekes G, van Ieperen-van Dijk AG et al (1995) Influence of culture period on the allergenic composition of *Pityrosporum orbiculare* extracts. Clin Exp All 25:1235–1245
16. Lintu P, Savolainen J, Kalimo K et al (1998) Stability of *Pityrosporum ovale* allergens during storage. Clin Exp All 28:486–490
17. Jensen-Jarolim E, Poulsen LK, With H et al (1992) Atopic dermatitis of the face, scalp, and neck: type I reaction to the yeast *Pityrosporum ovale*? J All Clin Immunol 89:44–51
18. Guého E, Midgley G, Guillot J (1996) The genus *Malassezia* with description of four new species. Antonie van Leeuwenhoek 69:337–355
19. Schmidt M, Zargari A, Holt P et al (1997) The complete cDNA sequence and expression of the first major allergenic protein of *Malassezia furfur*, Mal f 1. Eur J Biochem 246:181–185
20. Bos JD (1997) The skin as an organ of immunity. Clin Exp Immunol 107:3–5
21. Meyer T, Stockfleth E, Christophers E (2007) Immune response profiles in human skin. Br J Dermatol 157(suppl 2):1–7
22. Elias PM (2005) Stratum corneum defensive functions: an integrated view. J Invest Dermatol 125:183–200
23. Ganz T (2003) Defensins: antimicrobial peptides of innate immunity. Nat Rev Immunol 3:710–720
24. Niyonsaba F, Ogawa H (2005) Protective roles of the skin against infection: implication of naturally occurring human antimicrobial agents beta-defensins, cathelicidin LL-37 and lysozyme. J Dermatol Sci 40:157–168
25. Markiewski MM, Lambris JD (2007) The role of complement in inflammatory diseases from behind the scenes into the spotlight. Am J Pathol 171:715–727
26. Ray TL, Wuepper KD (1976) Activation of the alternative (properdin) pathway of complement by *Candida albicans* and related species. J Invest Dermatol 67:700–703
27. Webster GF, McArthur WP (1982) Activation of components of the alternative pathway of complement by *Propionibacterium acnes* cell wall carbohydrate. J Invest Dermatol 79:137–140
28. Belew PW, Rosenberg EW, Jennings BR (1980) Activation of the alternative pathway of complement by *Malassezia ovalis (Pityrosporum ovale)*. Mycopath 70:187–191
29. Shoham S, Levitz SM (2005) The immune response to fungal infections. Br J Haematol 129:569–582
30. Wagner DK, Sohnle PG (1997) Cutaneous defense mechanisms against fungi. Basic Clin Dermatol 12:161–189
31. Janssens S, Beyaert R (2003) Role of Toll-like receptors in pathogen recognition. Clin Microbiol Rev 16:637–646
32. Ben-Ami R, Lewis RE, Kontoyiannis DP (2008) Immunocompromised hosts: immunopharmacology of modern antifungals. Clin Infect Dis 47:226–235
33. Romani L (2004) Immunity to fungal infections. Nat Rev Immunol 4:1–13
34. Ashbee HR, Muir SR, Cunliffe WJ et al (1997) IgG subclasses specific to *Staphylococcus epidermidis* and *Propionibacterium acnes* in patients with acne vulgaris. Brit J Dermatol 136:730–733
35. Ingham E, Gowland G, Ward RM et al (1987) Antibodies to *Propionibacterium acnes* and *P. acnes* exocellular enzymes in the normal population at various ages and in patients with acne vulgaris. Br J Dermatol 116:805–812
36. Metze D, Kersten A, Jurecka W et al (1991) Immunoglobulins coat microorganisms of skin surface: a comparative immunohistochemical and ultrastructural study of cutaneous and oral microbial symbionts. J Invest Dermatol 96:439–445
37. Casadevall A, Feldmesser M, Pirofski LA (2002) Induced humoral immunity and vaccination against major human fungal pathogens. Curr Opin Microbiol 5:386–391
38. Martin-Scott J (1952) The *Pityrosporum ovale*. Brit J Dermatol 64:257–273
39. Sohnle PG, Collins-Lech C (1983) Activation of complement by *Pityrosporum orbiculare*. J Invest Dermatol 80:93–97

40. Suzuki T, Ohno N, Oshima Y et al (1998) Activation of the complement system, alternative and classical pathways, by *Malassezia furfur*. Pharm Pharmacol Lett 8:133–136
41. Scott DG, Cunliffe WJ, Gowland G (1979) Activation of complement-a mechanism for the inflammation in acne. Br J Dermatol 101:315–320
42. Faergemann J, Bergbrant IM, Dohse M et al (2001) Seborrhoeic dermatitis and *Pityrosporum (Malassezia)* folliculitis: characterization of inflammatory cells and mediators in the skin by immunohistochemistry. Brit J Dermatol 144:549–556
43. Piérard-Franchimont C, Arrese JE, Piérard GE (1995) Immunohistochemical aspects of the link between *Malassezia ovalis* and seborrhoeic dermatitis. J Eur Acad Dermatol Venereol 4:14–19
44. Lopez-Garcia B, Lee PH, Gallo RL (2006) Expression and potential function of cathelicidin antimicrobial peptides in dermatophytosis and tinea versicolor. J Antimicrob Chemother 57:877–882
45. Donnarumma G, Paoletti I, Buommino E et al (2004) *Malassezia furfur* induces the expression of beta-defensin-2 in human keratinocytes in a protein kinase C-dependent manner. Arch Dermatol Res 295:474–481
46. White A, Handler P, Smith et al (1992) Blood. In: Principles of biochemistry. McGraw Hill, Tokyo, pp 902–928
47. Saylers AA, Whitt DD (1994) Bacterial pathogenesis. A molecular approach. ASM Press, Washington, pp 16–17
48. Ardehali R, Shi L, Janatova J et al (2003) The inhibitory activity of serum to prevent bacterial adhesion is mainly due to apo-transferrin. J Biomed Mater Res A 66:21–28
49. Artis WM, Patrusky E, Rastinejad F et al (1983) Fungistatic mechanism of human transferrin for *Rhizopus oryzae* and *Trichophyton mentagrophytes*: alternative to simple iron deprivation. Infect Immun 41:1269–1278
50. King RD, Khan HA, Foye JC et al (1975) Transferrin, iron, and dermatophytes. I. Serum dermatophyte inhibitory component definitively identified as unsaturated transferrin. J Lab Clin Med 86:204–212
51. Shiraishi A, Arai T (1979) Antifungal activity of transferrin. Sabouraudia 17:79–83
52. Minn Y, Brummer E, Stevens DA (1997) Effect of iron on fluconazole activity against *Candida albicans* in presence of human serum or monocyte-derived macrophages. Mycopath 138:29–35
53. Grover DD, Brummer E, Stevens DA (1996) Study of the role of iron in the anticryptococcal activity of human serum and fluconazole. Mycopathol 133:71–77
54. Sridhar S, Ahluwalia M, Brummer E et al (2000) Characterization of an anticryptococcal protein isolated from human serum. Infect Immun 68:3787–3791
55. Sutcliffe MC, Savage AM, Alford RH (1980) Transferrin-dependent growth inhibition of yeast-phase *Histoplasma capsulatum* by human serum and lymph. J Infect Dis 142:209–219
56. Abate O, Zanatta R, Malisano T et al (2000) Canine serum protein patterns using high-resolution electrophoresis (HRE). Vet J 159:154–160
57. Graber HU, Pfister H, Martig J (1995) Increased concentrations of transferrin in the urine and serum of cattle with cardiomyopathy. Res Vet Sci 59:160–163
58. Jenkinson DM, Lloyd DH, Mabon RM (1979) The antigenic composition and source of soluble proteins on the surface of the skin of sheep. J Comp Pathol 89:43–50
59. Nakayashiki N (1990) Sweat protein components tested by SDS-polyacrylamide gel electrophoresis followed by immunoblotting. Tohoku J Exp Med 161:25–31
60. Lloyd DH, Dick WD, Jenkinson DM (1979) Location of the microflora in the skin of cattle. Br Vet J 135:519–526
61. Malcolm SA, Hughes TC (1980) The demonstration of bacteria on and within the stratum corneum using scanning electron microscopy. Br J Dermatol 102:267–275
62. Montes LF, Wilborn WH (1969) Location of bacterial skin flora. Br J Dermatol 81(Suppl 1):23–26

63. Williamson P, Kligman AM (1965) A new method for the quantitative investigation of cutaneous bacteria. J Invest Dermatol 45:498–503
64. King RD (1976) Nutritional immunity in cutaneous fungous infections. Int J Dermatol 15:358–359
65. Bond R, Kim Y, Lloyd DH (2005) Bovine and canine transferrin inhibit the growth of *Malassezia pachydermatis in vitro*. Med Mycol 43:447–451
66. Mauldin EA, Scott DW, Miller WH et al (1997) *Malassezia* dermatitis in the dog: a retrospective histopathological and immunopathological study of 86 cases (1990–95). Vet Dermatol 8:191–202
67. Richardson MD, Shankland GS (1991) Enhanced phagocytosis and intracellular killing of *Pityrosporum ovale* by human neutrophils after exposure to ketoconazole is correlated to changes of the yeast cell surface. Mycoses 34:29–33
68. Cech P, Lehrer RI (1984) Heterogeneity of human neutrophil phagolysosomes: functional consequences for candidacidal activity. Blood 64:147–151
69. Murphy JW (1991) Mechanisms of natural resistance to pathogenic fungi. Ann Rev Microbiol 45:509–538
70. Nazzaro-Porro M, Passi S (1978) Identification of tyrosinase inhibitors in cultures of *Pityrosporum*. J Invest Dermatol 71:205–208
71. Akamatsu H, Komura J, Asada Y et al (1991) Inhibitory effect of azelaic acid on neutrophil functions: a possible cause for its efficacy in treating pathogenetically unrelated diseases. Arch Dermatol Res 283:162–166
72. Fitton A, Goa KL (1991) Azelaic acid: a review of its pharmacological properties and therapeutic efficacy in acne and hyperpigmentary skin disorders. Drugs 41:780–798
73. Mittag H (1995) Fine structural investigation of *Malassezia furfur*. II. The envelope of the yeast cells. Mycoses 38:13–21
74. Ashbee HR, Evans EGV (2002) Immunology of diseases associated with *Malassezia* species. Clin Micro Rev 15:21–57
75. Krämer HJ, Kessler D, Hipler UC et al (2005) Pityriarubins, novel highly selective inhibitors of respiratory burst from cultures of the yeast *Malassezia furfur*: comparison with the bisindolylmaleimide arcyriarubin A. Chembiochem 6:2290–2297
76. Suzuki T, Ohno N, Ohshima Y et al (1998) Soluble mannan and beta-glucan inhibit the uptake of *Malassezia furfur* by human monocytic cell line, THP-1. FEMS Immunol Med Microbiol 21:223–230
77. Suzuki T, Tsuzuki A, Ohno N et al (2000) Enhancement of IL-8 production from human monocytic and granulocytic cell lines, THP-1 and HL-60, stimulated with *Malassezia furfur*. FEMS Immunol Med Microbiol 28:157–162
78. Bunse T, Mahrle G (1996) Soluble *Pityrosporum*-derived chemoattractant for polymorphonuclear leukocytes of psoriatic patients. Acta Derm Venereol 76:10–12
79. Takahashi M, Ushijima T, Ozaki Y (1984) Biological activity of *Pityrosporum*. I. Enhancement of resistance in mice stimulated by *Pityrosporum* against *Salmonella typhimurium*. Immunology 51:697–702
80. Takahashi M, Ushijima T, Ozaki Y (1986) Biological activity of *Pityrosporum*. II. Antitumor and immune stimulating effect of *Pityrosporum* in mice. J Nat Cancer Inst 77:1093–1097
81. Walters CE, Ashbee HR, Eady EA et al (1995) *Malassezia furfur* suppresses *in vitro* interleukin-6 release by peripheral blood mononuclear cells. J Invest Dermatol 105:492
82. Kesavan S, Walters CE, Holland KT et al (1998) The effects of *Malassezia* on pro-inflammatory cytokine production by human peripheral blood mononuclear cells *in vitro*. Med Mycol 36:97–106
83. Kesavan S, Holland KT, Ingham E (2000) The effects of lipid extraction on the immunomodulatory activity of *Malassezia* species *in vitro*. Med Mycol 38:239–247
84. Scheynius A, Faergemann J, Forsum U et al (1984) Phenotypic characterization *in situ* of inflammatory cells in pityriasis (tinea) versicolor. Acta Derm Venereol 64:473–479

85. Brasch J, Martens H, Sterry W (1993) Langerhans cell accumulation in chronic tinea pedis and pityriasis versicolor. Clin Exp Dermatol 18:329–332
86. Wroblewski N, Bar S, Mayser P (2005) Missing granulocytic infiltrate in pityriasis versicolor – Indication of specific anti-inflammatory activity of the pathogen? Mycoses Suppl 48: 66–71
87. Thomas DS, Ingham E, Bojar RA et al (2008) *in vitro* modulation of human kerationcyte pro- and anti-inflammatory cytokine production by the capsule of *Malassezia* spp. FEMS Immunol Med Microbiol 54:203–214
88. Pinkus H, Mehregan AH (1966) The primary histologic lesion of seborrhoeic dermatitis and psoriasis. J Invest Dermatol 46:109–116
89. Bergbrant IM, Johansson S, Robbins D et al (1991) An immunological study in patients with seborrhoeic dermatitis. Clin Exp Dermatol 16:331–338
90. Borton LK, Schwartz RA (1981) *Pityrosporum* folliculitis: a common acneiform condition of middle age. Ariz Med 38:598–601
91. Bufill JA, Lum LG, Caya JG et al (1988) *Pityrosporum* folliculitis after bone marrow transplantation. Clinical observations in five patients. Ann Int Med 108:560–563
92. Ford GP, Ive FA, Midgley G (1982) *Pityrosporum* folliculitis and ketoconazole. Brit J Dermatol 107:691–695
93. Weary PE, Russell CM, Butler HK et al (1969) Acneiform eruptions resulting from antibiotic administration. Arch Dermatol 100:179–183
94. Ferrandiz C, Ribera M, Barranco JC et al (1992) Eosinophilic pustular folliculitis in patients with acquired immunodeficiency syndrome. Inter J Dermatol 31:193–195
95. Potter BS, Burgoon CF, Johnson WC (1973) *Pityrosporum* folliculitis. Report of seven cases and review of the *Pityrosporum* organism relative to cutaneous disease. Arch Dermatol 107:388–391
96. Gross TL, Ihrke PJ, Walder EJ et al (2005) *Malassezia* dermatitis. In: Skin diseases of the dog and cat. Blackwell Science, Oxford
97. Guaguère E, Prélaud A (1996) A retrospective study of 54 dogs with *Malassezia pachydermatis* dermatitis: epidemiological, clinical, cytological and histopathological results. Pract Méd Chir An Comp 31:309–323
98. Mason KV, Evans AG (1991) Dermatitis associated with *Malassezia pachydermatis* in 11 dogs. J Am Anim Hosp Assoc 27:13–20
99. Mauldin EA, Morris DO, Goldschmidt MH (2002) Retrospective study: the presence of *Malassezia* in feline skin biopsies. A clinicopathological study. Vet Dermatol 13:7–13
100. Charles CR, Sire DJ, Johnson BL et al (1972) Hypopigmentation in tinea versicolor: a histochemical and electronmicroscopic study. Inter J Dermatol 12:48–58
101. El-Gothamy Z, Abdel-Fattah A, Ghaly AF (1975) Tinea versicolor hypopigmentation: histochemical and therapeutic studies. Inter J Dermatol 14:510–515
102. Karaoui R, Bou-Resli M, Al Zaid NS et al (1981) Tinea versicolor: ultrastructural studies on hypopigmented and hyperpigmented skin. Dermatologica 162:69–85
103. Allen HB, Charles CR, Johnson BL (1976) Hyperpigmented tinea versicolor. Arch Dermatol 112:1110–1112
104. Galadari I, El Komy M, Mousa A et al (1992) Tinea versicolor: histologic and ultrastructural investigation of pigmentary changes. Inter J Dermatol 31:253–256
105. Dotz WI, Henrikson DM, Yu GSM (1985) Tinea versicolor: a light and electron microscopic study of hyperpigmented skin. J Am Acad Dermatol 12:37–44
106. De Luca C, Picardo M, Breathnach A et al (1996) Lipoperoxidase activity of *Pityrosporum*: characterisation of by-products and possible role in pityriasis versicolor. Exp Dermatol 5:49–56
107. Nazzaro-Porro M, Passi S, Picardo M et al (1986) Lipoxygenase activity of *Pityrosporum in vitro* and *in vivo*. J Invest Dermatol 87:108–112
108. Krämer HJ, Podobinska M, Bartsch A et al (2005) Malassezin, a novel agonist of the aryl hydrocarbon receptor from the yeast *Malassezia furfur*, induces apoptosis in primary human melanocytes. Chembiochem 6:860–865

109. Aly R, Bibel DJ (1992) Adherence of skin microorganisms and the development of skin flora from birth. In: Noble WC (ed) The skin microflora and microbial skin disease. Cambridge University Press, Cambridge, pp 355–372
110. Kennedy MJ (1988) Adhesion and association mechanisms of *Candida albicans*. Curr Top Med Mycol 4:73–169
111. Tsuboi R, Ogawa H, Bramono K (1994) Pathogenesis of superficial mycoses. J Med Vet Mycol 32:91–104
112. Zurita J, Hay RJ (1987) Adherence of dermatophyte microconidia and arthroconidia to human keratinocytes *in vitro*. J Invest Dermatol 89:529–534
113. Calderone RA, Diamond R, Senet J-M et al (1994) Host cell-fungal cell interactions. J Med Vet Mycol 32:151–168
114. Dranginis AM, Rauceo JM, Coronado JE et al (2007) A biochemical guide to yeast adhesins: glycoproteins for social and antisocial occasions. Microbiol Mol Biol Rev 71:282–294
115. Borgers M, Cauwenbergh G, van de Ven MA et al (1987) Pityriasis versicolor and *Pityrosporum ovale*. Morphogenetic and ultrastructural considerations. Inter J Dermatol 26:586–589
116. Faergemann J, Aly R, Maibach HI (1983) Adherence of *Pityrosporum orbiculare* to human stratum corneum cells. Arch Dermatol Res 275:246–250
117. Bergbrant IM, Faergemann J (1994) Adherence of *Malassezia furfur* to human stratum corneum cells *in vitro*: a study of healthy individuals and patients with seborrhoeic dermatitis. Mycoses 37:217–219
118. Schechtman RC, Midgley G, Bingham JS et al (1995) Adherence of *Malassezia* isolates to human keratinocytes *in vitro* – a study of HIV-positive patients with seborrhoeic dermatitis. Brit J Dermatol 133:537–541
119. Bond R, Lloyd DH (1996) Factors affecting the adherence of *Malassezia pachydermatis* to canine corneocytes *in vitro*. Vet Dermatol 7:49–56
120. Bond R, Lloyd DH (1998) Studies on the role of carbohydrates in the adherence of *Malassezia pachydermatis* to canine corneocytes *in vitro*. Vet Dermatol 9:105–109
121. Bond R, Lloyd DH (1998) The relationship between population sizes of *Malassezia pachydermatis* in healthy dogs and in basset hounds with *M. pachydermatis*-associated seborrhoeic dermatitis and adherence to canine corneocytes *in vitro*. In: Kwochka KW, Willemse T, von Tscharner C (eds) Advances in veterinary dermatology, vol 3. Butterworth, Heineman, pp 283–289
122. Ray TL, Digre KB, Payne CD (1984) Adherence of *Candida* species to human epidermal corneocytes and buccal mucosal cells: correlation with cutaneous pathogenicity. J Invest Dermatol 83:37–41
123. Bailey A, Wadsworth E, Calderone R (1995) Adherence of *Candida albicans* to human buccal epithelial cells: host-induced protein synthesis and signalling events. Infect Immun 63:569–572
124. Merkel GJ, Scofield BA (1993) Conditions affecting the adherence of *Cryptococcus neoformans* to rat glial and lung cells *in vitro*. J Med Vet Mycol 31:55–64
125. Kimura LH, Pearsall NN (1978) Adherence of *Candida albicans* to human buccal epithelial cells. Infect Immun 21:64–68
126. King RD, Lee JC, Morris AL (1980) Adherence of *Candida albicans* and other *Candida* species to mucosal epithelial cells. Infect Immun 27:667–674
127. Collins-Lech C, Kalbfleisch JH, Franson TR et al (1984) Inhibition by sugars of *Candida albicans* adherence to human buccal mucosal cells and corneocytes *in vitro*. Infect Immun 46:831–834
128. Segal E, Lehrer N, Ofek I (1982) Adherence of *Candida albicans* to human vaginal epithelial cells: inhibition by amino sugars. Exp Cell Biol 50:13–17
129. Sandin RL, Rogers AL, Patterson RJ et al (1982) Evidence for mannose-mediated adherence of *Candida albicans* to human buccal cells *in vitro*. Infect Immun 35:79–85

130. Aly R, Shinefield HI, Strauss WG et al (1977) Bacterial adherence to nasal mucosal cells. Infect Immun 17:546–549
131. Bibel DJ, Aly R, Shinefield HR et al (1982) Importance of the keratinized epithelial cell in bacterial adherence. J Invest Dermatol 79:250–253
132. Kurono Y, Fujiyoshi T, Mogi G (1989) Secretory IgA and bacterial adherence to nasal mucosal cells. Ann Otol Rhinol Laryngol 98:273–277
133. Bond R, Lloyd DH (1997) Skin and mucosal populations of *Malassezia pachydermatis* in healthy and seborrhoeic Basset Hounds. Vet Dermatol 8:101–106
134. Power HT, Ihrke PJ, Stannard AA et al (1992) Use of etretinate for treatment of primary keratinization disorders (idiopathic seborrhea) in cocker spaniels, west highland white terriers, and basset hounds. J Am Vet Med Assoc 201:419–429
135. Tokura Y, Kobayashi M, Kabashima K (2008) Epidermal chemokines and modulation by antihistamines, antibiotics and antifungals. Exp Dermatol 17:81–90
136. Watanabe S, Kano R, Sato H et al (2001) The effects of *Malassezia* yeasts on cytokine production by human keratinocytes. J Invest Dermatol 116:769–773
137. Baroni A, Perfetto B, Paoletti I et al (2001) *Malassezia furfur* invasiveness in a keratinocyte cell line (HaCat): effects on cytoskeleton and on adhesion molecule and cytokine expression. Arch Dermatol Res 293:414–419
138. Ishibashi Y, Sugita T, Nishikawa A (2006) Cytokine secretion profile of human keratinocytes exposed to *Malassezia* yeasts. FEMS Immunol Med Microbiol 48:400–409
139. Sugita T, Suto H, Unno T et al (2001) Molecular analysis of *Malassezia* microflora on the skin of atopic dermatitis patients and healthy subjects. J Clin Microbiol 39:3486–3490
140. Baroni A, Orlando M, Donnarumma G et al (2006) Toll-like receptor 2 (TLR2) mediates intracellular signalling in human keratinocytes in response to *Malassezia furfur*. Arch Dermatol Res 297:280–288
141. Baroni A, Paoletti I, Ruocco E et al (2004) Possible role of *Malassezia furfur* in psoriasis: modulation of TGF b 1, integrin and HSP 70 expression in human keratinocytes and in the skin of psoriasis-affected patients. J Cutaneous Pathol 31:35–42
142. Chen TA, Halliwell RE, Hill PB (2002) Failure of extracts from *Malassezia pachydermatis* to stimulate canine keratinocyte proliferation *in vitro*. Vet Dermatol 13:323–329
143. von Tscharner C, Wyder MT, Busato A (1999) Proliferation characteristics of canine keratinocyte cultures infected with *Malassezia pachydermatis*. American Academy of Veterinary Dermatologists, Annual Meeting, Abstract 186
144. Chen TA, Halliwell REW, Shaw DJ et al (2004) Assessment of the ability of *Malassezia pachydermatis* to stimulate proliferation of canine keratinocytes *in vitro*. Am J Vet Res 65:787–796
145. Baroni A, Perfetto B, Paoletti I et al (2001) Uptake of *Malassezia furfur* by human dermal fibroblasts: effect of ketoconazole and cytoskeleton inhibitors. Arch Dermatol Res 293:407–413
146. Ashbee HR, Gunning J, Holland KT et al (1995) Titres of IgE specific to *Malassezia furfur* serovars A, B and C in patients with pityriasis versicolor and controls. J Invest Dermatol 105:492
147. Bergbrant IM, Faergemann J (1989) Seborrhoeic dermatitis and *Pityrosporum ovale*: a cultural and immunological study. Acta Derm Venereol 69:332–335
148. Faergemann J, Johansson S, Back O et al (1986) An immunologic and cultural study of *Pityrosporum* folliculitis. J Am Acad Dermatol 14:429–433
149. Kieffer M, Bergbrant IM, Faergemann J et al (1990) Immune reactions to *Pityrosporum ovale* in adult patients with atopic and seborrheic dermatitis. J Am Acad Dermatol 22:739–742
150. Silva V, Fischman O, de Camargo ZP (1997) Humoral immune response to *Malassezia furfur* in patients with pityriasis versicolor and seborrheic dermatitis. Mycopath 139:79–85
151. Squiquera L, Galimberti R, Morelli L et al (1994) Antibodies to proteins from *Pityrosporum ovale* in the sera from patients with psoriasis. Clin Exp Dermatol 19:289–293

152. Sternberg TH, Keddie FM (1961) Immunofluorescence studies in tinea versicolor. Arch Dermatol 84:161–165
153. Wu YC, Chen KT (1985) Humoral immunity in patients with tinea versicolor. J Dermatol 12:161–166
154. Alexander S (1968) Loss of hair and dandruff. Brit J Dermatol 79:549–552
155. Sohnle PG, Collins-Lech C, Huhta KE (1983) Class-specific antibodies in young and aged humans against organisms producing superficial fungal infections. Brit J Dermatol 108:69–76
156. Bergbrant IM, Faergemann J (1988) Variations of *Pityrosporum orbiculare* in middle-aged and elderly individuals. Acta Derm Venereol 68:537–540
157. Cunningham AC, Ingham E, Gowland G (1992) Humoral responses to *Malassezia furfur* serovars A, B and C in normal individuals of various ages. Brit J Dermatol 127:476–481
158. Faggi E, Pini G, Campisi E et al (1998) Anti-*Malassezia furfur* antibodies in the population. Mycoses 41:273–275
159. Saadatzadeh MR (1998) The immunology of the mycelial phase of *Malassezia*. PhD Thesis, University of Leeds
160. Bond R, Elwood CM, Littler RM et al (1998) Humoral and cell-mediated responses to *Malassezia pachydermatis* in healthy dogs and dogs with *Malassezia* dermatitis. Vet Record 143:381–384
161. Bond R, Saijonmaa-Koulumies LE, Lloyd DH (1995) Population sizes and frequency of *Malassezia pachydermatis* at skin and mucosal sites on healthy dogs. J Small Anim Pract 36:147–150
162. Nuttall TJ, Halliwell RE (2001) Serum antibodies to *Malassezia* yeasts in canine atopic dermatitis. Vet Dermatol 12:327–332
163. Bond R, Lloyd DH (2002) Immunoglobulin G responses to *Malassezia pachydermatis* in healthy dogs and dogs with *Malassezia* dermatitis. Vet Record 150:509–512
164. Chen TA, Halliwell REW (2002) Immunoglobulin G responses to *Malassezia pachydermatis* antigens in atopic and normal dogs. Adv Vet Dermatol 4:202–209
165. Habibah A, Catchpole B, Bond R (2005) Canine serum immunoreactivity to *M. pachydermatis in vitro* is influenced by the phase of yeast growth. Vet Dermatol 16:147–152
166. Chen TA, Halliwell RE, Pemberton AD et al (2002) Identification of major allergens of *Malassezia pachydermatis* in dogs with atopic dermatitis and *Malassezia* overgrowth. Vet Dermatol 13:141–150
167. Farver K, Morris DO, Shofer F et al (2005) Humoral measurement of type-1 hypersensitivity reactions to a commercial *Malassezia* allergen. Vet Dermatol 16:261–268
168. Koranda FC, Dehmel EA, Kahn G et al (1974) Cutaneous complications in immunosuppressed renal homograft recipients. J Am Med Assoc 229:419–424
169. Segal R, Shmueli D, Yussim A et al (1989) Renal transplant incidence of superficial fungal infection as related to immunosuppression therapy. Transplant Proc 21:2114–2116
170. Rhie S, Turcios R, Buckley H et al (2000) Clinical features and treatment of *Malassezia* folliculitis with fluconazole in orthotopic heart transplant recipients. J Heart Lung Transplant 19:215–219
171. Coldiron BM, Bergstresser PR (1989) Prevalence and clinical spectrum of skin disease inpatients infected with HIV. Arch Dermatol 125:357–361
172. Smith KJ, Skelton HG, Yeager J et al (1994) Cutaneous findings in HIV-1 positive patients: a 42 month prospective study. J Am Acad Dermatol 31:746–754
173. Vidal C, Girard PM, Dompmartin D et al (1990) Seborrhoeic dermatitis and HIV infection. J Am Acad Dermatol 23:1106–1110
174. Federlin K, Maini RN, Russell AS et al (1971) A micro-method for peripheral leucocyte migration in tuberculin sensitivity. J Clin Pathol 24:533–536
175. Cunningham AC (1991) An investigation of the immune status of man to *Malassezia* spp. PhD Thesis, University of Leeds
176. Sohnle PG, Collins-Lech C (1978) Cell-mediated immunity to *Pityrosporum orbiculare* in tinea versicolor. J Clin Invest 62:45–53

177. Sohnle PG, Collins-Lech C (1982) Analysis of the lymphocyte transformation response to *Pityrosporum orbiculare* in patients with tinea versicolor. Clin Exp Immunol 49:559–564
178. Wu C, Chen KT (1987) Lymphocyte proliferation to crude extract of *Pityrosporum* species and natural killer activity in tinea versicolor. J Med Assoc Thailand 70:45
179. Bergbrant IM, Andersson B, Faergemann J (1999) Cell-mediated immunity to *Malassezia furfur* in patients with seborrhoeic dermatitis and pityriasis versicolor. Clin Exp Dermatol 24:402–406
180. Saadatzadeh MR, Ashbee HR, Cunliffe WJ et al (2001) Cell-mediated immunity to the mycelial phase of *Malassezia* spp. in patients with pityriasis versicolor and controls. Brit J Dermatol 144:77–84
181. Farr PM, Krause LB, Marks JM et al (1985) Response of scalp psoriasis to oral ketoconazole. Lancet 2:921–922
182. Rosenberg EW, Belew PW (1982) Improvement of psoriasis of the scalp with ketoconazole. Arch Dermatol 118:370–371
183. Morris DO, Clayton DJ, Drobatz KJ et al (2002) Response to *Malassezia pachydermatis* by peripheral blood mononuclear cells from clinically normal and atopic dogs. Am J Vet Res 63:358–362
184. Bond R, Patterson-Kane JC, Lloyd DH (2004) Clinical, histopathological and immunological effects of exposure of canine skin to *Malassezia pachydermatis*. Med Mycol 42:165–175
185. Bhattacharyya T, Edward M, Cordery C et al (1998) Colonization of living skin equivalents by *Malassezia furfur*. Med Mycol 36:15–19
186. Holland DB, Bojar RA, Jeremy AH et al (2008) Microbial colonization of an *in vitro* model of a tissue engineered human skin equivalent – a novel approach. FEMS Microbiol Lett 279:110–115
187. Moore M (1940) *Malassezia furfur* the cause of tinea versicolor: cultivation of the organism and experimental production of the disease. Arch Dermatol 41:253–260
188. Drouhet E, Dompmartin D, Papachristou-Moraiti A et al (1980) Experimental dermatitis caused by *Pityrosporum ovale* and (or) *Pityrosporum orbiculare* in the guinea pig and the mouse. Sabouraudia 18:149–156
189. Van Cutsem J, Van Gerven F, Fransen J et al (1990) The *in vitro* antifungal activity of ketoconazole, zinc pyrithione, and selenium sulfide against *Pityrosporum* and their efficacy as a shampoo in the treatment of experimental pityrosporosis in guinea pigs. J Am Acad Dermatol 22:993–998
190. Faergemann J (1979) Experimental tinea versicolor in rabbits and humans with *Pityrosporum orbiculare*. J Invest Dermatol 72:326–329
191. Faergemann J, Fredriksson T (1981) Experimental infections in rabbits and humans with *Pityrosporum orbiculare* and *P. ovale*. J Invest Dermatol 77:314–318
192. Oble DA, Collett E, Hsieh M et al (2005) A novel T cell receptor transgenic animal model of seborrhoeic dermatitis-like skin disease. J Invest Dermatol 124:151–159
193. Rosenberg EW, Belew P, Bale G (1980) Effect of topical applications of heavy suspensions of killed *Malassezia ovalis* on rabbit skin. Mycopath 72:147–154
194. Guillot J, Chermette R, Guého E (1994) Prévalence du genre *Malassezia* chez les mammifères. J Mycol Méd 4:72–79
195. Gustafson BA (1955) Otitis externa in the dog. A bacteriological and experimental study. PhD Thesis, Royal Veterinary College, Stockholm
196. Mansfield PD, Boosinger TR, Attleberger MH (1990) Infectivity of *Malassezia pachydermatis* in the external ear canal of dogs. J Am Anim Hosp Assoc 26:97–100
197. Uchida Y, Mizutani M, Kubo T et al (1992) Otitis externa induced with *Malassezia pachydermatis* in dogs and the efficacy of pimaricin. J Vet Med Sci 54:611–614
198. Bond R, Rose JF, Ellis JW et al (1995) Comparison of two shampoos for treatment of *Malassezia pachydermatis*-associated seborrhoeic dermatitis in basset hounds. J Small Anim Pract 36:99–104

Pityriasis Versicolor and Other *Malassezia* Skin Diseases

6

Vicente Crespo Erchiga and Roderick J. Hay

Core Messages

- *Malassezia* yeasts have been implicated in the pathogenesis of a variety of different human diseases. These include the common skin infection, pityriasis versicolor (PV), along with systemic infection, particularly in neonates. Increasingly, though, they are associated with other conditions where the role of *Malassezia* in the pathogenesis is less clear and, in some cases, speculative. Examples include the common skin diseases, psoriasis and seborrhoeic dermatitis, as well as much rarer conditions, such as confluent and reticulate papillomatosis.
- PV is a superficial infection associated mainly with *M. globosa*. It is a non-life threatening condition that presents with scaling. Diagnosis depends on the demonstration of organisms by direct microscopy, and treatment is usually successfully accomplished either by topical or systemic imidazole therapy. Associated diseases include folliculitis, which is an infection associated with symptomatic hair follicle invasion by *Malassezia*, responding to antifungal therapy along with conditions, such as psoriasis and cephalic pustulosis where the connection between the organisms and the disease is less clearly established. In the case of onychomycosis, there is little evidence that *Malassezia* yeasts are implicated in nail plate invasion, although this may be a rare occurrence.

R. J. Hay (✉)
Professor of cutaneous infection, Skin Infection Clinic, Dermatology Department, Kings College Hospital, London SE5 9RS, UK
e-mail: roderick.hay@ifd.org

T. Boekhout et al. (eds.), *Malassezia and the Skin*
DOI: 10.1007/978-3-642-03615-6_6,

6.1 Pityriasis Versicolor

6.1.1 Introduction

Pityriasis versicolor (PV) is a skin infection caused by species of the genus *Malassezia*. Under the old taxonomic classification, the cause of this condition was ascribed to *M. furfur*, but this has now changed. With the modern molecular-genetically based taxonomy, in most cases it is thought to be due to *M. globosa,* although other species are likely to be involved in some cases, particularly those forms of PV that have atypical distributions. The disease itself has been given a number of different names, notably either PV or tinea versicolor, the latter being favored in the USA. The infection is common in most parts of the world, but more particularly in the tropics and subtropics, where it may affect a high proportion of the normal population. Generally asymptomatic, it is usually noticed by patients because of the obvious discoloration of the skin from which it derives its name – versicolor.

6.1.2 Epidemiology

PV is undoubtedly one of the commonest of the superficial mycoses. The epidemiology of the disease is poorly understood because only a few large studies that provide sufficient resolution have been documented, providing detailed figures on the prevalence of the skin disease. The prevalence is high in regions with hot and humid climates, and here, more than 15% of the population may be affected. In a recent paper, the overall prevalence of PV was found to be 15.52% in a fishing community in Venezuela [1]. In Malawi, this figure was calculated to be about 8% in a large unselected population [2] examined clinically for signs of skin disease. PV is much more common in certain age groups. Most studies have shown that it is commonest in children, adolescents and young adults in the tropics; it has also been found to be commoner in young women. For instance, in a recent study of over 800 Tanzanian school children, 26% had PV [3]. PV can also occur in infants under 1 year of age [4]. The disease is assumed to be more common in patients with pigmented skin, although this may be due to the higher chance of detection following pigmentary change.

This prevalence figure is usually much lower in temperate climates, although only a few studies are available. In a survey carried out in the USA between 1971 and 1974, PV was found in 0.8% of 28,000 persons [5]. In Sweden, the prevalence of PV was 1.1% among 20,296 people attending a general hospital in 1979 [6]. A subsequent study of 3,302 people, also carried out in Sweden in 1983, diagnosed PV in 0.5% of males and 0.3% of females [7]. Recently, a study on a representative sample of 1,024 young Italian sailors, found a 2.1% prevalence of PV. There was no association between the disease and sport practice, swimming or hyperhidrosis; the only significant association was a previous clinical history of PV, which supports the hypothesis that constitutional factors play a major role in the pathogenesis of the condition [8]. In comparison with other superficial mycoses, PV represents a high percentage of the total number of cases of dermatomycoses. In a recent study performed in a large hospital in Venezuela, PV accounted for 28.5% of all dermatomycoses [9], almost the same percentage as that found in the same year in Libya by Ellabib et al. (27.8%) [10]. In studies carried out in the Dermatology Departments of two University Hospitals in Southern Spain, PV only represented 10% of the total amount of dermatomycoses, though it is likely that the real frequency of the disease is much higher, because it is usually easily diagnosed by the general practitioners, and therefore only a small number of patients are referred to dermatologists (Crespo Erchiga and Delgado, unpublished data).

The disease occurs generally in otherwise healthy adult individuals. A number of factors influencing the appearing of PV have been described, but the cause or causes of the change of the *Malassezia* species from the saprophytic or yeast state to the parasitic or mycelial state remain unclear. To some extent, genetic factors may play a role, as the disease occurs more frequently among first-degree family members [11]. A positive family history among relatives is found more often than by chance, thus suggesting that there may be an inherited predisposition to acquire PV, but whether this is caused by a genetically determined host susceptibility factor, or by a greater susceptibility for heavy colonization by *Malassezia* species, is at present unknown. In a recent study from China, 21.1% patients had a positive family history of PV. There was also a higher rate of recurrence and longer duration in the patients with a positive family history. The heritability of PV in first-, second- and third-degree relatives was estimated to be 48.13, 40.11 and 27.20%, respectively [12].

There are also reports on the role of other factors, such as the use of oral anticoagulants [13] or hyperhidrosis [14], though this has not been confirmed by more recent studies [8]. Other endogenous factors, such as the use of systemic corticosteroids or immunosuppressive drugs may play a role as well [15]. For instance, two cases with particularly extensive PV in patients under treatment with anti-TNF drugs (etanercept) have recently been observed (Crespo, unpublished data), and neither patient reported a previous episode of the disease. Its relationship with other forms of therapeutic immunosuppression is less clear, but both cyclosporine and azathioprine have been reported as independent risk factors for PV in solid organ transplant patients [16]. It is also not clear whether PV is linked with other commonly used immunosuppressive agents, such as tacrolimus. Yet in terms of mechanism, the association with drugs that affect the function of T-lymphocyte mediated pathways suggest that reduced levels of T-cell function is one factor in the development of the disease. PV, however, has not been associated with other diseases in which CD4 lymphocyte immunosuppression is prominent, such as HIV-infection/AIDS. PV is not associated with pregnancy nor with diabetes mellitus.

Studies of the epidemiology of *Malassezia* yeasts on the skin have shown that in normal individuals there is an increased chance of hyphal formation in patients from tropical areas even without the presence of overt PV. Thus hyphal formation per sé does not indicate the presence of the disease, although it is possible that patients who show this change will develop PV (Deesi and G. Midgley, unpublished data). Similarly, hyphal transformation has been observed to occur in patients after renal transplantation when high doses of immunosuppressive drugs are administered [17]. Exposure to ultraviolet irradiation on sunbeds producing either low or high UVB emissions showed a reduction of yeast counts in users of both forms of sunbed [18].

The immunological responses in PV are difficult to interpretate, as different authors reported different or equivocal results [19]. Altogether, it seems likely that the pathogenesis of PV is dominated by local nonimmunological factors, such as high temperature and humidity, the occlusive role of clothes, and minor changes in the chemical composition of sebum. These factors are thought to influence a change in the morphological state of the yeasts that are usually present on healthy skin, into the parasitic form. One study showed that significantly more extractable skin surface amino acids, a shorter alkali neutralization time and reduced water spreading occurred in PV patients when compared to controls [20]. Likewise, *Malassezia* species require certain lipids as nutrients, such as myristate or palmitate, but as yet there is no evidence that inhibitory lipids or changes in lipid concentration occur that triggers PV in certain individuals. Higher yeast counts on lesional than nonlesional skin suggest though that local triggers are present to cause proliferation of *Malassezia* yeasts in PV.

The pathogenesis of depigmentation in the hypopigmented or white form of the disease (PV alba) also remains to be elucidated. Some authors have indicated a role for fungal metabolites that have a toxic effect on melanocytes. Examples of such possible toxins include dicarboxylic acids [21] and lipoperoxidase [22].

Recently, a trytophan-dependent metabolite of *M. furfur* has been described which, under UV light, exhibits yellow–green fluorescence. Among the tryptophan metabolites, a potent broad spectrum UV effect is shown by the lipophilic substances pityriacitrine and pityrialactone. A light-screening effect of these compounds might explain the absence of pigmentation in exposed areas in patients with PV [23]. However, pigment production was seen only amongst strains of *M. furfur* and some strains of *M. pachydermatis*, and recent epidemiological studies have shown that both species are not very common in PV lesions [19].

6.1.3 Etiology and Pathogenesis

In 1951, Gordon [24] was the first to propose that the rounded forms of *Malassezia*, which he referred as *P. orbiculare*, were the probable causative agent of PV. Since the taxonomic revision of 1996, a progressive number of studies have dealt with the etiology of PV from a new taxonomic perspective (Table 6.1).

In 1999, Crespo-Erchiga et al. [25, 26], after they carried out two parallel studies in Southern Spain on a limited number of PV patients, compared a similar number of samples from subjects with seborrhoeic dermatitis (SD) and from healthy skin, and showed, for the first time, that *M. globosa* was the predominant species in PV lesions (Fig. 6.1a, b).

These results were subsequently confirmed by the same authors in a study of 96 PV patients [27]. Most of the studies published over the following years, such as those of Nakabayashi et al. in Japan in 2000 [28], and Aspiroz et al. in Zaragoza (Spain) in 2002 [6], supported these results. Specifically in the last work, the percentages obtained for *M. globosa* and *M. sympodialis* were almost the same as those published by Crespo-Erchiga et al. in the above-mentioned studies. Aspiroz et al. also showed the presence of higher enzyme production, especially of lipases and esterases, in strains of *M. globosa* compared to other *Malassezia* species, a fact that may be a reason for an enhanced level of pathogenicity in

Table 6.1 Pityriasis versicolor: epidemiological studies since 1996

Patient numbers	*M. globosa*[a]	*M. sympodialis*[a]	*M. restricta*[a]	*M. slooffiae*[a]	*M. furfur*[a]	*M. obtusa*[a]	Medium[b]	Year	Refs
75	87	34	3	8	–	–	Dixon	1999	[26]
22	55	9	–	5	5	–	Dixon	2000	[28]
96	97	32	–	7	–	–	Dixon	2000	[27]
111	25	59	1	1	11	–	LNA	2001	[29]
79	90	40.5	–	1	–	–	LNA	2002	[6]
250	63.6	4.8	–	–	34	–	Dixon	2002	[31]
3	100	–	–	–	–	–	PCR	2002	[35]
11	47	27	13	–	13	–	Dixon	2003	[30]
94	53.3	9.3	–	4	25.3	8.1	Dixon	2004	[32]
65	24	–	–	26.8	70.7	–	Unknown	2004	[40]
222	65	5	–	–	13	–	Dixon, PCR	2004	[33]
150	5.3	4.6	–	5.3	42	0.6	Dixon	2005	[41]
95	2	11.4	–	–	77.8	8.3	Dixon	2006	[42]
100	97	34	1	3	2	–	Dixon	2006	[37]
71	90	3–4	–	3–4	3–4	–	Dixon	2006	[35]
90	63	14	–	4	10	8	Dixon	2007	[36]

[a]Expressed in %

[b]Dixon or Leeming and Notman (LNA) agar

PCR polymerase chain reaction

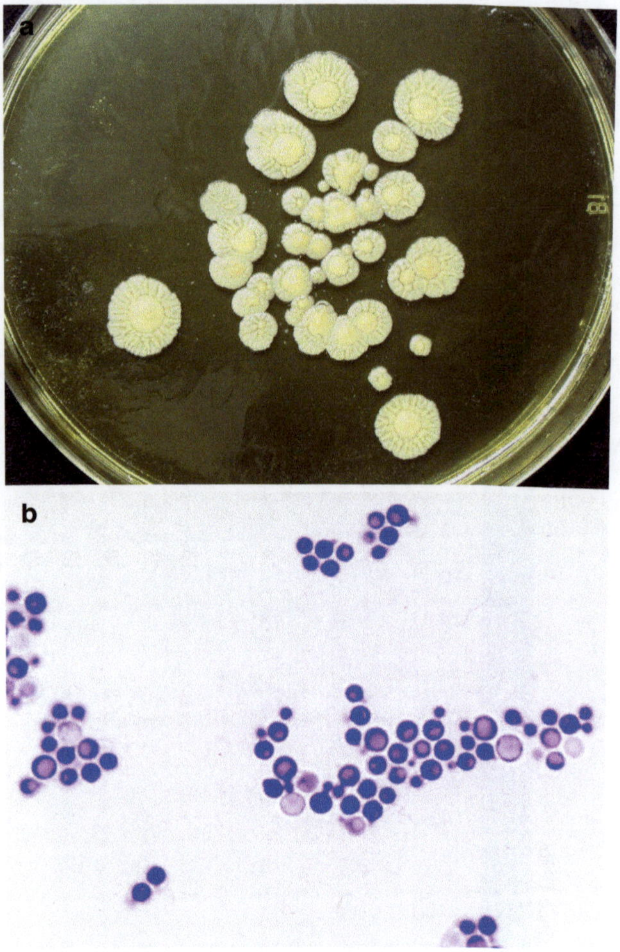

Fig. 6.1 (**a**) Culture of PV in mDixon medium showing abundant colonies of *Malassezia globosa*. (**b**) Typical spherical yeasts of *Malassezia globosa* in culture. Lactophenol cotton blue, ×1,000 magnification

human skin associated with this species. This observation provided support for a role of these potential virulence factors in the etiology of PV. The only study that showed different results was that of Gupta et al. [29] in Canada, where the predominant species involved in PV was found to be *M. sympodialis*, but it is important to note that the methodology and the culture medium used in this work was different from the other studies mentioned.

Hernandez et al., [30] found also that *M.globosa* was the predominant species in conventional cultures of PV samples from Mexico. Similar results were published by three other authors, namely Dutta et al. from India [31] who isolated *M. globosa* in 63.6% of cases, Tarazooie et al. [32] from Iran with this species occurring in 53.3% of cases, and Makni et al. [33] in Tunisia, who confirmed the predominance of *M. globosa* observed in Dixon medium cultures (65%) by means of molecular techniques [34]. In 2006, Gaitanis et al. [35] reported on the isolation of *Malassezia* spp. in patients with PV and SD using Dixon medium. From PV, they isolated *M. globosa* in 90% of cases, either alone (77%) or/and associated with *M. sympodialis*, *M. furfur* or *M. slooffiae* (13%). As these other species were isolated alone in less than 10% of cases, the authors did not record their respective percentages. Finally, in 2007, a study by Prohic et al. [36], carried out in Bosnia-Herzegovina, identified *M. globosa* as the main species in PV lesions (63%). Crespo-Erchiga and Delgado [37] analyzed 100 samples from PV patients from three different main Hospitals in Southern Spain that showed similar results, with *M. globosa* isolated in 97% of cases, followed by *M. sympodialis* (34%), *M. slooffiae* (3%), *M. furfur* (2%) and *M. restricta* (1%). Yeasts morphologically identical to those of *M. globosa*, but with pseudomycelium, were observed by direct microscopy in 98% of cases. Nakabayashi [38] used PCR identification of *Malassezia* in PV samples and found that *M. globosa* was the most commonly detected species (97%), followed by *M. restricta* (79%) and *M. sympodialis* (68%). Gaitanis et al. [39] used a DNA–PCR procedure that was applicable to lesional skin scales, and detected and identified only *M. globosa* from PV lesions.

In contrast, other studies, mainly carried out in areas with tropical or subtropical climates, showed a predominance of *M. furfur* in PV lesions. The first study was performed in Madagascar [40] in 2004, the second one in Panama [41], and the last one in Brazil [42]. In all three studies *M. furfur* was isolated as the predominant species, with respective percentages of 70.7, 42, and 77.8%. This last result support the hypothesis proposed 80 years ago by early investigators, such as Panja [43] or Castellani [44] and later by Midgley [45] and by Crespo-Erchiga and Delgado [19], that species different from *M. globosa*, which is mainly observed in Northern countries, might predominate in PV in other climates. Although more studies are necessary to either confirm or reject this hypothesis, there are other mycosis in which more than one species can be involved, such as chromoblastomycosis, where the same cutaneous lesions can be caused by at least three different fungi, namely *Fonsecaea pedrosoi, Phialophora verrucosa* and *Cladophialophora carrionii*, with a different geographical distribution. The clinical features of PV associated with yeasts seen on microscopy are described below.

In conclusion, the available data suggest that *M. globosa*, which is always present in its yeast state in healthy adult skin together with other *Malassezia* species, produces the clinical lesions of PV after developing into the mycelial state. The conditions that induce this transformation remain unclear, but it seems to be triggered mainly by changes in local conditions, such as heat, humidity, and sebum composition, but on an idiosyncratic basis. The possibility of the existence of more virulent strains of *M. globosa* should also be considered. Finally, the predominance of *M. furfur*, the sole other species that is able to develop pseudohyphae *in vitro*, in PV in some studies from tropical regions may lead to the concept of PV as a mycosis that can be caused by more than one fungal pathogen, depending on geographical distribution. This latter hypothesis, however, needs to be confirmed.

6.1.4 Clinical Presentation

PV usually presents with lightly scaling hypo- or hyperpigmented macules on the trunk, neck or face [19, 46] (Fig. 6.2a, b). The sites of involvement are normally most prominent in these areas, but it may also affect the proximal parts of the limbs, mainly the arms, and the scalp may also be affected. More widespread forms are also found.

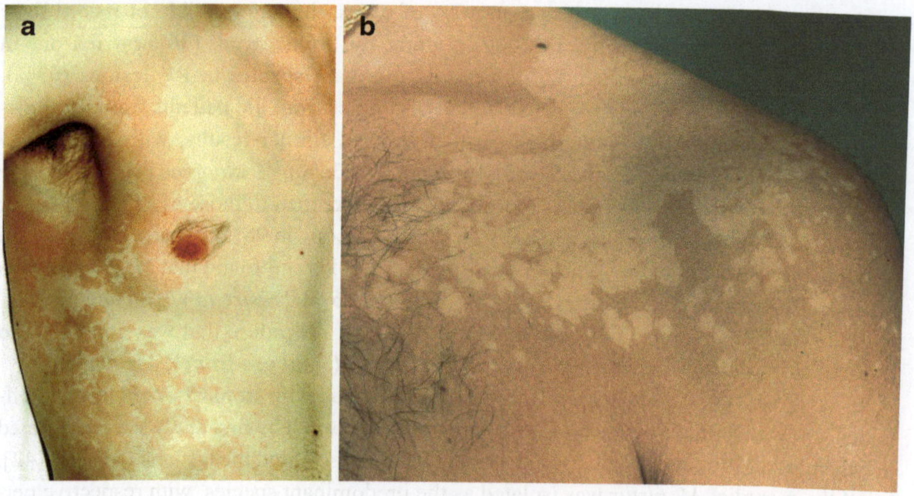

Fig. 6.2 Pityriasis versicolor. (**a**) Patient showing hyperpigmented lesions. (**b**) Patient showing hypopigmented lesions

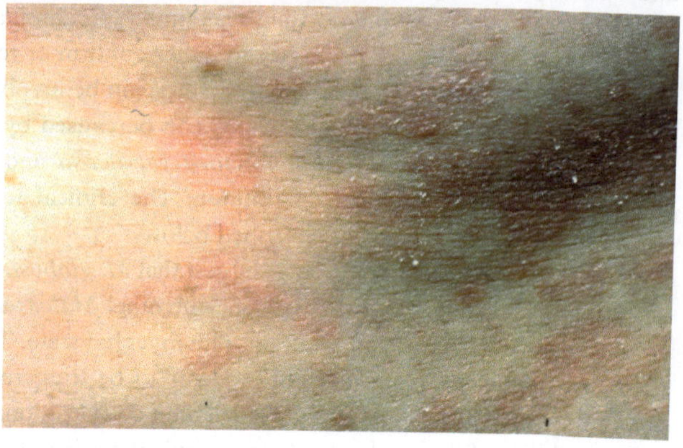

Fig. 6.3 Pityriasis versicolor. The erythematous form in a renal transplant recipient

Associated itching and irritation generally is minimal. Within these broad parameters, different forms of infection exist, sometimes with showing more erythema and less pigmentary change, and vice versa. With time, adjacent macules coalesce to produce large areas of pigmentary change and scaling. The scaling is often difficult to detect unless the skin is gently scratched. Often this also elicits a flare reaction. Patients who are immunosuppressed are more likely to have the erythematous forms in temperate areas (Fig. 6.3) and, similarly, these appear more common in familiar cases of infection without a clear history of sun exposure.

Most patients present with disease of gradual onset, which is scarcely noticed until the discoloration is more prominent. It is not clear why in some patients hyper pigmented changes develop, whereas in others these are hypopigmented. More extensive forms, usually hypopigmented, with a distribution affecting body sites, such as the face, neck and

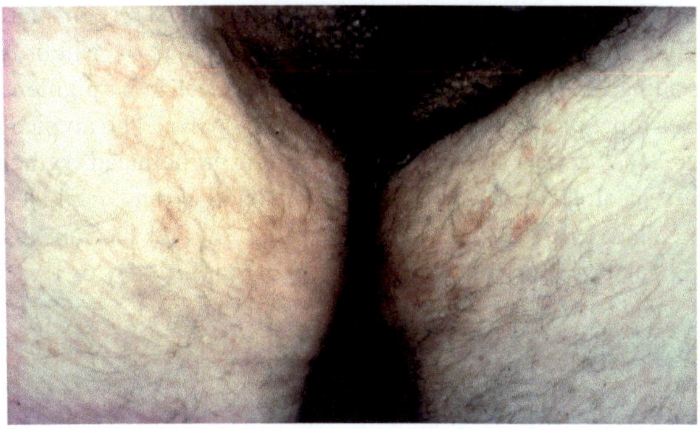

Fig. 6.4 Pityriasis versicolor. An atypical form affecting the groin area. Oval shaped yeasts predominate in direct microscopy.

arms as well as trunk, are often seen in the tropics, an observation that prompted Castellani to refer to the tropical form of PV as a distinct entity [44].

In addition, there are some distinctive and different variants of PV. One form is a type more often seen on the proximal limbs and below the waist area (Fig. 6.4). In Venezuela this accounted for about 13% of cases [47]. Typically scraping the lesions reveals ovoid yeast forms without hyphal formation rather than the usually observed round and hyphal forms [45]. This variant is seen more often in slightly older patients; in the above study the mean age of affected individuals was 40. The modern genetic typing of this condition has not been carried out so it is not clear which of the *Malassezia* yeasts is responsible for this type of infection or if this relates to the predominance of *M. furfur* as noted in some tropical surveys.

A second unusual form is a variety where lesions develop significant degrees of atrophy, sometimes resembling a form of anetoderma [48]. Here, there is clear loss of dermal substance with lesions appearing depressed in relation to the surrounding skin surface. These forms usually show the normal pattern of surface colonization with round yeasts and stubby hyphae; but there is usually less scaling or obvious pigmentary change. There is a variable degree of epidermal and dermal atrophy with reduction of the rete-ridges, subepidermal fibroplasia, pigment incontinence and degeneration of elastin fibers [48]. The cause is not clear and it is thought to be the end product of an inflammatory reaction leading to elastin degeneration, although similar changes are not seen with other fungi.

6.1.5 Differential Diagnosis

The differential diagnosis of PV includes a number of both common and rare conditions [19,49]. Usually in vitiligo there is much more severe hypopigmentation and no scales, with a stark cut off between hypopigmented and normally pigmented areas and pigment loss is generally total. There are also islets of pigmentation around hair follicles in vitiligo. PV occurs more prominent on certain body areas, such as the face and hands. In hot climates idiopathic guttate hypomelanosis needs to be distinguished, but here the hypopigmented areas are scattered in a tear drop fashion and do not become confluent. They occur also widely scattered on the trunk and limbs, and in these conditions scaling is absent.

Chloasma (melasma) on the face should also be considered. Here, the patches of hyperpigmentation are often irregular and prominent over light exposed areas. There is also no scaling. Other differentials include hypo - or hyperpigmented mycoses fungoides, pityriasis alba, pityriasis rotunda or pityriasis rosea. The atypical forms on the lower abdomen area may appear like erythrasma, but the pronounced erythema and confluence of the latter is typical. However, both erythrasma and PV may co-exist.

Tinea corporis, secondary syphilis, SD and the rare dermatosis called confluent and reticulate papillomatosis may also be confused with PV. The last two entities are also related to *Malassezia* species, and by direct microscopy, abundant yeasts can be observed in the scales, although always without hyphae, a feature that differentiates the doubtful cases.

6.1.6 Mycological Diagnosis

In most cases, the clinical diagnosis of PV is easy to confirm by an experienced dermatologist or laboratory technician using laboratory methods. Scrapings are generally taken with a scalpel, those with a solid blade/handle being the most useful. However, other methods have been used in the past, such a sellotape stripping or the use of acrylic glues. The materials obtained in this way are mounted directly on glass slides for microscopical investigation, usually using special stains that highlight the organisms (see below).

The diagnosis, however, should always be confirmed by direct examination with KOH and demonstration of pseudohyphae and blastoconidia in the typical "spaghetti and meatballs" pattern. The 10–20% KOH formulation can be improved by adding an equal part of Parker ink (permanent blue or black), which quickly stains the fungi in an easily visible blue color (Fig. 6.5) (even with the black ink). Some authors prefer the use of calcofluor white, which gives good results but it is necessary to examine the sample with an ultraviolet fluorescence microscope.

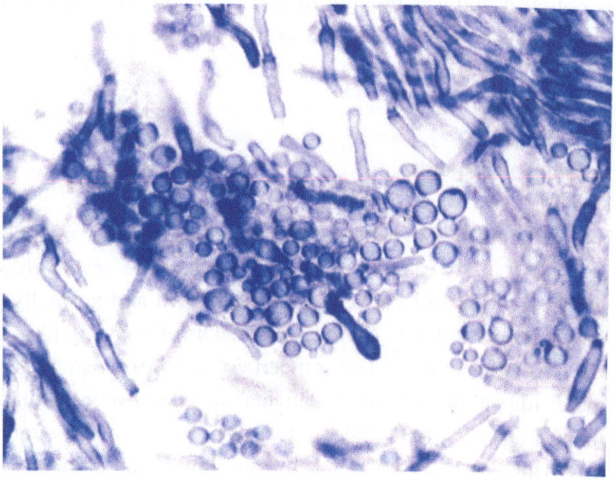

Fig. 6.5 Direct microscopy of PV. Note the presence of globose yeasts and pseudohyphae. KOH + black Parker ink, ×1,000 magnification

Also the Wood's light can be used to confirm the diagnosis, especially to detect subclinical patches of PV, but the yellow-green fluorescence visible using this method is positive in about one third of the cases only. This could be due to an infection by nonfluorescent species, as only *M. furfur* seems to produce fluorochromes [50]. Use of Wood's light depends critically on having darkness in the surrounding room, and with older equipment, it is important that the machine is warmed up before use as otherwise the intensity of the light produced is insufficient. Care should be taken to protect the patient's eyes, when the face is being irradiated with the light source.

Skin biopsy is rarely necessary for the diagnosis of PV, although the yeasts and hyphae present in the stratum corneum can be stained by Periodic acid-Schiff or methenamine silver. Similarly, culture is not necessary for routine diagnosis, but is indispensable in order to recognize the species involved in the disease, as well as those present in the normal skin microbiota. *Malassezia* yeasts are fastidious in their requirements for culture and, except *M. pachydermatis*, none of the other species grows in routine mycological media, like Sabouraud agar. The media used by most investigators in the field are Leeming and Notman's agar (LNA) and the modified Dixon agar (mDA). Colonies develop in 3–4 days and their macro and micromorphology are described in specialized papers [51, 52] (see also Chap. 2.1).

6.1.7 Treatment

Treatment of PV is generally straightforward. The main difficulty may be when the rash is very extensive [49, 53]. As PV may involve large areas of the trunk, ideally treatment should be easy to apply or, if oral, safe and of short duration. Most of the creams available for the management of PV are easy to apply and effective, but inappropriate treatments are still given at the primary care level. Whatever the treatment used, assessment of the extent of cure can be difficult as the organisms remain on the skin surface for a few weeks after cell death, and pigmentary changes remain after the cure of the infection and only resolve slowly over the following weeks. This is especially seen with PV associated with hypopigmentation.

Earlier treatments used for PV were 2.5 % selenium sulfide or 20% sodium hyposulphite given for 2–4 weeks [46]. An alternative is a mixture of one volume propylene glycol in one volume water [54]. This has also been used intermittently as long-term suppressive therapy to prevent relapse. These medications are well tolerated, although selenium sulfide occasionally stings or causes skin irritation. The effects of shorter treatment periods have not been studied. Nowadays, treatment usually depends on the use of azole antifungal preparations. All the imidazoles that have been tested were found to be effective. These include miconazole, econazole, clotrimazole, bifonazole, ketoconazole, sulconazole and tioconazole [55–58]. The antifungals are available as creams, lotions, gels, powders and shampoos. Adequate spreading of the applied topical agent is important to treat all diseased areas of the patients. Ketoconazole has the advantage that it is available as a shampoo, which is often used in this condition. Clinical trials have shown reasonable efficacy for this topical approach. Econazole shampoo is available in some areas. Generally, clinical trials have used 2–4 week periods of treatment. Evidence exists that shorter periods of application of ketoconazole shampoo are effective as well. Cyclopirox, which also has a shampoo formulation, and terbinafine are topically active against PV. Other agents, including calcineurin antagonists, such as tacrolimus or pimecrolimus, show *in vitro* efficacy against *Malassezia* yeasts [59]. Oral treatments that are effective against *Malassezia* include ketoconazole, itraconazole and fluconazole. Originally ketoconazole (200 mg) was used as a 7-day treatment, but it has been proposed to be used as a single dose treatment of 400 mg. Doubts remain whether this will be effective in tropical or hot environments [60,61], but short term treatments are possible.

The optimal treatment dose for itraconazole is 800–1,000 mg i.e., 4–5 days of 200 mg [62]. Studies of itraconazole were the first to point out the lag phase between apparent demise of the organisms as seen in electromicroscopy and their disappearance from the skin surface [63]. *M. globosa* has a thick cell wall, and cell ghosts with degeneration of the cytoplasm remain on the skin surface when the organism is not viable anymore. Shorter periods of treatment with itraconazole are not as effective, but have been shown to produce responses in some studies [64]. Fluconazole can be used, although it is generally given for 5–7 days. There is evidence that a single dose of fluconazole (200 mg) can produce significant cure rates, viz., 65% in one study with relapse rates of 35% [65], and that this may be more effective than a single dose of itraconazole [66].

Amongst the newer agents, no clinical data are available for voriconazole, albaconazole and posaconazole, although they appear to be active *in vitro* [67, 68]. Studies with pramiconazole in PV have shown that it is clinically effective, and *Malassezia* spp. are susceptible *in vitro* [69]. At the time of writing, this agent is not licensed for clinical use. Oral terbinafine does not achieve sufficient drug levels in the stratum corneum to destroy *Malassezia* although it is active topically.

6.1.8 Relapse and Prevention

As relapse is very common, particularly in patients living in the tropics, studies of the use of prophylactic treatment with propylene glycol in water have shown that this can be used to prevent relapses [54]. Others have tried periodically applied ketoconazole shampoo, but there is no clinical trial data on the latter approach.

6.2 Association with Other Skin Diseases

6.2.1 *Malassezia* (*Pityrosporum*) Folliculitis

Malassezia folliculitis (MF) is a benign disorder, characterized by follicular papules and pustules localized predominantly on the back and chest and upper arms. The typical lesion is described as a molluscoid, dome-shaped comedonal, papule, about 2–3 mm in

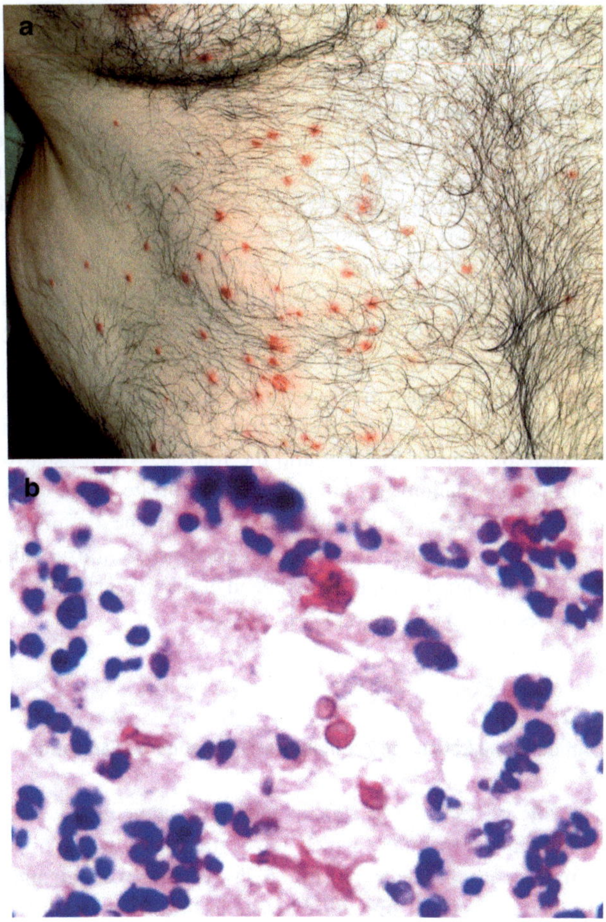

Fig. 6.6 (**a**) *Malassezia* folliculitis. (**b**) Histopathology of *Malassezia* folliculitis showing globose yeasts and pseudohyphae. PAS ×1,000 magnification

diameter. In severe cases, pustules, nodules and cysts can be present [15] (Fig. 6.6a). The condition is probably common, but often misdiagnosed as acne [70]. Pruritus and the absence of comedones and facial lesions allow one to distinguish both diseases, although pruritus may be minimal or absent in immunocompromised patients. MF is more frequent in tropical countries and during the summer in temperate regions. Occlusion seems to play an important role in this disorder [71], which has also been associated with antibiotic treatment (tetracyclines), corticosteroids and immunosuppression due to organ transplantation [72] and in Hodgkin disease [73]. In this last case, *M. globosa* was isolated in pure culture. A minor epidemic outbreak has been recently reported in an intensive care unit [74], and in heart transplant recipients receiving immunosuppressive treatment, 11 cases were described during a 4-month period (among 198 cardiac transplants) [75].

Circulating IgG antibodies against *P. ovale* are present in high titers [76]. Faergemann suggested that MF could be explained by an overgrowth of *Malassezia* in the follicle, and the inflammatory changes being due to products released, such as free fatty acids, and the complement-inducing activity of the yeast [77].

Direct microscopy of pustules shows the presence of abundant *Malassezia* yeasts in the pilosebaceous follicles, in the absence of other microorganisms. In a study published in 1985 [78], round yeast forms (cited as *P. orbiculare*) were observed and cultured in most of the cases. This suggests that, as in PV, *M. globosa* could be the predominant causative agent. In some of the heart transplant recipients referred to above, the culture was identified as either *M. furfur* or *M. pachydermatis*, but these authors did not apply the new *Malassezia* nomenclature [75]. In any case, colonization of hair follicles by *Malassezia* yeasts is not abnormal, and can be seen in acne vulgaris lesions, but here it is often mixed with staphylococci and propionibacteria.

The histological pattern of MF shows dilated hair follicles often full of keratinous material, with numerous budding yeasts that can be localized by staining with periodic acid-Schiff or methenamine silver techniques (Fig. 6.6b). In our experience, the yeasts observed in the biopsy of our patients were predominantly or exclusively of the globose type, supporting the hypothesis, expressed above, that *M. globosa* is likely to be the main species in the development of MF. Around the follicle, which usually shows a focal rupture of the follicular epithelium, an inflammatory infiltrate of lymphocytes, histiocytes and neutrophils can be observed in most cases [79, 80]. The diagnosis of MF is confirmed indirectly by the response, often dramatic, to appropriate antifungal therapy. Topical treatment is effective in some cases, but most need oral ketoconazole [71], itraconazole or fluconazole. Relapses are very common [70].

6.2.2
Confluent and Reticulate Papillomatosis of Gougerot-Carteaud

The role of *Malassezia* spp. in this rare cutaneous disorder, characterized by confluent hyperkeratotic papules, which have grayish-brown pigmentation and are often located symmetrically on the trunk (Fig. 6.7a), is controversial, and is based on the presence, in a number of cases, of lipophilic yeasts in the stratum corneum observed by direct microscopy and histopathology (Fig. 6.7b), and on the favorable response to topical or systemic antifungal therapy. In 1969, Roberts and Lachapelle [81] suggested that confluent and reticulate papillomatosis could represent a peculiar host reaction to heavy colonization by *P. orbiculare*. However, in the series reviewed by Nordby and Mitchell [82], only 5 out of 19 patients responded to topical treatment with selenium sulfide, and there are reports describing the complete recovery of the disease using minocycline [83], or retinoids [84], or azithromycin [85]. Furthermore, two cases were successfully treated with topical vitamin D analogs, either calcipotriol [86] or tacalcitol [87].

Out of three cases, *M. furfur* was isolated from one and *M. sympodialis* from the other two. In all cases, direct microscopy showed the presence of abundant yeast cells that were globose and oval in shape, but without pseudohyphae (V. Crespo Erchiga, unpublished data). It must be pointed out that, due to the hyperkeratotic nature of the lesions, the yeasts could be observed only after prolonged digestion with potassium hydroxide. Only one of the patients, with lesions on the upper trunk, was cleared completely after treatment with oral itraconazole and topical selenium sulfide.

Recently, Stein et al. [88] described three teenage siblings with confluent and reticulated papillomatosis, two of them having tinea versicolor prior to, or in association with,

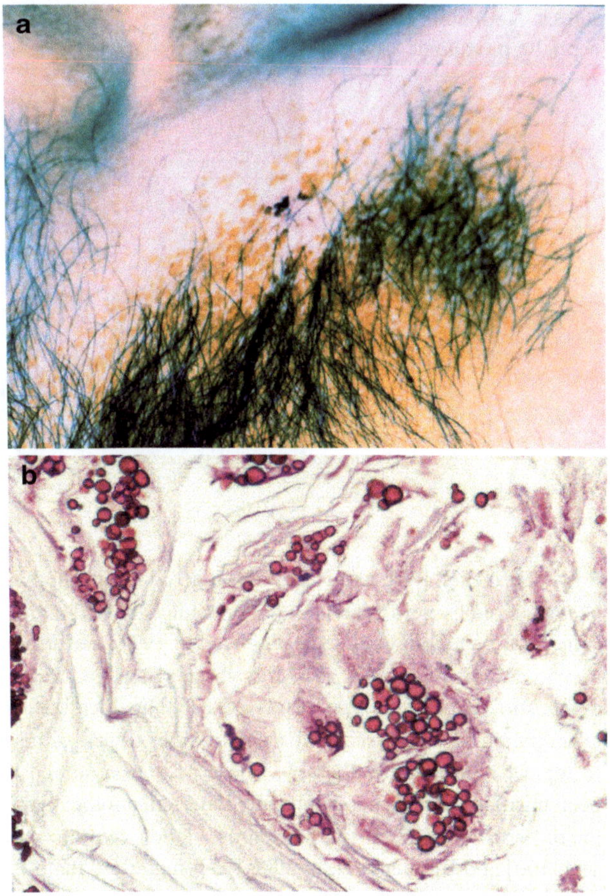

Fig. 6.7 (**a**) Confluent and reticulate papillomatosis. (**b**) Globose *Malassezia* yeasts in the stratum corneum in a case of confluent and reticulate papillomatosis. PAS ×1,000 magnification

papillomatosis. The pityriasis cleared with antifungal medication, but the papillomatosis did not, and this was treated successfully with minocycline. The authors suggested that these cases support the hypothesis of an aetiological relationship between both diseases, and that there may also be a familial predisposition to their development.

In conclusion, it is possible that this rare dermatological disease represents more than one pathological condition, and that *Malassezia* yeasts are involved in the pathogenesis of only some cases, perhaps those with a clinical pattern similar to that of PV. It could also be a basic disorder of keratinisation, determined by an abnormal response to *Malassezia* yeasts, or to bacteria or their products.

6.2.3 Neonatal Cephalic Pustulosis

Common neonatal cephalic pustulosis is a benign entity, formerly named neonatal acne, which is frequently identified in neonates. The incidence was estimated to be 10% by

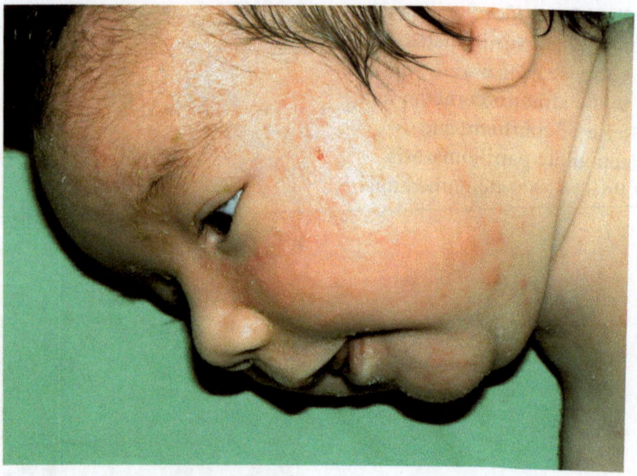

Fig. 6.8 Neonatal cephalic pustulosis

Rapelanoro et al. [89], and 3% in hospitalized infants by Niamba et al. [90]. The disease is characterized by nonfollicular papulopustules located on face and neck of newborn infants (Fig. 6.8). Aractigni et al. [91] proposed in 1991, that it represents an inflammatory reaction to colonization by *Malassezia* yeasts. In 1996, Rapelanoro et al. [89] proposed as diagnostic criteria, the age of onset, cephalic localization, positive direct microscopy examination for *Malassezia*, absence of other causes of neonatal pustulosis, and response to topical ketoconazole therapy. These authors assumed that this condition could have been misdiagnosed as neonatal acne. Later on, another report by Pont et al. [92] described the presence of *M. furfur* in the lesions, while two more recent studies by Niamba et al. in 1998 [90] and Bernier et al. in 2000 [93], suggested that *M. sympodialis* could be the triggering agent of this rash.

The saprophytic role of these yeasts in the skin of healthy adults and children makes the causative role of *Malassezia* spp. in this disease, a matter of discussion. It is well known that the degree of *Malassezia* colonization increases with age. Koseki and Takahasi [94] found *Malassezia* in 50% of healthy newborns, and in 80% of infants after 7 days of life. Borderon et al. [95] found *Malassezia* in 13, 30 and 77% of premature neonates younger than 15, 16–30 days and 31–90 days, respectively. In a recent study by Niamba et al. [90], progressive skin colonization by *Malassezia* was found to be correlated with age, beginning after 2 weeks of age, and yet, the mean age of onset of cephalic pustulosis was 12 days. Ayhan et al. in 2007 [96] found no correlation between the severity of the disease and the isolation of *Malassezia* yeasts. Skin colonization of patients with neonatal cephalic pustulosis was not higher than colonization of healthy newborns. The authors concluded that *Malassezia* is neither a direct causative factor in neonatal cephalic pustulosis, nor is *Malassezia* colonization related to the development or clinical stage of the disease. More extensive studies are needed to establish a possible role of *Malassezia* yeasts in this clinical entity. Up to now, *Malassezia* spp. may only be one of a number of factors causing neonatal cephalic pustulosis.

6.2.4 Otitis

Recently, Chai et al. in Australia [97] described a case of malignant otitis externa caused by *M. sympodialis*. The patient, a man with a poorly controlled type II diabetes mellitus, developed a severe infection, which started with erythema and cellulitis around the external auditory canal, and progressed to marked destruction of the temporal bone and the formation of an adjacent large soft tissue mass below the skull base. The infection showed total resolution after therapy with intravenous amphotericin B and fluconazole.

In contrast to the rarity of this disease in humans, ear infections due to *M. pachydermatis* are common in animals, particularly in dogs [98]. Some cases have recently been reported in which *M. furfur* and *M. obtusa* were isolated [99], and in cats, the disease was found to be associated with *M. sympodialis* [100].

6.2.5 Onychomycosis

A few cases of onychomycosis involving *Malassezia* species have been reported [101–103]. The presence of *Malassezia* yeasts in nail material, a well known, though infrequent finding, is probably caused by subungual colonization by members of the cutaneous microbiota, which makes the relevance of such cases uncertain. *Malassezia* yeasts of the oval form have been observed occasionally in patients with distal fingernail onycholysis, and were found to be associated with *Candida* spp. (V. Crespo Erchiga, personal observation). *M. furfur* was identified from an AIDS patient with infected nails (E. Guého, unpublished result) and in a series of patients with onychomycosis in Brazil [103]. This species is also able to form hair nodules, [104], and in this case the identification was confirmed by E. Guého. *M. furfur* seems to be the only *Malassezia* species able to survive in such unusual conditions. Nails are not a good source of lipids, and so, there is little probability that *Malassezia* could really infect them.

A role of *Malassezia* yeasts, as opportunistic agents of onychomycosis, cannot, however, be ruled out completely. Recently, Ertam et al. [105] reported a case of onychomycosis in an immunosuppressed liver transplant recipient. Abundant *Malassezia* yeasts were seen by direct microscopy, and *M. furfur* was isolated in the absence of *Candida* yeasts or dermatophytes. Complete resolution was achieved after topical therapy with cyclopirox-olamine nail solution. Likewise, Chowdhary [106] reported a case of a child, with the clinical features of onychomycosis and histopathological evidence of invasion by typical *Malassezia* yeasts on three occasions, and that responded to oral ketoconazole.

6.2.6 Balanitis

PV is occasionally found on the penile skin [107–109]. It is characterized by fine scaling macules either slightly erythematous or hypopigmented. The depigmentation, more visible

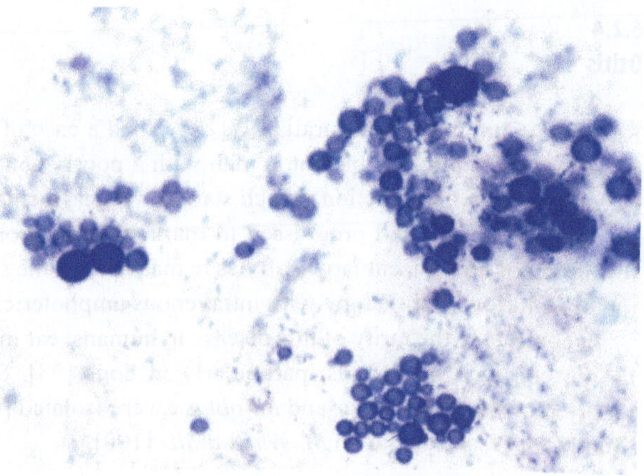

Fig. 6.9 Direct microscopy in a case of balanitis showing abundant globose yeasts. KOH+Parker blue ink ×1,000 magnification

in black patients, must be differentiated from vitiligo, which is an alternative cause of depigmentation in the anogenital region. The diagnosis is easy as, like in PV, direct microscopy shows the typical presence of *Malassezia* yeasts and pseudohyphae. A recent study by Mayser et al. [110] showed asymptomatic colonization of the glans penis by *Malassezia* spp., predominantly *M. sympodialis* and *M. globosa*, in more than 49% of adult men. In contrast, *Malassezia* yeasts were not found in vaginal smears. The authors suggest that *Malassezia* yeasts belong to the resident microflora of the male genital region, especially in the prepuce and the glans penis, where sebaceous glands, i.e., Tyson glands occur abundantly. These results were confirmed in another study carried out in circumcised adult men by Aridogan et al. [111], but here the incidence of *Malassezia* colonization was much lower (7%). The most common species was *M. furfur*, followed by *M. globosa, M. obtusa* and *M. slooffiae*, while *M. sympodialis* was not isolated. In a similar study by Iskit et al. [112], *Malassezia* colonization was demonstrated in 18% of circumcised and 43% of uncircumcised children. The predominant species in this report was *M. globosa*, followed by *M. furfur* and *M. slooffiae*. These yeasts could be involved in some forms of unspecific balanitis, SD or psoriasis. Mayser recommended the use of lipid-containing media in cases of recurrent, unspecific balanitis, where other infectious agents like *Candida* have been excluded. One case of chronic, mild balanitis, with negative culture for *Candida* spp., and where the presence of abundant *M. globosa* yeasts, but without hyphae, was observed by direct microscopy, was cured after treatment with clotrimazole cream (Crespo-Erchiga, personal observation) (Fig. 6.9).

6.2.7 Keratitis

Recently, Suzuki et al. [113] reported a case of keratitis caused by *M. restricta*. The species identification was made by a PCR technique. The corneal ulcer disappeared after treatment with topical pimaricin and miconazole ointment and oral itraconazole. The clinical findings resembled keratitis caused by filamentous fungi.

6.2.8 Psoriasis

The role of *Malassezia* yeasts in psoriasis is still undetermined. Historically, Rivolta made the first association of lipophilic yeasts with psoriasis in 1873 [114]. He reported on the presence of round budding cells in the epidermis of a patient with psoriasis, and named the fungus, *Cryptococcus psoriasis*.

In recent years, several reports have associated lipophilic yeasts with the development of skin lesions in psoriasis, especially when it involves the scalp region. Lober et al. [115] were able to induce psoriasis-like lesions in ten patients with psoriasis and ten controls, by using patch tests with *Malassezia*. Their hypothesis was that psoriasis was produced through activation of the alternative pathway of complement. Rosenberg et al. [116] and Farr et al. [117] reported the improvement of scalp psoriasis with oral and topical ketoconazole therapy. Also Rosenberg et al. [118] described the association between scalp psoriasis and the presence of oval yeasts that could possibly correspond to *M. restricta*. Elewski [119] reported, the development of guttate psoriasis in sites of MF during erythromycin therapy, and that the lesions cleared after further therapy with ketoconazole. Recently, a case of guttate psoriasis associated with PV alba has been published by Narang et al. [120]. In this case, direct microscopy of PV lesions showed the typical "spaghetti and meatballs" appearance, and the skin biopsy from psoriatic lesions revealed evidence of both fungal elements and psoriasis. Both lesions cleared after fluconazole therapy. Gupta et al. [121] observed that in patients with psoriasis and SD, *M. globosa* was the species most frequently isolated, with equal frequency from lesions in the scalp, forehead and trunk. A later and interesting study by Prohic [122] reported significant differences in the distribution of *Malassezia* yeasts between psoriatic and healthy scalp skin, and confirmed that *M. globosa,* in its yeast phase, is the predominant species (55%) in psoriatic patients, followed by *M. slooffiae* (18%) and *M. restricta* (10%). This last species is the commonest isolated from healthy scalp skin. In another analysis on the presence of *Malassezia* yeasts in the scales of psoriatic patients, Takahata et al. [123] used a real time PCR assay and concluded that *M. restricta* was the predominant species. Colonization by this yeast of the scales from all body sites studied, was approximately five times higher than colonization by *M. globosa*.

Other researchers have investigated the immune reactions of patients with psoriasis, showing that these individuals have immunological responses to both *Malassezia* yeasts and to proteins derived from them. It has been demonstrated that antibodies to the yeasts are present in serum from patients, but not from control subjects [124]. In 2002, Kanda et al. [125] reported that *Malassezia* yeasts induce Th-1 and Th-2 related cytokines, chemokines and prostaglandin E2 production in peripheral blood mononuclear cells from patients with psoriasis. Baroni et al. [126] performed a western blot analysis on human keratinocytes treated or untreated with *M. furfur,* and on biopsies from healthy and psoriatic patients. They concluded that *M. furfur* can induce the over-production of molecules involved in cell migration and hyperproliferation, thereby favoring the exacerbation of psoriasis.

In conclusion, there is no convincing scientific evidence of the role of *Malassezia* species in psoriasis. As with other skin diseases, further studies are needed to clarify the relationship between these yeasts and psoriasis in an adequate number of subjects.

References

1. Acosta Quintero ME, Cazorla Perfetti DJ (2004) Clinical-epidemiological aspects of pityriasis versicolor (PV) in a fishing community of the semiarid region in Falcon State, Venezuela. Rev Iberoam Micol 21:191–194
2. Ponnighaus JM, Fine PE, Saul J (1996) The epidemiology of pityriasis versicolor in Malawi, Africa. Mycoses 39:467–470
3. Ferié J, Dinkela A, Mbata M et al (2006) Skin disorders among school children in rural Tanzania and an assessment of therapeutic needs. Trop Doct 36:219–221
4. Jena DK, Sengupta S, Dwari BC et al (2005) Pityriasis versicolor in the pediatric age group. Indian J Dermatol Venereol Leprol 71:259–261
5. Johnson ML, Johnson KG, Engel A (1984) Prevalence, morbidity and cost of dermatologic diseases. J Am Acad Dermatol 11:930–936
6. Aspiroz C, Ara M, Varea M et al (2002) Isolation of *Malassezia globosa* and *M. sympodialis* from patients with pityriasis versicolor in Spain. Mycopathologia 154:111–117
7. Hellgren L, Vincent J (1983) The incidence of tinea versicolor in central Sweden. J Med Microbiol 16:501–502
8. Ingordo V, Naldi L, Colecchia B et al (2003) Prevalence of pityriasis versicolor in young Italian sailors. Br J Dermatol 149:1270–1272
9. Cermeño JR, Hernandez I, Godoy G et al (2005) Mycoses at Hospital Universitario "Ruiz y Paez", Ciudad Bolivar, Venezuela 2002. Invest Clin 46:37–42
10. Ellabib MS, Khalifa Z, Kavanagh K (2002) Dermatophytes and other fungi associated with skin mycoses in Tripoli Libya. Mycoses 45:101
11. Hafez M, el Shamy S (1985) Genetic susceptibility in pityriasis versicolor. Dermatol 171:86–88
12. He SM, Du WD, Yang S et al (2008) The genetic epidemiology of tinea versicolor in China. Mycoses 51:55–62
13. Borelli D, Jacobs PH, Nall L (1991) Tinea versicolor: epidemiologic, clinical and therapeutic aspects. J Am Acad Dermatol 25:300–305
14. Burke RC (1961) Tinea versicolor: susceptibility factors and experimental infection in human beings. J Invest Dermatol 36:389–401
15. Gupta AK, Batra R, Bluhm R et al (2004) Skin diseases associated with *Malassezia* species. J Am Acad Dermatol 51:785–798
16. Gülec AT, Demirbilek M, Seckin D et al (2003) Superficial fungal infections in 102 renal transplant recipients: a case-control study. J Am Acad Dermatol 49:187–192
17. Roberts SOB (1969) *Pityrosporum orbiculare*: incidence and distribution on clinically normal skin. Br J Dermatol 81:264–269
18. Rivers JK, Norris PG, Murphy GM et al (1989) UVA sunbeds: tanning, photoprotection, acute adverse effects and immunological changes. Br J Dermatol 120:767–777
19. Crespo-Erchiga V, Delgado Florencio V (2002) *Malassezia* species in skin diseases. Curr Opin Infect Dis 15:133–142
20. Patel SD, Noble WC (1993) Analyses of skin surface lipid in patients with microbially associated skin disease. Clin Exp Dermatol 18:405–409

21. Nazzaro Porro M, Passi S, Picardo M (1986) Lipoxygenase activity of *Pityrosporum in vitro* and *in vivo*. J Invest Dermatol 87:108–112
22. De Luca C, Picardo M, Breathnach A et al (1996) Lipoperoxidase activity of *Pityrosporum*: characterisation of by-products and possible role in pityriasis versicolor. Exp Dermatol 5:49–56
23. Thoma W, Kramer HJ, Mayser P (2005) Pityriasis versicolor alba. J Eur Acad Dermatol Venereol 19:147–152
24. Gordon MA (1951) Lipophilic yeast organism associated with tinea versicolor. J Invest Dermatol 17:267–272
25. Crespo-Erchiga V, Ojeda A, Vera A et al (1999) Aislamiento e identificacion de *Malassezia* spp en pitiriasis versicolor, dermatitis seborreica y piel sana. Rev Iberoam Micol 16:S16–21
26. Crespo-Erchiga V, Ojeda A, Vera A et al (1999) Mycology of pityriasis versicolor. J Mycol Méd 9:143–148
27. Crespo-Erchiga V, Ojeda A, Vera A et al (2000) *Malassezia globosa* as the causative agent of pityriasis versicolor. Br J Dermatol 143:799–803
28. Nakabayashi A, Sei Y, Guillot J (2000) Identification of *Malassezia* species isolated from patients with seborrhoeic dermatitis, atopic dermatitis, pityriasis versicolor and normal subjects. Med Mycol 38:337–341
29. Gupta AK, Kohli Y, Faergemann J et al (2001) Epidemiology of *Malassezia* yeasts associated with pityriasis versicolor in Ontario, Canada. Med Mycol 39:199–206
30. Hernandez F, Mendez LJ, Bazan E et al (2003) Especies de *Malassezia* asociadas a diversas dermatosis y a piel sana en población mexicana. Rev Iberoam Micol 20:141–144
31. Dutta S, Bajaj AK, Basu S et al (2002) Pityriasis versicolor: socio-economic and clinico-mycologic study in India. Int J Dermatol 41:823–824
32. Tarazooie B, Kordbacheh P, Zaini F et al (2004) Study of the distribution of *Malassezia* species in patients with pityriasis versicolor and healthy individuals in Tehran, Iran. BMC Dermatol 4:5–8
33. Makni F, Ben Salah I, Affes M et al (2004) Identification des espèces de *Malassezia* isolées des lésions de pityriasis versicolor et des sujets sains dans la région de Sfax (Tunisie). Abst FSMM J Mycol Méd 14:159
34. Ben Salah S, Affes M, Makni F et al (2005) Les levures du genre *Malassezia*: identification moléculaire par nested-PCR des souches isolées à partir des différentes affections dermiques. Abst FSMM. J Mycol Méd 15:175
35. Gaitanis G, Velegraki A, Alexopoulos EC et al (2006) Distribution of *Malassezia* species in pityriasis versicolor and seborrhoeic dermatitis in Greece. Typing of the mayor pityriasis versicolor isolate M. globosa. Br J Dermatol 154:854–859
36. Prohic A, Ozagovic L (2007) *Malassezia* species isolated from lesional and non-lesional skin in patients with pityriasis versicolor. Mycoses 50:58–63
37. Crespo-Erchiga V, Delgado Florencio V (2006) *Malassezia* yeasts and pityriasis versicolor. Curr Opin Infect Dis 19:139–147
38. Nakabayashi A (2002) Identification of *Malassezia* associated dermatoses. Jap J Med Mycol 43:65–68
39. Gaitanis G, Velegraki A, Frangoulis E et al (2002) Identification of *Malassezia* species from patient skin scales by PCR-RFLP. Clin Microbiol Infect 8:162–173
40. Razanakolona I, Rakotozandrindrainy N, Razafimahefa J et al (2004) Pityriasis versicolor à Antananarivo: première étude sur l´identification d´espèces de *Malassezia* responsables. Abst FSMM J Mycol Méd 14:152
41. De Quinzada M (2005) Estudio de las especies de *Malassezia* relacionadas con la patología cutánea pitiriasis versicolor en Panamá. Thesis Doctoral, Panamá
42. Miranda KC, Rodrigues de Araujo C, Soares AJ et al (2006) Identificaçâo de espécies de *Malassezia* em pacientes com pitiríase versicolor em Goiania-GO. Rev Soc Bras Med Trop 39:582–583
43. Panja G (1928) The *Malassezia* of the skin, their cultivation, morphology and species. Trans 7th Congress Far East Assoc Trop Med 2:442–453

44. Castellani A (1935) Pityriasis versicolor tropicalis. Proc R Soc Med 28:1558
45. Midgley G (2000) The lipophilic yeasts: state of the art and prospects. Med Mycol 38(suppl 1): 9–16
46. Roberts SOB (1969) Pityriasis versicolor: a clinical and mycological investigation. Br J Dermatol 81:315–326
47. Borelli D (1985) Pityriasis versicolor due to *Malassezia ovalis*. Mycopathol 89:147–153
48. Crowson AN, Magro CM (2003) Atrophying tinea versicolor: a clinical and histological study of 12 patients. Intern J Dermatol 42:928–932
49. Savin R (1996) Diagnosis and treatment of tinea versicolor. J Family Pract 43:127–132
50. Schwartz RA (2004) Superficial fungal infections. Lancet 364:1173–1182
51. Crespo-Erchiga V, Guého E (2005) Superficial diseases caused by *Malassezia* species. In: Merz WG, Hay RJ (eds) Medical mycology. Topley and Wilson's microbiology and microbial infections, 10th edn. Arnold, London, pp 202–216
52. Guého E, Midgley G, Guillot J (1996) The genus *Malassezia* with description of four new species. Antonie van Leeuwenhoek 69:337–355
53. Gupta AK, Kogan N, Batra R (2005) Pityriasis versicolor: a review of pharmacological treatment options. Expert Opin Pharmacother 6:165–178
54. Faergemann J, Fredriksson T (1980) Propylene glycol in the treatment of pityriasis versicolor. Acta Derm Venereol (Stockh) 60:92–93
55. Katsambas A, Rigopoulos D, Antoniou C et al (1996) Econazole 1% shampoo versus selenium in the treatment of tinea versicolor: a single-blind randomized clinical study. Int J Dermatol 35:667–668
56. Mellen LA, Vallee J, Feldman SR et al (2004) Treatment of pityriasis versicolor in the United States. J Dermatol Treatment 15:189–192
57. Svejgaard E (1973) Double blind trial of miconazole in dermatomycosis. Acta Derm Venereol (Stockh) 53:497–499
58. Tanenbaum L, Anderson C, Rosenberg MJ et al (1984) 1% sulconazole cream v 2% miconazole cream in the treatment of tinea versicolor. A double-blind, multicenter study. Arch Dermatol 120:216–219
59. Sugita T, Tajima M, Ito T et al (2005) Antifungal activities of tacrolimus and azole agents against the eleven currently accepted *Malassezia* species. J Clin Microbiol 43:2824–2829
60. Hay RJ, Midgley G (1984) Short course ketoconazole therapy in pityriasis versicolor. Clin Exp Dermatol 9:571–573
61. Jacobs PH (1985) Evolution in the treatment of pityriasis versicolor. In: Meinhof W (ed) Oral therapy in dermatomycoses: a step forward. Medical Publishing Foundation, Oxford, pp 107–110
62. Delescluse J (1990) Itraconazole in tinea versicolor: a review. J Am Acad Dermatol 23:551–554
63. Galimberti RL, Villalba I, Galarza S et al (1987) Itraconazole in pityriasis versicolor: ultrastructural changes in *Malassezia furfur* produced during treatment. Rev Infect Dis 9(suppl 1):S134–S138
64. Kose O, Bülent Tastan H, Riza Gur A et al (2002) Comparison of a single 400 mg dose versus a 7-day 200 mg daily dose of itraconazole in the treatment of tinea versicolor. J Dermatol Treat 13:73–76
65. Partap R, Kaur I, Chakrabarti A et al (2004) Single-dose fluconazole versus itraconazole in pityriasis versicolor. Dermatology 208:55–59
66. Farschian M, Yaghoobi R, Samadi K (2002) Fluconazole versus ketoconazole in the treatment of tinea versicolor. J Dermatol Treat 13:73–76
67. Garau M, Pereiro M Jr, del Palacio A (2003) *In vitro* susceptibilities of *Malassezia* species to a new triazole, albaconazole (UR-9825), and other antifungal compounds. Antimicrob Agents Chemother 47:2342–2344
68. Velegraki A, Alexopoulos EC, Kritikou S et al (2004) Use of fatty acid RPMI 1640 media for testing susceptibilities of eight *Malassezia* species to the new triazole posaconazole and

to six established antifungal agents by a modified NCCLS M27–A2 microdilution method and Etest. J Clin Microbiol 42:3589–3593
69. Faergemann J, Ausma J, Vandeplassche L et al (2007) The efficacy of oral treatment with pramiconazole in pityriasis versicolor: a phase II a trial. Br J Dermatol 156:1385–1388
70. Levy A, Feuilhade de Chauvin M, Dubertret L et al (2007) Folliculites à *Malassezia*. Caractéristiques et réponses thérapeutiques chez 26 malades. Ann Dermatol Vénéréol 134:823–828
71. Crespo A, Crespo V, Sanz A (1990) Folliculitis por *Malassezia*. Dos casos tratados con ketoconazol. Rev Iberoam Micol 7:119–121
72. Alves EV, Martins JE, Ribeiro EB et al (2000) *Pityrosporum* folliculitis: renal transplantation case report. J Dermatol 27:49–51
73. Garau M, Relaño MT, Molina L et al (2003) Foliculitis en paciente con enfermedad de Hodgkin. Enferm Infec Microbiol Clin 21:101–102
74. Archer-Dubon C, Icaza-Chivez ME, Orozco-Topete R et al (1999) An epidemic outbreak of *Malassezia* folliculitis in three adult patients in an intensive care unit: a previously unrecognized nosocomial infection. Int J Dermatol 38:453–456
75. Rhie S, Turcios R, Buckley H et al (2000) Clinical features and treatment of *Malassezia* folliculitis with fluconazole in orthotopic heart transplant recipients. J Heart Lung Transplant 19:215–219
76. Guého E, Faergemann J, Lyman C et al (1994) *Malassezia* and *Trichosporon*: two emerging pathogenic basidiomycetous yeast-like fungi. J Med Vet Mycol 32:367–378
77. Faergemann J (1999) *Pityrosporum* species as a cause of allergy and infection. Allergy 54:413–419
78. Bäck O, Fergemann J, Hornqvist R (1985) *Pityrosporum* folliculitis: a common disease of the young and middle aged. J Am Acad Dermatol 12:56–61
79. Faergemann J, Johansson S, Back O et al (1986) An inmunologic and cultural study of *Pityrosporum* folliculitis. J Am Acad Dermatol 14:429–433
80. Sina B, Kauffman CL, Samorodin CS (1995) Intrafollicular mucin deposits in *Pityrosporum* folliculitis. J Am Acad Dermatol 32:807–809
81. Roberts SOB, Lachapelle JM (1969) Confluent and reticulate papillomatosis (Gougerot-Carteaud) and *Pityrosporum orbiculare*. Br J dermatol 81:841–845
82. Nordby CA, Mitchell AJ (1986) Confluent and reticulated papillomatosis responsive to selenium sulfide. Int J Dermatol 25:194–199
83. Shimizu S, Han-Yaku H (1997) Confluent and reticulated papillomatosis responsive to Minocycline. Dermatol 194:59–61
84. Baalbaki SA, Malak JA, Al-Khars MA (1993) Confluent and reticulated papillomatosis: treatment with etretinate. Arch Dermatol 129:961–963
85. Raja Babu KK, Snehal S, Sudha Vani D (2000) Confluent and reticulate papillomatosis: successful treatment with azithromycin. Br J Dermatol 142:1252–1253
86. Gülec AT, Seçkin D (1999) Confluent and reticulate papillomatosis: treatment with topical calcipotriol. Br J Dermatol 141:1150–1151
87. Ginarte M, Fabeiro JM, Toribio J (2002) Confluent and reticulate papillomatosis (Gougerot-Carteaud) successfully treated with tacalcitol. J Dermatol Treatment 13:27–30
88. Stein JA, Shin HT, Chang MW (2005) Confluent and reticulated papillomatosis associated with tinea versicolor in three siblings. Ped Dermatol 22:331–333
89. Rapelanoro R, Mortureux P, Couprie B et al (1996) Neonatal *Malassezia furfur* pustulosis. Arch Dermatol 132:190–193
90. Niamba P, Weill FX, Sarlangue J et al (1998) Is common neonatal cephalic pustulosis (neonatal acne) triggered by *Malassezia sympodialis*? Arch Dermatol 134:995–998
91. Aractingi S, Cdranel S, Reygagne P et al (1991) Pustulose néonatale induite par *Malassezia furfur*. Ann Dermatol Vénéréol 118:856–858
92. Pont V, Grau C, Del Pino M et al (1998) Pustulosis neonatal por *Malassezia furfur*. Actas Dermosifiliogr 89:199–205

93. Bernier V, Weill FX, Hirigoyen V et al (2002) Skin colonization by *Malassezia* species in neonates. A prospective study and relationship with neonatal cephalic pustulosis. Arch Dermatol 138:215–218
94. Koseki S, Takahashi S (1988) Serial observation on the colonization of *Pityrosporum orbiculare (ovale)* on the facial skin surface of newborn infants. Jpn J Med Mycol 29: 209–215
95. Borderon JC, Langier J, Vaillant MC (1989) Colonisation du nouveu-né par *Malassezia furfur*. Bull Soc Fr Mycol Méd 1:129–132
96. Ayhan M, Sancak B, Karaduman A et al (2007) Colonization of neonate skin by *Malassezia* species: Relatioship with neonatal cephalic pustulosis. J Am Acad Dermatol 57:1012–1018
97. Chai FC, Auret K, Christiansen K et al (2000) Malignant otitis externa caused by *Malassezia sympodialis*. Head Neck 22:87–89
98. Guillot J, Bond R (1999) *Malassezia pachydermatis*: a review. Med Mycol 37:295–306
99. Crespo MJ, Abarca ML, Cabañes FJ (2000) Atypical lipid-dependent *Malassezia* species isolated from dogs with otitis externa. J Clin Microbiol 38:2383–2385
100. Crespo MJ, Abarca ML, Cabanes FJ (2000) Otitis externa associated with *Malassezia sympodialis* in two cats. J Clin Microbiol 38:1263–1266
101. Crozier WJ, Wise KA (1993) Onicomicosis due to *Pityrosporum*. Austral J Dermatol 34:109–112
102. Escobar ML, Carmona-Fonseca J, Santamaria L (1999) Onicomicosis por *Malassezia*. Rev Iberoam Micol 16:225–229
103. Silva V, Moreno GA, Zaror L et al (1997) Isolation of *Malassezia furfur* from patients with onicomicosis. J Med Vet Mycol 35:73–74
104. Lopes JO (1994) Nodular infection of the hair caused by *Malassezia furfur*. Mycopathologia 125:149–152
105. Ertam I, Aytimur D, Alper S (2007) *Malassezia furfur* onychomycosis in an immunosuppressed liver transplant recipient. Indian J Dermatol Venereol Leprol 73:425–426
106. Chowdary A, Randhawa HS, Sharma S et al (2005) *Malassezia furfur* in a case of onychomycosis: colonizer or etiologic agent. Med Mycol 43:87–90
107. Blumenthal H (1971) Tinea versicolor of the penis. Arch Dermatol 103:461–462
108. Nia AK, Smith EL (1979) Pityriasis versicolor of the glans penis. Br J Venereol Dis 55:230
109. Smith EL (1978) Pityriasis versicolor of the penis. Br J Vener Dis 54:441
110. Mayser P, Schütz H, Schuppe C et al (2001) Frequency and spectrum of *Malassezia* yeasts in the area of the prepuce and glans penis. BJU Int 8:554–558
111. Aridogan A, Ilkit M, Izol V et al (2005) *Malassezia* and *Candida* colonization on glans penis of circumcised men. Mycoses 48:352–356
112. Iskit S, Ilkit M, Turc-Bicer A et al (2006) Effect of circumcision on genital colonization of *Malassezia* spp. in a pediatric population. Med Mycol 44:113–117
113. Suzuki T, Hori N, Miyake T et al (2007) Keratitis caused by a rare fungus, *Malassezia restricta*. Jpn J Ophtalmol 51:292–294
114. Rivolta S (1873) In: Di Giulio Speirani F (ed) Parasiti vegetali, 1st edn. Torino, pp 469–470
115. Lober CW, Belew PW, Rosemberg EW et al (1982) Patch test with killed sonicated microflora in patients with psoriasis. Arch Dermatol 118:322–325
116. Rosenberg EW, Belew PW (1982) Improvement of psoriasis of the scalp with ketoconazole. Arch Dermatol 118:370–371
117. Farr PM, Krause LB, Marks JM et al (1985) Response of scalp psoriasis to oral ketoconazole. Lancet 26:921–922
118. Rosenberg EW, Noah PW, Skinner RB et al (1989) Microbial associations of 167 patientswith psoriasis. Acta Derm Venereol Suppl (Stockh) 146:72–74
119. Elewski B (1990) Does *Pityrosporum ovale* have a role in psoriasis? Arch Dermatol 126:1111–1112

120. Narang T, Dogra S, Kaur I et al (2007) *Malassezia* and psoriasis: Koebner´s phenomenon or direct causation? JEADV 21:1111–1112
121. Gupta AK, Kohli Y, Summerbell RC et al (2001) Quantitative culture of *Malassezia* species from different body sites of individuals with or without dermatoses. Med Mycol 39:243–251
122. Prohic A (2003) Identification of *Malassezia* species isolated from scalp skin of patients with psoriasis and healthy subjects. Acta Dermatol Venereol Croat 11:1–8
123. Takahata Y, Sugita T, Hiruma M et al (2007) Quantitative analysis of *Malassezia* in the scale of patients with psoriasis using a real-time polymerase chain reaction assay. Br J Dermatol 157:670–673
124. Squiquera L, Galimberti R, Morelli L et al (1994) Antibodies to proteins from *Pityrosporum ovale* in the sera from patients with psoriasis. Clin Exp Dermatol 19:289–293
125. Kanda N, Tani K, Enomoto U et al (2002) The skin fungus-induced Th-1 and Th-2 related cytokine, chemokine and psostaglandin E2 production in peripheral blood mononuclear cells from patients with atopic dermatitis and psoriasis vulgaris. Clin Exp Allerg 32:1243–1250
126. Baroni A, Paoletti I, Ruocco E et al (2004) Possible role of *Malassezia furfur* in psoriasis: modulation of TGF-beta1, integrin, and HSP70 expression in human keratinocytes and in the skin of psoriasis-affected patients. J Cut Pathol 31:35–42

Malassezia Yeasts in Seborrheic and Atopic Eczemas

7

George Gaitanis, Peter Mayser, Annika Scheynius and Reto Crameri

Core Messages

- *Malassezia* yeasts are implicated in the pathogenesis of two common diseases: seborrheic and atopic eczema. Seborrheic eczema (SE) affects areas of the body with an increased density of sebaceous glands (scalp, mid eyebrow, nasolabial folds, sternum and interscapular regions) and body folds). *M. globosa* and *M. restricta* are more commonly isolated in lesional skin, but this is also true for the isolation rate from healthy skin. In lesions of SE, a nonimmunogenic irritant reaction is recognized by immunocytochemistry. SE-specific molecular subtypes of *M. globosa*, *M. restricta*, and *M. furfur* have been recognized. Production of potent aryl hydrocarbon receptor ligands by *Malassezia* yeasts could offer new insight in the pathogenesis of SE through modulation of the immune system. The persistent and recurring nature of SE is highlighted through the existence of multiple, but not yet curative topical and systemic therapies.
- Atopic eczema (AE) has variable clinical presentation of skin lesions depending on the age of the patient. Three types of AE are recognized: "extrinsic," "intrinsic," and one-third group of patients with extrinsic or intrinsic AE showing IgE-mediated sensitization to self-antigens. *Malassezia* yeasts have been shown to act as allergens in AE, and thirteen allergens have been cloned, characterized, and produced as recombinant proteins from *Malassezia* species. Some of these recombinant allergens show a high degree of sequence identity to homologous human self antigens. The human proteins are capable of inducing positive skin prick and atopy patch tests in patients sensitized to the *Malassezia* proteins, indicating a role of IgE-mediated autoreactivity in the pathogenesis of AE in a subset of patients. The recombinant allergens will contribute to improve the diagnosis of sensitization to the yeast in AE patients. A correct patient stratification according to specific sensitization patterns could open novel possibilities for an immunotherapeutic treatment of a subset of AE patients in the future.

G. Gaitanis (✉)
Department of Skin and Venereal Diseases, University Hospital of Ioannina, University of Ioannina, Medical School, S. Niarchou Avenue, 45110, Ioannina, Greece
e-mail: ggaitanis@med.uoa.gr

T. Boekhout et al. (eds.), *Malassezia and the Skin*
DOI: 10.1007/978-3-642-03616-3, © Springer Verlag Berlin Heidelberg 2010

7.1 Seborrheic Eczema/Dermatitis

George Gaitanis and Peter Mayser

7.1.1 Definition

Seborrheic eczema (SE) or seborrheic dermatitis (SD) is a skin disease that can be readily recognized in its typical form when it involves areas of the scalp, eyebrows, paranasal folds (Fig. 7.1), chest, back, armpits, and genitals that are rich in sebaceous glands [1–4]. Diagnosis might constitute a clinical problem when SE lesions appear on the face of patients with psoriasis (sebopsoriasis).

The initial description of this condition was made by Unna [5], who named it SE. The association of SE with *Malassezia* yeasts was accepted during the late nineteenth and up to the middle of the twentieth century. However, at that time, the observed epidermal hyperproliferative state of SE fostered the belief that SE is a condition with increased epidermal cell turnover, such as psoriasis, and *Malassezia* yeasts were considered merely innocuous colonizers [1]. The focus on the pathogenic action of *Malassezia* in SE returned

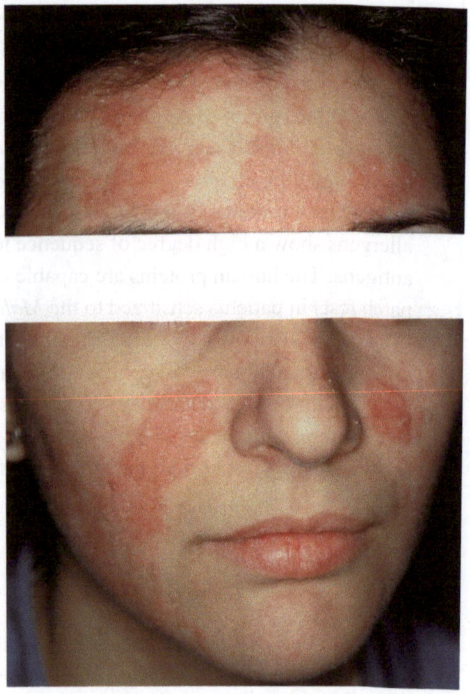

Fig. 7.1 Eczematous plaques of seborrheic eczema (SE) in the face and the sternum area in a middle-aged female patient. Redness and scaling is evident as well as postinflammatory hyperpigmentation in areas were the lesions have subsided (courtesy of Eveline Guého)

during the 1980s as the only common property of the treatment regimens that were used to treat SE was their antifungal activity [6].

The gradual change in the contemporary beliefs regarding the cause of SE is depicted in the established reference textbook "Rook's textbook of dermatology." In the third edition (1979) the authors state that "many attempts have been made to relate SE to infections with bacteria or with *Pityrosporum ovale*", but they conclude that "proof of this widely held hypothesis is totally lacking" [7]. Seven years later, in the fourth edition [8], enough evidence had accumulated. The review article by Shuster [6] analyzed with scrutinized analytically published data that supported an etiological association between *Malassezia* yeasts and SE. Thus, in the sixth edition of Rook's textbook of dermatology [9], it was highlighted that Koch's postulates have been fulfilled in the case of *Malassezia* being associated with SE [10]. Although in the last edition of this textbook [2], the association between *Malassezia* and SE is reconfirmed, attention is called to the considerable gap in the pathobiology that links the yeast with the disease state.

7.1.2 Seborrheic Eczema Clinical Morphology

The typical morphology of SE includes greasy or yellow loosely adherent scales located in the scalp, retroauricular areas, nasolabial folds, presternal and interscapular regions, and the flexures. There are several morphological patterns, which are shown in Table 7.1 [2, 4, 9]. SE eczema occurs commonly with an overall prevalence of 11.6% in the United States [3]. Moreover, 2.6% of men and 3.0% of women of a representative sample examined had clinically significant disease as determined by the dermatologist [3]. Some authors stated that SE is much more common among the elderly [11]. Forms of SE that are severe and resistant to treatment have been associated with neuroleptic induced Parkinson's disease [12], but the pathophysiological association is poorly understood yet.

Table 7.1 The clinical forms of seborrheic eczema/dermatitis

Infantile seborrheic eczema (ISE)
Scalp (cradle cup)
Trunk
Adult seborrheic eczema (ASE)
Face (may include blepharitis and conjunctivitis)
Trunk
Petaloid
Pityriasiform
Flexural
Eczematous plaques
Follicular
Generalized (may appear as erythroderma)

7.1.3 Seborrheic Eczema and *Malassezia*

Currently available data demonstrate that the pathogenesis of SE is complex, but it could be explained only after *Malassezia* yeasts are perceived as an integral part of the skin homeostasis comprising the sebaceous gland and the surrounding epidermis. The ability of *Malassezia* yeasts to induce flares of SE is probably strain- and not species-specific. The most common species implicated in SE are *M. globosa* and *M. restricta,* followed by *M. sympodialis, M. furfur*, and *M. slooffiae* [13, 14]. Yet, there is evidence for the involvement of certain subtypes of *M. globosa*, *M. restricta* [14], and *M. furfur* [15] in SE. *M. globosa* SE associated subtypes were differentiated by intergenic spacer sequences into seven groups: four including *M. globosa* strains from SE lesions, two groups with mixed strains from healthy and SE skin, and a distinct group with *M. globosa* strains from healthy skin [14]. Likewise, *M. restricta* isolates from SE clustered in a single group while strains from healthy and diseased skin formed a separate cluster [14].

The implication of the immune system in SE is crucial; yet, no specific immune response to *Malassezia* yeasts is elicited [16]. Immunocytochemistry of skin biopsies from SE lesions, the healthy skin of SE patients and from healthy controls showed that this disease is characterized by a nonimmunogenic irritant reaction of the skin immune system in the epidermis and around the follicles [17]. Inflammation markers, viz. (IL-1α, IL-1β, IL-2, IL-4, IL-6, IL-10, IL-12, IFN-γ, and TNF-α) are significantly increased in SE compared with those in skin from healthy volunteers. Comparison of SE skin with adjacent clinically healthy skin, however, did not show a statistical significant difference in the aforementioned markers. In accordance to these findings, *M. globosa, M. restricta, M. furfur, M. sympodialis, M. slooffiae,* and *M. obtusa* have been shown to induce the production of the proinflammatory cytokines IL-1α, IL-6, and the antiinflammatory IL-10 from normal human keratinocytes [18]. Factors that affected the production of these cytokines were (a) the presence or absence of the lipid capsule that consists part of the cell wall of *Malassezia* yeasts [19], (b) the culture growth phase (exponential vs. stationary), and (c) yeast cell viability. The proinflammatory cytokines were upregulated after the removal of the lipid capsule, whereas the antiinflammatory IL-10 was downregulated [20]. Contrary to the immunohistochemistry findings [17], IL-8 production was increased following stimulation of natural healthy keratinocytes with acapsular *Malassezia* yeasts belonging to the six lipophilic species, but TNF-α production was not enhanced by either capsular or acapsular *Malassezia* forms. In a previous experiment [21], TNF-α production was increased along with IL-1β, IL-6, and IL-8 in human normal keratinocytes co-cultured with *M. sympodialis, M. slooffiae,* and *M. pachydermatis,* whereas *M. furfur* failed to enhance the production of the aforementioned cytokines. Ishibashi et al. [22] did not detect any increases in Th1 (IL-1α, IL-1β, IL-2, and TNF-α) cytokine mRNA by microarrays applied on keratinocytes exposed to *M. globosa* and *M. restricta* cells. Yet, this outcome could be attributed to the short period of exposure time (2 and 4 h) of keratinocytes to *M. globosa* and *M. restricta* cells, as the primary objective of the investigators was to assess mRNA for Th2 cytokines (IL-4, IL-5, IL-10, and IL-13). However, this study showed that *M. globosa* and *M. restricta* stimulated the production of different inflammatory mediators. Specifically, *M. globosa* stimulated IL-3, IL-5, IL-6, IL-7, IL-10, IL-13, and GM-CSF (granulocyte monocyte-colony stimulation factor) and *M. restricta* IL-4 and Leptin. In accordance with the recent epidemiological data that demonstrate the existence of more than one *Malassezia* species on SE skin [14], development of this skin disease could be the result of the quantitative expression and secretion of a broad panel of cytokines occurring through *Malassezia* stimulation of keratinocytes. On these grounds, the therapeutic infusion in a patient of recombinant IL-2 caused the appearance of SE in a dose dependent manner [23]. Lesions, however, developed in the known seborrheic regions and not in a more wide distribution, as would be expected after systemic administration of an inciting agent.

These data cannot explain the observation that clinically healthy skin proximally to SE lesions demonstrates equivalent cytokine and immune cell infiltrating profiles to lesional skin [17]. Thus, the existence of other currently unknown, host-specific factors that contribute to SE appearance is highly probable. The effect of *Malassezia* on the skin immune system and the development of SE is not restricted to cytokine production. *M. furfur* has the ability to up-regulate Toll-like receptor 2 and stimulate the expression of Human-β-defensins 2 (HBD-2) and HBD-3 as well as IL-8 when cocultured with human keratinocytes [24, 25]. HBD-2 has a potent antimicrobial activity against gram-negative bacteria and *Candida albicans*, and it is tempting to speculate that *M. furfur* cells could manipulate the skin innate immune system by stimulating epidermal keratinocytes to produce HBD-2 for direct antimicrobial action and IL-8 for attraction of neutrophils to purge antagonistic microorganisms. Thus, exploiting the defense mechanisms of their host would be a valuable means for keeping away antagonistic microorganisms and maintaining the skin flora equilibrium. As we know from psoriasis and atopic eczema (AE) studies, the skin immune system has an effect on the microbiota. In psoriasis, the upregulation of the inflammatory response with expression of antimicrobial peptides results in absence of infections in psoriatic skin despite colonization with *Staphylococcus aureus* [26]. On the other hand, AE is characterized by a Th 2 immune response with decreased production of antimicrobial peptides, allowing survival of a plethora of nonpathogenic and pathogenic microorganisms such as *S. aureus* [27]. Thus, SE could be underlied by a nonspecific, direct or indirect activation of the immune system by the presence of *Malassezia* cells or their products (e.g., secreted enzymes or metabolites).

Indication of the ability of *Malassezia* yeast cells to further interact with the skin immune system in SE is the increased production of potent aryl hydrocarbon receptor (AHR) ligands [Indole (3,2 b) carbazole (ICZ), indirubin, and malassezin] by *M. furfur* isolates when grown in L-tryptophan agar [28]. *M. furfur* strains from SE constitutively produced higher amounts of ICZ, indirubin, and malassezin than *M. furfur* strains from healthy skin [28]. Although the extent of implication of these substances in SE development is far from clear, they could possibly reflect the response of *Malassezia* yeasts to alterations in the skin microenvironment in an effort to modulate the immune reaction. Differentiation of the recently described Th17 cells is dependent on IL-6 and transforming growth factor-beta (TGF-β), while it is modulated by AHR activation [29]. Th17 cells are considered to play an important role in autoinflammatory phenomena, such as experimental autoimmune encephalomyelitis and inclusion of the potent AHR ligand 6-formyloindol [3,2-b]carbazole in the antigenic emulsion injected in mice accelerated the onset of encephalomyelitis [30]. Furthermore, Th17 cytokines modulate psoriatic keratinocytes to express and secrete large amounts of β-defensins [31], while tolerance of *Malassezia* cells on the skin can be achieved by AHR activation in dendritic cells [32]. Their subsequent effects on regulatory T-cell survival and function in Balb/c mice allograft-specific tolerance of pancreatic islet cells was achieved through AHR activation of this pathway [32]. The effect of constitutive AHR activation on the immune system has been focused mainly on dioxin because of its detrimental environmental effects [33]. However, unremitting *in vivo* production of the aforementioned AHR ligands by *Malassezia* can have an impact on the surrounding epidermal cells. In fact, constitutive activation of the AHR receptor in transgenic mice resulted in inflammatory skin lesions [34], while part of the damage of ultraviolet light in keratinocytes is considered to be mediated through AHR activation by L-tryptophan photoproducts, such as 6-formyloindol [3,2-b]carbazole [35]. Actually, SE has been reported to appear after intense psoralen ultraviolet light therapy [36] and in high-altitude guides who are exposed to ambient UV light [37].

Interestingly, SE is not the result of increased sebum production [38], but it is rather associated with fluctuations in sebum production as demonstrated by the epidemiology of the disease. In newborns, SE in the form of infantile SE is quite common, especially during the first trimester when sebaceous glands are active under the influence of maternal androgens, and the skin is colonized with *Malassezia* yeasts among other microbes [39]. In Australia, infantile SE had a prevalence of 10% in a cohort of preschool-aged children and was higher in infants during the first 3 months of life (71.7%) [40]. SE gradually declined by the end of the first year of life, as sebaceous gland activity minimized devoid of maternal androgen stimulation. It is well known that SE recurs with the advent of adulthood [2], but it is more common after the age of 50 [11] when sebaceous gland activity gradually diminishes [11, 41]. Thus, at least in part, SE could be the result of selection of more pathogenic *Malassezia* strains that rapidly expand to cover a biocenosis [42] during infant life and the teens. Alternatively, it could be the result of a selection of metabolically more active strains, which struggle for survival in an environment with compromised homeostatic mechanisms due to aging and reduction of available nutrients, as sebum production decreases in the aging skin [11]. The well-known SE flares occurring during the autumn following the higher sebum secretion rates of the summer are in accordance to this [43], as is the observation that women with SE have decreased sebum excretion rates compared with controls [38]. Likewise, in animals, pathogenic strains of *M. pachydermatis* respond to β-endorphin stimulation with an increased production of phospholipase [44, 45]. β-Endorphin increased sebum production in humans [46], and the existence of such a pathway, i.e., β-endorphin-increased phospholipase production, in the anthropophilic *Malassezia* species could imply better utilization of the sebaceous lipids by the yeasts. A side effect of the aberrant production of *Malassezia* phosholipases could be the removal of epidermal lipids, the compromise of epidermal barrier function and the subsequent development of SE as sebaceous gland activity decreases. In *C. albicans*, phospholipase production constitutes an established virulence factor [47]. Evidence accumulated for the significance of phospholipase as a virulence factor in *M. pachydermatis,* but the significance of these enzymes for the anthropophilic *Malassezia* species remains to be elucidated [48]. Recently, the existence of a functional μ-opioid receptor in pathogenic and nonpathogenic strains of *M. pachydermatis* has been demonstrated [49]. Given the extensive distribution of G-protein-coupled receptors in fungi and their role in sensing various environmental parameters [50], it is highly possible that the anthropophilic *Malassezia* species do possess an extensive repertoire of these receptors. Comparison of their cellular effects between strains isolated from healthy skin and SE-affected skin could highlight aspects of SE pathogenesis.

Certain *M. globosa* IGS sequence subtypes [14] as well as *M. furfur* genotypes with specific microsatellite fingerprints [15] have been associated with SD. Thus, the selective production of bioactive indoles by *M. furfur* strains isolated from SE, such as the antimicrobial substance indirubin [51], could benefit the yeast by controlling the growth of antagonistic bacteria. Consistent to this proposed mechanism is the emergence of more pathogenic strains of *Malassezia* yeasts during SE flares. In addition, the possibility that mating of *Malassezia* species under nutrient deprivation, which may happen during aging, may give rise to *Malassezia* strains that are better adapted for survival cannot be ruled out, as recent evidence shows that *M. globosa* species possess a mating-type locus [52].

In conclusion, current data show the existence of diverse, complex mechanisms in the development of SE. Systematic investigation of virulence factors in *Malassezia* yeasts and associated susceptibility traits in SE will highlight the significance of this skin commensal and pathogen in skin homeostasis and disease development.

7.1.4 Seborrheic Eczema and HIV/AIDS

Seborrheic eczema occurs more commonly in HIV/AIDS patients, as it has been reported to range between 20 and 40% of HIV-1 seropositive patients and between 40 and 80% in those with AIDS [53]. SE is more severe and more resistant to therapy in HIV/AIDS patients and advances with the stage of the disease [54]. HIV/AIDS affects the response of keratinocytes to stress signals [55], while it also has detrimental effects on Langerhans cells and their cross-talk with CD-4 memory lymphocytes resulting in virus replication and destruction of both subsets of immune cells, thus destabilizing the skin immune system [56]. Although HIV/AIDS is associated with the development of multiple skin diseases, SE has been considered the most common [53]. The significance of *Malassezia* yeasts in the development of HIV/AIDS associated SE is under debate, as some studies noted an increase in *Malassezia* yeasts on the skin of HIV/AIDS positive non-SE volunteers [57] and on the healthy skin of HIV-positive SE patients [58] compared with controls. Decreased *Malassezia* counts from HIV-infected patients [54, 59] questioning the implication of *Malassezia* have, however, been reported in this subset of SE [58]. As pointed out earlier, the *Malassezia* strains implicated and not just the yeast load [58] may be significant for the development of SE in HIV/AIDS patients. In the declining HIV/AIDS skin immune environment, the corresponding microbiota is in constant disequilibrium due to the development of multiple, treatment resistant viral, bacterial, and fungal dermatoses [60]. Thus, survival on a compromised skin environment is a constant challenge, forcing the selection of more survival-suitable *Malassezia* strains by the expression of relevant traits.

7.1.5 Infantile Seborrheic Eczema

Infantile seborrheic eczema (ISE) is a term that is commonly used to describe a distinctive eczematous or psoriasiform eruption with a predilection for the scalp and the proximal flexures, and one that has a favorable prognosis compared with AE [39, 61].

It commonly appears between the second week of life up to the sixth month and can involve the face, scalp, trunk, and napkin area individually or in any combination. Pruritus is usually absent, although it is difficult to evaluate this symptom in the neonate. The clinical appearance is that of SE with erythema, scales, and crusts. The lesions might coalesce, especially in the face and flexures, but they are more distinct on the trunk.

As the clinical appearance of ISE is not specific, it might be easily confused with infantile AE and infantile psoriasis. Thus, various studies have considered ISE as a predilection factor for the development of psoriasis or AE [39]. However, it is quite difficult to unequivocally conclude on this as all three conditions share clinical characteristics in this age group. Furthermore, as they occur quite commonly, it is possible that they can coexist in the same infant.

Culture studies have reported variable results regarding the isolation of *Malassezia* yeasts from lesional or control skin. In one study, the same rate of positive cultures for *Malassezia* was shown during ISE, and after 1 year when the lesions had subsided [20]. Unfortunately, the authors did not use the same sampling plates throughout the study, and the results may have been biased by this. Other studies reported increased isolation rates from neonates with ISE when compared with healthy controls [62, 63] or infants with AE [63].

In conclusion, as mentioned earlier for adult SE, it is difficult to unequivocally differentiate a typical ISE from AE or infantile psoriasis as they share clinical characteristics [64], and pathophysiological associations cannot be determined unequivocally by epidemiological studies alone.

7.1.6 Therapy

The multitude of therapeutic modalities proposed for SE demonstrates the persistent and relapsing nature of this disease. The common underlying action mechanism of most compounds used is their antifungal property [6]. Most cases of SE can be controlled with application of topical agents during flares of the disease. In mild cases, applying an antidandruff shampoo preparation on the skin lesions could demonstrate efficacy [65], but application of topical antiinflammatory, antifungal, or combinational medications might be necessary to control SE. Potent steroids for the treatment of SE should be restricted to short periods to avoid side effects such as atrophy and telangiectasias, or the development of perioral dermatitis. The use of topical calcineurin inhibitors (viz., tacrolimus, pimercolimus) in SE is devoid of the steroid local side effects, but it is associated with stinging during the first days of their use and patients should be informed accordingly. Topical antifungal azoles are a safe approach for long-term management of SE, and they can be used in combination with mild steroids for rapid control of SE flares. Below, we describe the different agents used in the treatment of SE in more detail.

7.1.6.1 Antifungals

For topical antifungal agents, controlled studies exist for ketoconazole, bifonazole, and ciclopiroxolamine. In the largest double-blinded ketoconazole trial, involving 1,162 people with mild to severe disease affecting multiple regions of the body, treatment was determined to be successful on the basis of a global assessment score at 4 weeks in 56% of patients who received ketoconazole foam twice daily, when compared with 42% who received placebo foam ($P<0.001$). Similar results were obtained comparing the drug with placebo in cream formulations [66]. Comparing 2% ketoconazole in a gel formulation, used once daily, with placebo in 459 subjects with moderate-to-severe disease in different areas of the body, the skin of 25% of subjects with active treatment was considered cleared or almost cleared at day 28 in comparison with 14% assigned to placebo ($P=0.001$) [67].

Intermittent use of ketoconazole can maintain remission. In one study, 312 patients with scalp lesions in whom dermatitis had initially cleared with twice-weekly shampoo containing 2% ketoconazole were subsequently enrolled in a 6-month placebo-controlled prophylactic trial; relapse rates were 47% among patients using placebo, 31% among patients using ketoconazole shampoo once every other week, and 19% among patients using the active treatment once weekly [68]. In a randomized trial with bifonazole involving 100 patients, the skin of 43% of patients using 1% bifonazole cream once daily, when compared with 23% of those using placebo, was shown to be almost clear after 4 weeks according to a global improvement scale [69]. In addition, bifonazole shampoo used three times weekly has also been shown to result in significantly greater improvement in scalp lesions than placebo [70].

In a randomized trial comparing ciclopiroxolamine shampoo, used once or twice weekly, with placebo in 949 patients with scalp lesions, rates of clearance over a 4-week period were 45 and 58% with the once-weekly and twice-weekly active treatments, respectively, when compared with 32% with placebo ($P<0.001$ for both comparisons with placebo). Among 428 patients with a response who were subsequently randomly assigned to ciclopiroxolamine prophylaxis once weekly, ciclopiroxolamine prophylaxis every 2 weeks, or placebo for 4 months, relapse rates were 15, 22, and 35%, respectively [71].

Limited data are available for comparisons of different antifungal agents. In a trial involving 303 patients with facial SE, ciclopiroxolamine cream administered twice daily for 28 days, followed by once-daily use for an additional 28 days, resulted in significantly higher rates of remission than the use of ketoconazole foaming gel used twice weekly for the first 28 days and then once a week (57% vs. 44% at 56 days in an intention-to-treat analysis, $P=0.03$). However, these results are difficult to interpret because of the much lower frequency of application for ketoconazole than for ciclopiroxolamine [72]. Local tolerance as well as overall acceptability was better with ciclopiroxolamine than with ketoconazole [73].

7.1.6.2
Topical Corticosteroids

Several randomized trials have compared short-term topical corticosteroids, including in approximate order of increasing potency, hydrocortisone, betamethasone dipropionate, clobetasol 17-butyrate, and clobetasol dipropionate, with topical antifungal agents [74, 75].

These trials have shown either no significant difference or a small difference in favor of the antifungal agent, but the largest trial included only 72 people [75].

7.1.6.3
Selenium Sulfide Preparations

In a randomized trial involving 246 patients with moderate-to-severe dandruff, 2.5% selenium sulfide shampoo, 2% ketoconazole shampoo, and placebo were compared. All shampoos were used twice weekly. Reductions in the score for dandruff at week four were 67% with selenium sulfide, 73% with ketoconazole, and 44% with placebo; the reductions were significantly larger with both medicated shampoos than with placebo [76].

7.1.6.4
Topical Lithium Salts

The mechanism of action of topical lithium succinate and lithium gluconate is poorly understood. They are effective alternative agents for the treatment of SE in areas other than the scalp. In a crossover, placebo-controlled trial of lithium succinate involving two 4-week treatment periods separated by a two-week washout period, twice-daily lithium succinate ointment was associated with significantly greater reductions in erythema, scaling, and the percentage of the area of the skin involved [77]. In a small, randomized trial involving 12 patients, lithium succinate was significantly more effective than placebo for the treatment of lesions in HIV-positive patients [78]. In 129 patients with facial lesions, application twice-daily of lithium gluconate was shown to be more effective than placebo in an 8-week trial, and it was shown to be superior to 2% ketoconazole in an 8-week noninferiority trial involving 288 patients with facial lesions; in the latter study, complete-remission rates were 52% with the use of lithium gluconate and 30% with the use of ketoconazole, but ketoconazole was applied only twice weekly [72, 79].

7.1.6.5
Topical Calcineurin Inhibitors

Calcineurin inhibitors prevent T-cell activation by downregulating the activity of type 1 and 2 T-helper cells. In a randomized trial involving 96 patients with moderate-to-severe facial SE (mean age, 59.6 years, 88.5% male), the mean change from baseline to 4 weeks in the total target-area score with twice-daily 1% pimecrolimus was significantly greater than with placebo in a per-protocol analysis, but not in an intention-to-treat analysis [80]. Two small, randomized trials did not show significant differences between pimecrolimus and topical corticosteroids, but these trials had limited statistical power [81, 82]. The effects of an intermittent or prophylactic use of calcineurin inhibitors were not yet tested in controlled studies.

7.1.6.6
Phototherapy

Although UV-light might provoke SE (see above) [36], ultraviolet B phototherapy is sometimes considered as an option for extensive or recalcitrant SD, but it has not been studied in randomized trials [83, 84].

7.1.6.7
Systemic Antifungal Therapy

Fluconazole: In a randomized trial involving 63 patients with mild-to-moderate SE, a single weekly dose of 300 mg of fluconazole was no better than placebo after 2 weeks [85].

Terbinafine: Scapparro and collaborators [86] performed a randomized placebo-controlled study with either oral terbinafine 250 mg o.d. or placebo (moisturizing ointment) twice daily for 4 weeks. Terbinafine significantly reduced the mean global clinical score as well as individual erythema, scaling, and itching scores. The good clinical response was maintained for 8 weeks after cessation of therapy [86]. In contrast, oral terbinafine (at a dose of 250 mg/day for 4 weeks) in a randomized, placebo-controlled trial [87] involving 174 patients was not found to be better than placebo in patients with lesions predominantly involving exposed areas of the skin, such as the face. A difference was noted in patients with lesions predominantly involving areas of the skin that were not exposed, such as the scalp, sternum, and interscapular areas.

Itraconazole: In a study by Baysal et al. [88] 32 patients applied 1% hydrocortisone cream twice daily for 1 month in parallel with itraconazole 200 mg/day during the first week of the first month, then followed by itraconazole 200 mg/day of the first days of the following 11 months and 2 months of no treatment. There was a statistically significant decrease in the mean severity score at the first, 12th, and 14th month. An open clinical study in 29 patients with oral itraconazole (200 mg/day for 7 days) and consecutive usage of 200 mg/day for the first 2 days of the following 2 months in patients with SE resulted in clinical improvement in 23 patients [89]. Compared with baseline values, the mean SD total score as well as individual erythema and desquamation were significantly reduced (Wilcoxon's signed test-two tailed) ($P<0.0001$) [89]. In an open noncomparative study 60 patients with moderate-to-severe SE were treated with oral itraconazole, initially 200 mg/day for a week, followed by a maintenance therapy of a single dose of 200 mg every 2 weeks for a total of 18 weeks. At the end of the initial treatment, significant improvement was reported in three clinical parameters: erythema, scaling, and itching. Maintenance therapy led to only slight further improvement [90]. A novel promising *per os* agent for the management of SE is pramiconazole, which accomplished control of SE within 3 days and maintained the result for 4 weeks after a single 200-mg dose [91]. The development of this drug has currently been suspended because of changes in the ownership status [92].

In future, understanding the particular mechanisms that lead to the development of SE lesions would lead to targeted therapies that would offer cure to this common and persistent skin condition.

7.2 *Malassezia* in Atopic Eczema/Dermatitis

Annika Scheynius and Reto Crameri

7.2.1 Atopic Eczema (Atopic Dermatitis)

Atopic eczema (AE), or atopic dermatitis (AD), is a chronic relapsing inflammatory skin disease, and its prevalence has doubled or tripled in industrialized countries during the past three decades with around 15–30% of children and 2–10% of adults being affected nowadays [93]. The disease is characterized by severely itchy, red, dry, and crusted skin with the distribution varying with age. This skin disorder has a great impact on quality of life at all ages; it interferes with work, play, and nightly rest, and affects family functions and relations. In childhood, it is often the first symptom of allergic disease that often progresses to allergic rhino-conjunctivitis and/or asthma at a later stage.

There are different subgroups of patients with AE. Approximately 80% of adult patients with symptoms of AE have immediate type skin reactions to environmental allergens (positive skin-prick test, SPT), allergen-specific IgE in serum, and elevated total serum-IgE levels [94]. This form of AE is often called "extrinsic," since it is assumed that allergens are involved in induction and maintenance of the disease. In contrast, patients with the "intrinsic" form of AE, also denoted "nonatopic eczema," have normal serum IgE levels and allergen reactivity has not been connected to their symptoms [94]. The two subgroups have an indistinguishable clinical picture. Still, another subgroup of patients with AE seems to have an additional autoimmune IgE-mediated reactivity against self antigens in addition to that against exogenous allergens [95, 96]. Thus, it is important to recognize the different subgroups of AE patients to identify underlying disease mechanisms and offer the patients proper advice and treatment.

The pathogenesis of AE is likely to result from a combination of a defective skin barrier and inappropriate immune responses with contribution of both genetic and environmental factors [93, 94]. A number of genes seem to play a role in the disease where some are related to the barrier function of the epidermis and others are involved in immunological responses favoring the development of IgE-mediated sensitization with an allergen-specific Th2 polarization [93, 94]. MicroRNAs are short noncoding RNAs that are important transcriptional and posttranscriptional regulators of gene expression and function [97]. We recently found that AE skin has a specific microRNA expression profile compared with healthy human skin and lesional skin from patients with psoriasis [98], suggesting functional implications that might represent potential therapeutic targets.

Environmental triggers can be different life style factors [99], stress, allergens, and microorganisms [94]. The gene–environment interactions during sensitization or allergen challenge may induce a permanently altered genetic state persisting during life, so-called epigenetic alteration [100]. Whether epigenetic mechanisms play a role in the development of allergic diseases is an open area of research [101].

7.2.2 Malassezia and Atopic Eczema

Interestingly, microorganisms have been demonstrated not only to cause skin infections but also to trigger allergic symptoms by acting as allergens. Among these, microorganisms are species that belong to the lipophilic yeast genus *Malassezia* that are also part of the normal human cutaneous microbiota. Clemmensen and Hjort [102] were the first who reported an association between *Malassezia* and AE. They described that treatment with the antifungal drug ketoconazole improved AE, especially in adult patients with a head and neck distribution of the eczema and a positive skin-prick test (SPT) to *Malassezia* extract. Later, studies supported their findings [103] and also found that treatment with ketoconazole lowered the levels of total serum IgE and *M. sympodialis*-specific IgE [103, 104]. Allergen-specific IgE and/or atopy patch test (APT) reactivity to *Malassezia* can be detected in 50% of adult patients with AE [103], but rarely in other allergic diseases [105], indicating a specific link between AE and *Malassezia* in a subset of patients. We have therefore chosen *M. sympodialis*, one of the most common species colonizing the skin of both healthy individuals and patients with AE [13, 103, 106] as a model to deepen our understanding of how a member of the normal skin microbiota can interact with the innate (natural) and adaptive (acquired) immune system and possibly contribute to the pathogenesis of AE.

7.2.3 Microbe and Allergen Interaction with Our Immune System

The initial contact among microbes, allergens, and our defense barriers, including the innate and adaptive immune system, is crucial. Various types of cells such as keratinocytes, eosinophils, mast cells, dendritic cells (DCs), lymphocytes, and NK-cells are involved in the pathogenic mechanisms of AE [93, 94]. Antigen-presenting cells (APC) will decide whether Th1- or Th2-like or Th17 effector cells or regulatory T-cells (Treg) will differentiate, resulting in immunity or tolerance. The most potent APCs in the body are the DCs [107] that respond to different foreign environmental stimuli. At the same time, they have the capacity to sample self antigens and environmental proteins continuously to create immune tolerance. This dual capacity gives them power to dictate immune responses and has evoked hope that DCs can be manipulated *in vitro* or *in vivo* to improve immunotherapy. It is thus vital to understand how DCs exert their control over the quality and magnitude of the immune response. How do antigen-presenting DCs recognize allergens? This is a delicate task for the DCs since allergens are in general harmless proteins that we should not react against. We know that by expressing different Toll-like receptors (TLRs), DCs can recognize different pathogen-associated molecules (PAMPs), but the mechanisms for allergen recognition are so far largely unknown.

7.2.4 Allergens in *Malassezia* Species

Without doubt recombinant allergens are the ideal reagents to improve diagnosis [108, 109], to study the mechanisms involved in the pathophysiology [110], and they may contribute in the future to the therapy of allergic diseases [111, 112]. This specially applies for fungal-related allergic complications because fungi display extremely complex allergen repertoires [113], hampering the production of standardized, consistent extract preparations [114]. Therefore, it is not astonishing that the outcome of serologic or skin test studies performed with different fungal extracts yielded partly discordant results [115]. Currently, thirteen allergens from *Malassezia* species are reported by the official allergen nomenclature list (www.allergen.org), three derived from strain *M. furfur* TIMM 2782, designated Mala f 2 – Mala f 4, and ten from *M. sympodialis* strain ATCC 42132, Mala s 1, and Mala s 5- Mala s 13 (Table 7.2). Although this repertoire is far from complete [116] and might also be confusing in terms of taxonomic nomenclature, because strain ATCC 42132, originally isolated by Dr. Jan Faergemann (Gothenburg, Sweden) was first designated

Table 7.2 Cloned *Malassezia* allergens

Allergens[a]	Size (kDa)	Function/sequence similarity	Strain	Acc. No	References
Mala s 1[b]	36	Unknown	ATTCC 42132	X96486	[121]
Mala f 2	20	Peroxisomal protein	TIMM 2782	ABO11804	[122]
Mala f 3	21	Peroxisomal protein	TIMM 2782	ABO11805	[122]
Mala f 4	35	Malate dehydrogenase	TIMM 2782	AFO84828	[123]
Mala s 5[b]	18	Peroxisomal protein	ATTCC 42132	AJO11955	[132]
Mala s 6[b]	17	Cyclophilin	ATTCC 42132	AJO11956	[132]
Mala s 7[b]	16	Unknown	ATTCC 42132	AJO11957	[133]
Mala s 8[b]	18	Unknown	ATTCC 42132	AJO11958	[133]
Mala s 9[b]	14	Unknown	ATTCC 42132	AJO11959	[133]
Mala s 10	86	Heat shock protein	ATTCC 42132	AJ428052	[135]
Mala s 11	22	MnSOD	ATTCC 42132	AJ548421	[135]
Mala s 12	67	GMC oxidoreductase	ATTCC 42132	AJ871960	[136]
Mala s 13	12	Thioredoxin	ATTCC 42132	AJ937743	[119]

[a] *Malassezia* allergens are referred to the first four letters in the genus (Mala) instead of only the first three, which is normally the case for allergens, in order to avoid confusion with apple (*Malus* = Mal) allergens; Mala s (*Malassezia sympodialis*); Mala f (*Malassezia furfur*)

[b] Originally designated as Mal f 1, Mal f 5–Mal f 9, respectively

Pityrosporum orbiculare, later *M. furfur*, and finally *M. sympodialis* (after sequencing by E. Guého, J. Faergemann personal communication), it is the best investigated allergen repertoire from a structural point of view. The crystal structure of Mala s 1 [117], Mala s 6 [118] and Mala s 13 [119] have been solved recently at high resolution, providing interesting insides into the structural world of allergens. Mala s 6, a cyclophilin, and Mala s 13, a thioredoxin, belong to the class of phylogenetically highly conserved proteins and are members of so-called pan-allergen families [118–120]. These proteins, together with Mala s 11, a manganese-dependent superoxide dismutase, share a high degree of sequence identity to the corresponding human enzymes as well and might play an essential role in the pathogenesis of AE [96] as discussed further.

In contrast, Mala s 1 shows a 6-fold β-propeller structure representing a new fold not known so far in any of the crystallized allergens [117]. Like Asp f 1 and Alt a 1, the major allergens of *Aspergillus fumigatus* [124] and *Alternaria alternata* [125], Mala s 1 seems to be a major species-specific fungal allergen without sequence homology to known proteins. These allergens appear to be highly conserved across a few species or species of a genus as shown by the lack of orthologous proteins in the fungal kingdom obtained by comparative genomic studies involving 22 fully sequenced fungal genomes [126]. The putative active site of Mala s 1 overlaps structurally to putative active sites in the potential homologues Q4P4P8 and Tri 14, from the maize fungal pathogen *Ustilago maydis* and the wheat fungal pathogen *Gibberella zeae*, respectively [127, 128]. Tri14 is a protein of the mycotoxin trichothecene synthesis gene cluster largely responsible for the pathogenicity of *G. zeae* [129], and it has been suggested to be involved in host–cell recognition or in cell-wall processes involved in plant cell invasion by the parasite. The conserved residues within the potential active sites might suggest similar functions for Mala s 1 and for the two related proteins of the plant fungal parasites. Interestingly, Mala s 1 is localized in the cell wall and exposed to the yeast cell surface [130]. The genomes of the *Malassezia* species *M. globosa* and *M. restricta* were recently sequenced [52, 131]. The resemblance between *Malassezia* and *U. maydis* is not limited to Mala s 1 and Q4P4P8, since DNA sequence comparison indicated that *U. maydis* is the fungus most closely related to *Malassezia* among all fungi with complete genome sequences [52].

Mala f 2 and Mala f 3 represent peroxisomal proteins reported as cross-reactive allergens derived from many fungal species [126, 132, 133]. Mala f 4 codes for a malate dehydrogenase, a known allergen structure that is reported also as a latex allergen [134], whereas Mala s 5 represents and additional peroxisomal protein [135]. Interestingly Mala s 7, Mala s 8, and Mala s 9, first detected as partial sequences encoding IgE-binding proteins from a phage surface-displayed *M. sympodialis* cDNA library [135], encode proteins of unknown function without sequence homology to known allergens or to other known proteins [136]. Although we should be careful in defining a sequence as unique because the repertoire of environmental IgE-binding proteins is far from complete, we can assume that Mala s 7, 8, and 9 are rare, if not unique IgE-binding structures, as otherwise homologous sequences would have been found among the many sequenced yeast genomes [137]. The last three *M. sympodialis* allergens to be mentioned, Mala s 10, Mala s 12, and Mala s 13, encode a heat-shock protein [138], a glucose-methanol-choline (GMC) oxidoreductase [139], and a thioredoxin [119] respectively. Heat-shock proteins and thioredoxins are well known as IgE-binding structures [119, 140], whereas Mala s 12 homologous allergens have not been reported so far [139]. Although many more allergens can be expected to be produced by *Malassezia* species as demonstrated by high throughput screening of phage surface displayed cDNA libraries [116], the available recombinant allergens have shown to be useful reagents for a highly specific detection of *Malassezia*-sensitized AE patients [105, 141–143], for the understanding of cross-reactivity at the molecular level [118, 119], and for the investigation of the pathological background played by allergens in AE (see further).

7.2.5 Host–Microbe Interactions in AE

Several factors, along with the genetic background, contribute to a dysfunctional skin barrier and unwanted immune responses in AE. The defective skin barrier in AE patients, both in lesional and nonlesional skin, fail to provide sufficient protection against microbes and allergens and causes increased transepidermal water loss due to a reduced content of water-retaining ceramides [94]. Moreover, the host interaction with microorganisms can stimulate the production of allergens. We recently showed that the release of *M. sympodialis* allergens is indeed significantly higher at pH 6, reflecting the higher pH of skin from patients with AE compared with that at pH 5.5, which is the normal skin pH [144].

Antimicrobial peptides, such as LL-37 and human β-defensin-2 (HBD-2), are produced in the skin as a first line of defense against bacteria, fungi, and some viruses [145]. *M. furfur* induces upregulation of HBD-2 in normal cultured human keratinocytes, which thus limits the uptake of further yeast cells and reflects a regulated innate host defense [25]. This defense mechanism seems to be hampered in patients with AE, since it has been described that patients with AE are lacking appropriate induction of these antimicrobial peptides compared with patients with psoriasis, also suggesting an explanation for the frequent infections with *Staphylococcus aureus* in the skin of AE patients [27]. The expression of the antimicrobial protein LL-37 has also been found to be enhanced both at the protein and transcriptional level in lesional skin of patients with AE and in other types of eczema, where LL-37 was always observed at the place for epidermal injury indicating that epidermal damage plays an important role for the expression [146]. Besides antimicrobial peptides, *Malassezia* yeasts can induce the production of distinct Th2-type cytokines from human keratinocytes, thereby contributing to a Th2-biased immune response [22].

The DC subtypes Langerhans' cells and inflammatory dendritic epidermal cells (IDECs) from AE patients express much higher levels of the high-affinity IgE receptor, FcεRI, [147] and, hence, are candidates to contribute to allergic inflammation in the skin. We have previously demonstrated that immature human monocyte-derived DCs (MDDCs) can efficiently bind and rapidly internalize *M. sympodialis* as well as allergenic components from the yeast [148]. This process is associated with maturation of the MDDCs, induction of lymphocyte proliferation and of a Th2-like immune response [149], suggesting that sensitization to *M. sympodialis* can be mediated by DCs in the skin. MDDCs in healthy individuals possess the ability to distinguish between Mala s 11 and the human homologue manganese superoxide dismutase (hMnSOD) [150] (see Sect. 7.2.6). One hypothesis is that DCs in AE patients are able to present self antigens thus breaking tolerance resulting in autoreactivity contrary to DCs in healthy individuals. cDNA microarray analyzes have indicated differences in the gene expression profiles with an overproduction of proinflammatory molecules in MDDCs from patients with AE compared with MDDCs from healthy individuals cocultured with *M. sympodialis* [151]. Furthermore, *M. sympodialis* triggers the innate immune response differently in patients with AE than in healthy individuals with a prominent secretion of LL-37 by MDDCs from patients with severe AE [152]. Whether these changes in DCs behavior upon contact with *M. sympodialis* reflect genetic differences in AE patients or are the result of phenotypic changes due to their sensitization to *Malassezia* remains to be clarified.

Peripheral blood mononuclear cells (PBMC) from both healthy individuals and patients with AE proliferate *in vitro* when stimulated with *M. sympodialis*, although with a significantly higher response in patients with AE [103, 141, 153, 154]. *M. sympodialis* reactive T cell clones generated from lesional skin show a Th2/Th0-like cytokine profile [153], and there is a relationship between circulating T cells with a Th2-like cytokine profile and positive APT reactions to *M. sympodialis* in AE patients [141]. These observations suggest that *Malassezia* sensitized AE patients have acquired a T cell reactivity that may contribute to maintain the skin inflammation.

Natural killer (NK) or NKT cells and DCs can mutually influence each other's respective activity, shaping the ensuing adaptive immune response [155]. We provided the first evidence that NK cells and DCs do interact in the skin *in vivo* and that *M. sympodialis* stimulates this interaction in patients with AE [156]. Furthermore, *M. sympodialis* enhances NK cell-induced DC maturation in healthy subjects [157]. Thus, NK cells and DCs seem to cooperate in regulating immune responses in the skin. We thus hypothesize that NK and/or NKT cells can selectively eliminate DCs that have picked up *Malassezia* cells before they activate the immune system, a function that might be impaired in AE.

The dominating symptom in AE is severe itch that provokes scratching and increased inflammation. Mast cells most likely play a central role in this vicious circle. Fungal products, such as zymosan, can activate mast cells through TLR2 [158] and crosslinking of the high-affinity IgE receptor, FcεRI, on mast cells leads to the release of potent inflammatory mediators, such as histamine, proteases, chemotactic factors, cytokines, and arachidonic acid metabolites [159]. Recent data show that *M. sympodialis* can activate mouse mast cells to release cysteinyl leukotriens, enhance the mast cell IgE response, modulate MAPK activation, and by signaling through the TLR-2/MyD88 pathway alter the IL-6 production [160]. These effects of *M. sympodialis* on mast cells may contribute to the inflammation and result in AE.

7.2.6 Cross-Reactivity Between *Malassezia* Allergens and Self-Proteins

Obviously molecular interactions are involved in every kind of pathogenesis. They are, however, in most cases largely unknown. This applies specially to multifactorial diseases such as AE, where interactions between genetic predisposition, mechanical barriers, and environmental factors are contributing to the development of a complex clinical syndrome [93]. However, there is accumulating circumstantial evidence showing that *M. sympodialis* allergens are indeed involved in the pathogenesis of at least a subset of AE patients. First, sensitization to *M. sympodialis* is almost specific for AE [105], and second, application of *M. sympodialis* extract to nonlesional skin areas of AE patients sensitized to the yeast is sufficient to induce an eczematous reaction [141]. Extracts are of course complex mixtures of undefined composition and as such not adequate reagents to elucidate the mechanisms underlying a disease. It has been shown, however, that the use of recombinant MnSOD as highly pure allergen preparation is sufficient to elicit positive SPT and APT in AE patients sensitized to this particular molecular structure [161]. Interestingly, MnSOD belongs to the class of phylogenetically highly conserved proteins [162]. MnSODs share, together with cyclophilins and thioredoxins, that have also been identified as relevant *M. sympodialis* allergens, a high degree of sequence identity to the corresponding human enzymes [118, 119]. There is no doubt that the human proteins can elicit positive SPT and specific T cell proliferation in AE patients sensitized to the corresponding environmental allergen [96], indicating an autoantigenic T cell-mediated pathogenesis based on molecular mimicry in a subset of AE patients. Interestingly, MnSOD, cyclophilin, and thioredoxin are stress-inducible enzymes, which can be upregulated by mechanical stress (e.g., scratching). In the case of MnSOD, it has been shown that the enzyme is upregulated in lesional skin areas of AE patients [161]. Release of the autoallergen at the side of inflammation and the presence of crossreacting allergen-specific IgE antibodies might be a logical explanation for the exacerbation of AE in the absence of exogenous allergen exposure [162, 163]. IgE and IgE-mediated reactions are situated at the end of the cascade resulting in allergic symptoms that might be explained by antigen–antibody interactions, which induce the release of toxic mediators (Fig. 7.2). The pathogenic events leading to sensitization are situated more early in the cascade and are at present poorly understood.

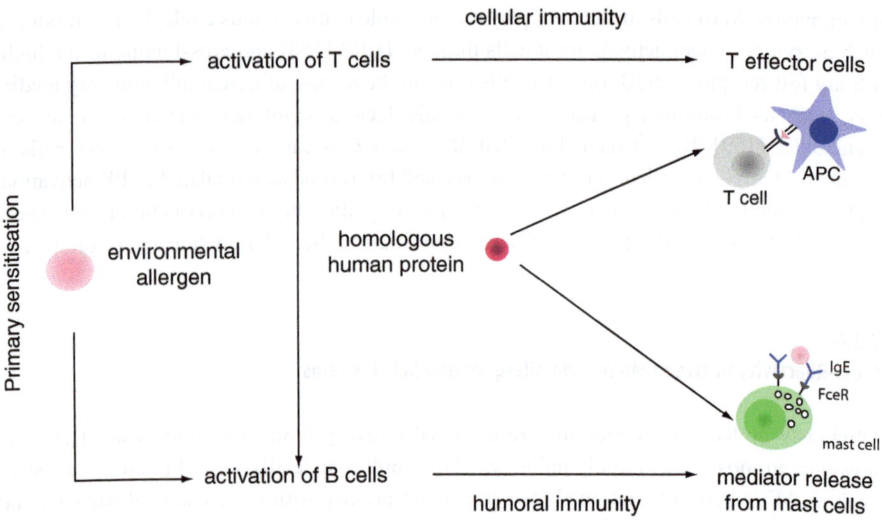

Fig. 7.2 IgE-mediated cross-reactivity between environmental allergens and homologous human proteins. Patients become sensitized to an environmental allergen which induces and establishes a Th2-mediated allergic response resulting in the production of allergen-specific IgE. Based on molecular mimicry human proteins homologous to the environmental allergen can stimulate allergen-specific Th2 cells and cross-link IgE on effector cells thus inducing allergy symptoms

7.2.7 Diagnosis of Sensitization to *Malassezia*

The diagnosis of AE is still mainly based on a clinical history of symptoms and signs fulfilling well-defined diagnostic criteria [94]. Assessment of severity and the extent of the eczema are achieved by the standardized SCORAD index [164]. Serological tests of allergen-specific IgE along with SPTs and APTs are important parts of the diagnosis to identify triggering allergens [94]. It is now possible to diagnose allergen specific IgE in serum or plasma to *Malassezia* extracts with ImmunoCAP (Phadia Diagnostics AB) [103, 142]. Although *Malassezia* ssp. share cross-reactive allergenic determinants to a great extent, there are also species-specific allergens [165]. This is the reason why the use of extract from only one species is not sufficient to detect all patients sensitized to *Malassezia* [166]. There remain, however, problems of reproducibility associated with the preparation of crude (fungal) allergen extracts [167]. These might be eliminated through the use of standardized mixtures of well-characterized recombinant allergens that also can increase specificity and sensitivity in diagnosis [109, 143]. With our recombinant *M. sympodialis* allergens (Table 7.2), we have created unique tools to improve the diagnosis of AE in patients sensitized to the yeast. We have also developed an APT method [141, 168, 169] and an *in vitro* lymphocyte proliferation assay in addition to an ELISpot method to detect T cell-mediated reactivity against *M. sympodialis* [141, 153, 154]. The ELISpot method has shown that there is a correlation between a positive APT reaction to *M. sympodialis* and the number of IL-4 and IL-5 producing PBMC, whereas the allergen-specific IgE levels did not predict the APT outcome, and no correlation was found between IgE levels and the APT scores [170]. Thus, the ELISpot method might provide an alternative *in vitro* diagnostic method and serve as an important diagnostic tool, which can reveal a previously overlooked impact of *M. sympodialis* hypersensitivity in certain subgroups of AE patients without detectable allergen specific serum IgE [168].

7.2.8 Future Therapy

Restoring the epidermal-barrier function [171] and avoiding IgE-sensitization are major targets for the prevention of AE. As for many other allergic diseases, the therapy of AE is currently mainly limited to symptomatic forms of treatment. However, the only therapy able to cure IgE-mediated allergic diseases is the allergen-specific immunotherapy (SIT). SIT works very well, especially in insect venom allergy [172], but also in patients that are monosensitized to a single allergen source, such as single-species pollen [173]. One of the reasons for this success is that insect venoms, as well as most pollen, are relatively simple allergen mixtures that can be easily standardized. In contrast, fungal extracts are extremely complex allergen mixtures and, therefore, difficult to standardize [114, 115]. Consequently, the few studies reporting on fungal allergen-specific immunotherapy have been disappointing [174]. From this point of view, there is less hope that *Malassezia* extracts will be useful to cure AE in *Malassezia*-sensitized patients. However, the availability of a large number of *M. sympodialis* recombinant allergens will allow a patient stratification in terms of sensitization patterns. This may also allow a component-resolved treatment of these patients with mixtures of perfectly standardized single recombinant allergens satisfying the requirements for pharmacological use in the future.

7.2.9 Conclusions

There is a need to improve the diagnostic methods to identify different subgroups of AE patients and to dissect underlying pathogenic mechanisms. Our recombinant *M. sympodialis* allergens will be useful tools in this respect, as well as to determine the future role of tailor-made immunotherapy for well-defined patients with AE sensitized to *Malassezia*. By characterizing different molecular structures of the yeast, we will also gain more information about the yeast itself and specific host–microbe interactions.

Acknowledgments GG wishes to express his appreciation to I.D. Bassukas for inspiring discussions during the preparation of Chap. 7.1, and A. Velegraki for time-honored, unrestricted mentoring. The work of Chap. 7.2 was supported by the Swedish Research Council, the Swedish Asthma and Allergy Association's Research Foundation, and through the regional agreement on medical training and clinical research (ALF) between Stockholm County Council and the Karolinska Institutet. Work at SIAF was supported by the Swiss National Science Foundation Grant No. 31-310000-114634/1 and by the OPO-Foundation, Zürich.

References

1. Hay RJ, Graham-Brown RAC (1997) Dandruff and seborrheic dermatitis: causes and management. Clin Exp Dermatol 22:3–6
2. Holden CA, Berth-Jones J (2004) Eczema, lichenification, prurigo and erythroderma. In: Burns T, Breathnach S, Cox N, et al Rook's textbook of dermatology, 7th edn. Blackwell, London, pp 17.1–17.55
3. Naldi L, Rebora A (2009) Seborrheic dermatitis. N Engl J Med 360:387–396
4. Plewig G, Jansen T (1999) Seborrheic dermatitis. In: Freedberg IM, Eisen AZ, Wolff K, et al (eds) Fitzpatrick's dermatology in general medicine. McGraw-Hill, New York, pp 1482–1489
5. Unna PG (1887) Seborrheic eczema. J Cut Dis 5:449–453
6. Shuster S (1984) The aetiology of dandruff and the mode of action of therapeutic agents. Br J Dermatol 111:235–242
7. Rook A, Wilkinson DS (1979) Eczema, lichen simplex and prurigo. In: Rook A, Wilkinson DS, Ebling FJG (eds) Rook's textbook of dermatology, 3rd edn. Blackwell, London, pp 308–312
8. Burton J, Rook A, Wilkinson DS (1986) Eczema, lichen simplex, erythroderma and prurigo. In: Rook A, Wilkinson DS, Ebling FJG (eds) Rook's textbook of dermatology, 4th edn. Blackwell, pp 375–381
9. Burton J, Holden CA (1999) Eczema, lichenification and prurigo. In: Champion RH, Burton JL, Burns DA et al (eds) Rook/Wilkinson/Ebling textbook of dermatology, 6th edn. Blackwell, pp 638–643

10. Shuster S, Blatchford N (1988) Seborrheic dermatitis and dandruff-a fungal disease. Royal Soc Med Serv 132:1–54
11. Kligman A (1979) Perspectives and problems in cutaneous gerontology. J Invest Dermatol 73:39–46
12. Binder RL, Jonelis FJ (1983) Seborrheic dermatitis in neuroleptic induced Parkinsonism. Arch Dermatol 119:473–475
13. Ahsbee HR (2007) Update on the genus *Malassezia.* Med Mycol 45:287–303
14. Tajima M, Sugita T, Nishikawa A et al (2008) Molecular analysis of *Malassezia* microflora in seborrheic dermatitis patients: comparison with other diseases and healthy subjects. J Invest Dermatol 128:345–351
15. Gaitanis G, Velegraki A, Alexopoulos EC et al (2009) *Malassezia furfur* fingerprints as possible markers for human phylogeography. ISME J 3:498–502
16. Parry ME, Sharpe GR (1998) Seborrheic dermatitis is not caused by an altered immune response to *Malassezia* yeast. Br J Dermatol 139:254–263
17. Faergemann J, Bergbrant IM, Dohsé M et al (2001) Seborrhoeic dermatitis and *Pityrosporum* (*Malassezia*) folliculitis: characterization of inflammatory cells and mediators in the skin by immunohistochemistry. Br J Dermatol 144:549–556
18. Thomas DS, Ingham E, Bojar RA et al (2008) *In vitro* modulation of human keratinocyte pro- and anti-inflammatory cytokine production by the capsule of *Malassezia* species. FEMS Immunol Med Microbiol 54:203–214
19. Mittag H (1995) Fine structural investigation of *Malassezia furfur.*II. The envelope of the yeast cells. Mycoses 38:13–21
20. Tolleson A, Frithz A, Stenlund K (1997) *Malassezia furfur* in infantile seborrheic dermatitis. Ped Dermatol 14:424–425
21. Watanabe S, Kano R, Sato H et al (2001) The effects of *Malassezia* yeasts on cytokine production by human keratinocytes. J Invest Dermatol 116:769–773
22. Ishibashi Y, Sugita T, Nishikawa A (2006) Cytokine secretion profile of human keratinocytes exposed to *Malassezia* yeasts. FEMS Immunol Med Microbiol 48:400–409
23. Yamamoto T, Tsuboi R (2008) Interleukin-2-induced seborrheic dermatitis-like eruption. J Eur Acad Dermatol Venereol 22:235–262
24. Baroni A, Orlando M, Donnarumma G et al (2006) Toll-like receptor 2 (TLR2) mediates intracellular signalling in human keratinocytes in response to *Malassezia furfur*. Arch Dermatol Res 297:280–288
25. Donnarumma G, Paoletti I, Buommino E et al (2004) *Malassezia furfur* induces the expression of beta-defensin-2 in human keratinocytes in a protein kinase C-dependent manner. Arch Dermatol Res 295:474–481
26. Marples RR, Heaton CL, Kligman AM (1973) *Staphylococcus aureus* in psoriasis. Arch Dermatol 107:568–570
27. Ong PY, Ohtake T, Brandt C et al (2002) Endogenous antimicrobial peptides and skin infections in atopic dermatitis. N Engl J Med 347:1151–1160
28. Gaitanis G, Magiatis P, Stathopoulou K et al (2008) AHR ligands, malassezin, and indolo[3, 2-b]carbazole are selectively produced by *Malassezia furfur* strains isolated from seborrheic dermatitis. J Invest Dermatol 128:1620–1625
29. Veldhoen M, Hirota K, Christensen J et al (2009) Natural agonists for aryl hydrocarbon receptor in culture medium are essential for optimal differentiation of Th17 T cells. J Exp Med 206:43–49
30. Veldhoen M, Hirota K, Westendorf AM et al (2008) The aryl hydrocarbon receptor links TH17-cell-mediated autoimmunity to environmental toxins. Nature 453:106–109
31. Jansen PA, Rodijk-Olthuis D, Hollox EJ et al (2009) Beta-defensin-2 protein is a serum biomarker for disease activity in psoriasis and reaches biologically relevant concentrations in lesional skin. PLoS ONE 4:e4725

32. Hauben E, Gregori S, Draghici E et al (2008) Activation of the aryl hydrocarbon receptor promotes allograft-specific tolerance through direct and dendritic cell-mediated effects on regulatory T cells. Blood 112:1214–1222
33. Kerkvliet NI (2002) Recent advances in understanding the mechanism of TCDD immunotoxicity. Int Immunopharmacol 2:277–291
34. Tauchi M, Hida A, Negishi T et al (2005) Constitutive expression of aryl hydrocarbon receptor in keratinocytes causes inflammatory skin lesions. Mol Cell Biol 25:9360–9368
35. Fritsche E, Schäfer C, Calles C et al (2007) Lightening up the UV response by identification of the arylhydrocarbon receptor as a cytoplasmatic target for ultraviolet B radiation. Proc Natl Acad Sci USA 104:8851–8856
36. Tegner E (1983) Seborrhoeic dermatitis of the face induced by PUVA treatment. Acta Derm Venereol 63:335–339
37. Moehrle M, Dennenmoser B, Schlagenhauff B et al (2000) High prevalence of seborrhoeic dermatitis on the face and scalp in mountain guides. Dermatology 201:146–147
38. Burton J, Pye R (1983) Seborrhea is not a feature of seborrheic dermatitis. Br Med J 286:1169–1170
39. Atherton DJ, Gennery AR, Cant AJ (2004) The neonate. In: Burns T, Breathnach S, Cox N, Griffiths C (eds) Rook's textbook of dermatology, 7th edn. Blackwell, Oxford, pp 1422–1429
40. Foley P, Zuo Y, Plunkett A et al (2003) The frequency of commons skin conditions in preschool-aged children in Australia. Arch Dermatol 139:318–322
41. Cotterill J, Cunliffe W, Williamson B et al (1972) Age and sex variation in skin surface lipid composition and sebum excretion rate. Br J Dermatol 87:333–340
42. Gobulev W (2006) Antagonistic interactions among yeasts. In: Rosa C, Peter G (eds) Biodiversity and ecophysiology of yeasts. Springer, Berlin, pp 197–219
43. Youn SW, Na JI, Choi SY et al (2005) Regional and seasonal variations in facial sebum secretions: a proposal for the definition of combination skin type. Skin Res Technol 11:189–195
44. Cafarchia C, Otranto D (2004) Association between phospholipase production by *Malassezia pachydermatis* and skin lesions. J Clin Microbiol 42:4868–4869
45. Cafarchia C, Dell'Aquila ME, Capelli G et al (2007) Role of beta-endorphin on phospholipase production in *Malassezia pachydermatis* in dogs: new insights into the pathogenesis of this yeast. Med Mycol 45:11–15
46. Zouboulis CC, Baron JM, Böhm M et al (2008) Frontiers in sebaceous gland biology and pathology. Exp Dermatol 17:542–551
47. Kumar CP, Kumar SS, Menon T (2006) Phospholipase and proteinase activities of clinical isolates of *Candida* from immunocompromised patients. Mycopathologia 161:213–218
48. Juntachai W, Oura T, Murayama SY et al (2008) The lipolytic enzymes activities of *Malassezia* species. Med Mycol 16:1–8
49. Cafarchia C, Dell'Aquila ME, Traversa D et al (2009) Expression of the micro-opioid receptor on *Malassezia pachydermatis* and its effect on modulating phospholipase production. Med Mycol 19:1–6
50. Xue C, Hsueh YP, Heitman J (2008) Magnificent seven: roles of G protein-coupled receptors in extracellular sensing in fungi. FEMS Microbiol Rev 32:1010–1032
51. Giakoumaki D, Stathopoulou K, Melliou E et al (2008) Identification of indirubin as a metabolite of *Malassezia furfur* strains isolated from diseased skin. Planta Medica 74:PB166
52. Xu J, Saunders CW, Hu P et al (2007) Dandruff-associated *Malassezia* genomes reveal convergent and divergent virulence traits shared with plant and human fungal pathogens. Proc Natl Acad Sci U S A 104:18730–18735
53. Tschachler E, Bergstresser P, Stingl G (1996) HIV-related skin diseases. Lancet 348:659–663
54. Muñoz-Pérez MA, Rodriguez-Pichardo A, Camacho F et al (1998) Dermatological findings correlated with CD4 lymphocyte counts in a prospective 3 year study of 1161 patients with

human immunodeficiency virus disease predominantly acquired through intravenous drug abuse. Br J Dermatol 139:33–39

55. Puig L, Fernandez-Figueras T, Ferrandiz C et al (1995) Epidermal expression of 65 and 72 kD heat shock proteins in psoriasis and AIDS-associated psoriasiform dermatitis. J Am Acad Dermatol 33:985–989
56. Pope M, Betjes MGC, Romani N et al (1994) Conjugates of dendritic cells and memory T-lymphocytes from skin facilitate productive infection with HIV-1. Cell 78:389–398
57. Pechere M, Krischer J, Remondat C et al (1999) *Malassezia* spp carriage in patients with seborrheic dermatitis. J Dermatol 26:558–561
58. Pechere M, Saurat J-H (1997) *Malassezia* yeast density in HIV-positive individuals. Br J Dermatol 136:138–139
59. Wikler JR, Nieboer C, Willemze R (1992) Quantitative skin cultures of *Pityrosporum* yeasts in patients seropositive for the human immunodeficiency virus with and without seborrheic dermatitis. J Am Acad Dermatol 27:37–39
60. Jing W, Ismail R (1999) Mucocutaneous manifestations of HIV infection: a retrospective analysis of 145 cases in a Chinese population in Malaysia. Int J Dermatol 38:457–463
61. Yates VM, Kerr RE, MacKie RM (1983) Early diagnosis of infantile seborrhoeic dermatitis and atopic dermatitis–clinical features. Br J Dermatol 108:633–638
62. Broberg A, Faergemann J (1989) Infantile seborrheic dermatitis and *Pityrosporum ovale*. Br J Dermatol 120:359–362
63. Ruiz-Maldonado R, Lopez-Matinez R, Perez Chavarria E et al (1989) *Pityrosporum ovale* in infantile seborrheic dermatitis. Ped Dermatol 6:16–20
64. Moises-Alfaro CB, Caceres-Rios HW, Rueda M et al (2002) Are infantile seborrheic and atopic dermatitis clinical variants of the same disease? Int J Dermatol 41:349–351
65. Johnson BA, Nunley JR (2000) Treatment of seborrheic dermatitis. Am Fam Physician 61:2703–2710
66. Elewski BE, Abramovits W, Kempers S et al (2007) A novel foam formulation of ketoconazole 2% for the treatment of seborrheic dermatitis on multiple body regions. J Drugs Dermatol 6:1001–1008
67. Elewski B, Ling MR, Phillips TJ (2006) Efficacy and safety of a new once-daily ketoconazole 2% gel in the treatment of seborrheic dermatitis: a phase III trial. J Drugs Dermatol 5:646–650
68. Peter RU, Richarz-Barthauer U (1995) Successful treatment and prophylaxis of scalp seborrhoeic dermatitis and dandruff with 2% ketoconazole shampoo: results of a multicentre, double-blind, placebo-controlled trial. Br J Dermatol 132:441–445
69. Zienicke H, Korting HC, Braun-Falco O et al (1993) Comparative efficacy and safety of bifonazole 1% cream and the corresponding base preparation in the treatment of seborrhoeic dermatitis. Mycoses 36:325–331
70. Segal R, David M, Ingber A et al (1992) Treatment with bifonazole shampoo for seborrhea and seborrheic dermatitis: a randomized, double-blind study. Acta Derm Venereol 72:454–455
71. Shuster S, Meynadier J, Kerl H et al (2005) Treatment and prophylaxis of seborrehic dermatitis of the scalp with antipityrosporal 1% ciclopirox shampoo. Arch Dermatol 141:47–52
72. Chosidow O, Maurette C, Dupuy P (2003) Randomized, open-labeled, non-inferiority study between ciclopiroxolamine 1% cream and ketoconazole 2% foaming gel in mild to moderate facial seborrheic dermatitis. Dermatology 206:233–240
73. Unholzer A, Varigos G, Nicholls D et al (2002) Ciclopiroxolamine cream for treating seborrheic dermatitis: a double-blind parallel group comparison. Infection 30:373–376
74. Faergemann J (1986) Seborrhoeic dermatitis and *Pityrosporum orbiculare*: treatment of seborrhoeic dermatitis of the scalp with miconazole-hydrocortisone (Daktacort), miconazole and hydrocortisone. Br J Dermatol 114:695–700

75. Stratigos JD, Antoniou C, Katsambas A et al (1988) Ketoconazole 2% cream versus hydrocortisone 1% cream in the treatment of seborrheic dermatitis: a double-blind comparative study. J Am Acad Dermatol 19:850–853
76. Danby FW, Maddin WS, Margesson LJ et al (1993) A randomized, double-blind, placebo controlled trial of ketoconazole 2% shampoo versus selenium sulfide 2.5% shampoo in the treatment of moderate to severe dandruff. J Am Acad Dermatol 29:1008–1012
77. Efalith Multicenter Trial Group (1992) A double-blind, placebo-controlled, multicenter trial of lithium succinate ointment in the treatment of seborrheic dermatitis. J Am Acad Dermatol 26:452–457
78. Langtry JA, Rowland Payne CM, Staughton RC et al (1997) Topical lithium succinate ointment (Efalith) in the treatment of AIDS-related seborrheic dermatitis. Clin Exp Dermatol 22:216–219
79. Dreno B, Chosidow O, Revuz J et al (2003) Lithium gluconate 8% vs ketoconazole 2% in the treatment of seborrhoeic dermatitis: a multicentre, randomized study. Br J Dermatol 148:1230–1236
80. Warshaw EM, Wohlhuter RJ, Liu A et al (2007) Results of a randomized, double-blind, vehicle-controlled efficacy trial of pimecrolimus cream 1% for the treatment of moderate to severe facial seborrheic dermatitis. J Am Acad Dermatol 57:257–264
81. Firooz A, Solhpour A, Gorouhi F et al (2006) Pimecrolimus cream, 1%, vs hydrocortisone acetate cream, 1%, in the treatment of facial seborrheic dermatitis: a randomized, investigator-blind, clinical trial. Arch Dermatol 142:1066–1067
82. Rigopoulos D, Ioannides D, Kalogeromitros D et al (2004) Pimecrolimus cream 1% versus. betamethasone 17-valerate 0.1% cream in the treatment of seborrhoeic dermatitis: a randomized open-label clinical trial. Br J Dermatol 151:1071–1075
83. Gambichler T, Breuckmann F, Boms S et al (2005) Narrowband UVB phototherapy in skin conditions beyond psoriasis. J Am Acad Dermatol 52:660–670
84. Pirkhammer D, Seeber A, Hönigsmann H et al (2000) Narrow-band ultraviolet B (ATL-01) phototherapy is an effective and safe treatment option for patients with severe seborrheic dermatitis. Br J Dermatol 143:964–968
85. Cömert A, Bekiroglu N, Gürbüz O et al (2007) Efficacy of oral fluconazole in the treatment of seborrheic dermatitis: a placebo-controlled study. Am J Clin Dermatol 8:235–238
86. Scaparro E, Quardi G, Virno G et al (2001) Evaluation of the efficacy and tolerability of oral terbinafine (Daskil®) in patients with seborrhoeic dermatitis. A multicentre, randomized, investigator-blinded, placebo-controlled trial. Br J Dermatol 144:854–857
87. Vena GA, Micali G, Santoianni P et al (2005) Oral terbinafine in the treatment of multi-site seborrheic dermatitis: a multicenter, double-blind placebo-controlled study. Int J Immunopathol Pharmacol 18:745–753
88. Baysal V, Yildirim M, Ozcanli C et al (2004) Itraconazole in the treatment of seborrheic dermatitis: a new treatment modality. Int J Dermatol 43:63–66
89. Kose O, Erbil H, Gur AR (2005) Oral itraconazole for the treatment of seborrhoeic dermatitis: an open, noncomparative trial. J Eur Acad Dermatol Venereol 19:172–175
90. Shemer A, Kaplan B, Nathansohn N et al (2008) Treatment of moderate to severe facial seborrheic dermatitis with itraconazole: an open non-comparative study. Isr Med Assoc J 10:417–418
91. Piérard GE, Ausma J, Henry F et al (2007) A pilot study on seborrheic dermatitis using pramiconazole as a potent oral anti-*Malassezia* agent. Dermatology 214:162–169
92. Geria AN, Scheinfeld NS (2008) Pramiconazole, a triazole compound for the treatment of fungal infections. IDrugs 11:661–670
93. Bieber T (2008) Mechanisms of disease. Atopic dermatitis. N Engl J Med 358:1483–1494
94. Akdis CA, Akdis M, Bieber T et al (2006) Diagnosis and treatment of atopic dermatitis in children and adults: European Academy of Allergology and Clinical Immunology/American

Academy of Allergy, Asthma and Immunology/ PRACTALL Consensus Report. J Allergy Clin Immunol 118:152–169
95. Altrichter S, Kriehuber E, Moser J et al (2008) Serum IgE autoantibodies target keratinocytes in patients with atopic dermatitis. J Invest Dermatol 128:2232–2239
96. Zeller S, Glaser AG, Vilhelmsson M et al (2008) Immunoglobulin E-mediated reactivity to self antigens: a controversial issue. Int Arch Allergy Immunol 145:87–93
97. Miska AE (2005) How microRNAs control cell division, differentiation and death. Curr Opin Gen Develop 15:563–568
98. Sonkoly E, Wei T, Janson PCJ et al (2007) MicroRNAs: novel regulators involved in the pathogenesis of psoriasis? PloS ONE 11:e610
99. Alfvén T, Braun-Fahrländer C, Brunekreef B et al (2006) Allergic diseases and atopic sensitisation in children related to farming and anthroposophic lifestyle in five European countries – the PARSIFAL study. Allergy 61:414–421
100. Vercelli D (2004) Genetics, epigenetics, and the environment: switching, buffering, releasing. J Allergy Clin Immunol 113:381–386
101. Vuillermin PJ, Ponsonby AL, Saffery R et al (2009) Microbial exposure, interferon gamma gene demethylation in naïve t-cells, and the risk of allergy disease. Allergy 64:343–353
102. Clemmensen OJ, Hjort N (1983) Treatment of dermatitis of the head and neck with ketoconazole in patients with type I hypersensitivity for *Pityrosporum orbiculare*. Semin Dermatol 2:26–29
103. Scheynius A, Johansson C, Buentke E et al (2002) Atopic eczema/dermatitis syndrome and *Malassezia*. Int Arch Allergy Immunol 127:161–169
104. Bäck O, Scheynius A, Johansson SG (1995) Ketoconazole in atopic dermatitis: therapeutic response is correlated with decrease in serum IgE. Arch Dermatol Res 287:448–451
105. Casagrande-Fischer B, Flückiger S, Linder MT et al (2006) Sensitization to the yeast *Malassezia sympodialis* is specific for extrinsic and intrinsic atopic eczema. J Invest Dermatol 126:2414–2421
106. Sandström Falk MH, Tengvall Linder M, Johansson C et al (2005) The prevalence of *Malassezia* yeasts in patients with atopic dermatitis, seborrhoeic dermatitis and healthy controls. Acta Derm Venereol 85:17–23
107. Steinman RM (2007) Dendritic cells: understanding immunogenicity. Eur J Immunol 37(suppl 1):S53–S60
108. Hiller R, Laffer S, Harwanegg C et al (2002) Microarrayed allergen molecules: diagnostic gatekeepers for allergy treatment. FASEB J 16:414–416
109. Schmid-Grendelmeier P, Crameri R (2001) Recombinant allergens for skin testing. Int Arch Allergy Immunol 125:96–111
110. Crameri R, Rhyner C (2007) Impact of native, recombinant, and cross-reactive allergens on humoral and T-cell-mediated immune responses. Immunol Allergy Clin North Am 27:65–68
111. Crameri R (2007) Allergy vaccines: dreams and reality. Exp Rev Vaccines 6:991–999
112. Valenta R, Niederberger V (2007) Recombinant allergens for immunotherapy. J Allergy Clin Immunol 119:826–830
113. Kodzius R, Rhyner C, Konthur Z et al (2003) Rapid identification of allergen-encoding cDNA clones by phage display and high-density arrays. Comb Chem High Throughput Screen 6:147–154
114. Vailes L, Sridhara S, Cromwell O et al (2001) Quantitation of the major fungal allergens, Alt a 1 and Asp f 1, in commercial allergenic products. J Allergy Clin Immunol 107:641–646
115. Crameri R, Weichel M, Flückiger S et al (2006) Fungal allergies: a yet unsolved problem. Chem Immunol Allergy 91:121–133
116. Crameri R, Kodzius R, Konthur Z et al (2001) Tapping allergen repertoires by advanced cloning technologies. Int Arch Allergy Immunol 124:43–47
117. Vilhelmsson M, Zargari A, Crameri R et al (2007) Crystal structure of the major *Malassezia sympodialis* allergen Mala s 1 reveals a beta-propeller fold: a novel fold among allergens. J Mol Biol 369:1079–1086

118. Glaser GA, Limacher A, Flückiger S et al (2006) Analysis of the cross-reactivity and of the 1.5 A crystal structure of the *Malassezia sympodialis* Mala s 6 allergen, a member of the cyclophilin pan-allergen family. Biochem J 396:41–49
119. Limacher A, Glaser AG, Meier C et al (2007) Cross-reactivity and 1.4 A crystal structure of *Malassezia sympodialis* thioredoxin (Mala s 13), a member of a new pan-allergen family. J Immunol 178:389–396
120. Flückiger S, Fijten H, Whitley P et al (2002) Cyclophilins, a new family of cross-reactive allergens. Eur J Immunol 32:10–17
121. Schmidt M, Zargari A, Holt P et al (1997) The complete cDNA sequence and expression of the first major allergenic protein of *Malassezia furfur*, Mal f 1. Eur J Biochem 246:181–185
122. Yasueda H, Hashida-Okado T, Saito A et al (1998) Identification and cloning of two novel allergens from the lipophilic yeast, *Malassezia furfur*. Biochem Biophys Res Comm 248:240–244
123. Onishi Y, Kuroda M, Yasueda H et al (1999) Two-dimensional electrophoresis of *Malassezia* allergens for atopic dermatitis and isolation of Mal f 4 homologs with mitochondrial malate dehydrogenase. Eur J Biochem 261:148–154
124. Moser M, Crameri R, Menz G et al (1992) Cloning and expression of recombinant *Aspergillus fumigatus* allergen I/a (rAsp f I/a) with IgE binding and type I skin test activity. J Immunol 149:454–460
125. De Vouge MW, Thaker AJ, Curran IH et al (1996) Isolation and expression of a cDNA clone encoding an *Alternaria alternata* Alt a 1 subunit. Int Arch Allergy Immunol 111:385–395
126. Bowyer P, Fraczek M, Denning DW (2006) Comparative genomics of fungal allergens and epitopes shows widespread distribution of closely related allergen and epitope orthologues. BMC Genomics 7:251
127. Martinez-Espinoza AD, Garcia-Pedrajas MD, Gold SE (2002) The Ustilaginales as plant pests and model systems. Fungal Genet Biol 35:1–20
128. Schisler DA, Khan NI, Boehm MJ (2002) Biological control of *Fusarium* head blight of wheat and deoxynivalenol levels in grain via use of microbial antagonists. Adv Exp Med Biol 504:53–69
129. Dyer RB, Plattner RD, Kendra DF et al (2005) *Fusarium graminearum* TRI14 is required for high virulence and DON production on wheat but not for DON synthesis in vitro. J Agric Food Chem 53:9281–9287
130. Zargari A, Emilson A, Halldén G et al (1997) Cell surface expression of two major yeast allergens in the *Pityrosporum* genus. Clin Exp Allergy 27:584–592
131. Dawson TL Jr (2007) *Malassezia globosa* and *restricta*: breakthrough understanding of the etiology and treatment of dandruff and seborrheic dermatitis through whole-genome analysis. J Investig Dermatol Symp Proc 12:15–19
132. Hemmann S, Blaser K, Crameri R (1997) Allergens of *Aspergillus fumigatus* and *Candida boidinii* share IgE-binding epitopes. Am J Respir Crit Care Med 156:1956–1962
133. Weichel M, Flückiger S, Crameri R (2002) Molecular characterization of mould allergens involved in respiratory complications. Recent Res Devel Resp Critical Care Med 2:29–45
134. Posch A, Chen Z, Dunn MJ et al (1997) Latex allergen database. Electrophoresis 18:2803–2810
135. Lindborg M, Magnusson CG, Zargari A et al (1999) Selective cloning of allergens from the skin colonizing yeast *Malassezia furfur* by phage surface display technology. J Invest Dermatol 113:156–161
136. Rasool O, Zargari A, Almqvist J et al (2000) Cloning, characterization and expression of complete coding sequences of three IgE binding *Malassezia furfur* allergens, Mala f 7, Mala f 8 and Mala f 9. Eur J Biochem 267:4355–4361
137. Wolfe KH (2006) Comparative genomics and genome evolution in yeasts. Philos Trans R Soc Lond B Biol Sci 361:403–412

138. Andersson A, Rasool O, Schmidt M et al (2004) Cloning, expression and characterization of two new IgE-binding proteins from the yeast *Malassezia sympodialis* with sequence similarities to heat shock proteins and manganese superoxide dismutase. Eur J Biochem 271: 1885–1894
139. Zargari A, Selander C, Rasool O et al (2007) Mala s 12 is a major allergen in patients with atopic eczema and has sequence similarities to the GMC oxidoreductase family. Allergy 62:695–703
140. Shen DH, Han SH (1998) Characterization of allergens of *Penicillium* and *Aspergillus* species. J Microbiol Immunol Infect 32:141–145
141. Johansson C, Eshaghi H, Tengvall Linder M et al (2002) Positive atopy patch test reaction to *Malassezia furfur* in atopic dermatitis patients correlates with a T h elper 2-like peripheral blood mononuclear cells response. J Invest Dermatol 118:1044–1051
142. Schmid-Grendelmeier P, Scheynius A, Crameri R (2006) The role of sensitization to *Malassezia sympodialis* in atopic eczema. Chem Immunol Allergy 91:98–109
143. Zargari A, Eshaghi H, Bäck O et al (2001) Serum IgE reactivity to *Malassezia furfur* extract and recombinant *M. furfur* allergens in patients with atopic dermatitis. Acta Derm Venereol 81:418–422
144. Selander C, Zargari A, Möllby R et al (2006) Higher pH level, corresponding to that on the skin of patients with atopic eczema, stimulates the release of *Malassezia sympodialis* allergens. Allergy 61:1002–1008
145. Ganz T (2003) Defensins: antimicrobial peptides of innate immunity. Nat Rev Immunol 3:710–720
146. Ballardini N, Johansson C, Lilja G et al (2009) Enhanced expression of the antimicrobial peptide LL-37 in lesional skin of adults with atopic eczema. Br J Dermatol Mar 20:40–47
147. Novak N, Bieber T, Kraft S (2004) Immunoglobulin E-bearing antigen presenting cells in atopic dermatitis. Curr Allergy Asthma Rep 4:263–269
148. Buentke E, Zargari A, Heffler LC et al (2000) Uptake of the yeast *Malassezia furfur* and its allergenic components by immature $CD1a^+$ dendritic cells. Clin Exp Allergy 30:1759–1770
149. Buentke E, Heffler LC, Wallin RP et al (2001) The allergenic yeast *Malassezia furfur* induces maturation of human dendritic cells. Clin Exp Allergy 31:1583–1593
150. Vilhelmsson M, Johansson C, Jacobsson-Ekman G et al (2007) The *Malassezia sympodialis* allergen Mala s 11 induces human dendritic cell maturation in contrast to its human homologue manganese superoxide dismutase. Int Arch Allergy Immunol 143:155–162
151. Gabrielsson S, Buentke E, Liedén A et al (2004) *Malassezia sympodialis* stimulation differently affects gene expression in dendritic cells from atopic dermatitis patients and healthy individuals. Acta Derm Venereol 84:339–345
152. Agerberth B, Buentke B, Bergman P et al (2006) *Malassezia sympodialis* differently affects the expression of LL-37 in dendritic cells from atopic eczema patients and healthy individuals. Allergy 61:422–430
153. Tengvall Linder M, Johansson C, Zargari A et al (1996) Detection of *Pityrosporum orbiculare* reactive T cells from skin and blood in atopic dermatitis and characterization of their cytokine profiles. Clin Exp Allergy 26:1286–1297
154. Tengvall Linder M, Johansson C, Bengtsson A et al (1998) *Pityrosporum orbiculare*-reactive T-cell lines in atopic dermatitis patients and healthy individuals. Scand J Immunol 47:152–158
155. Walzer T, Dalod M, Robbins SH et al (2005) Natural-killer cells and dendritic cells: "l'union fait la force". Blood 106:2252–2258
156. Buentke E, Heffler LC, Wilson JL et al (2002) Natural killer and dendritic cell contact in lesional atopic dermatitis skin *Malassezia*-influenced cell interaction. J Invest Dermatol 119:850–857
157. Buentke E, D'Amoto M, Scheynius A (2004) *Malassezia* enhances natural killer cell-induced dendritic cell maturation. Scand J Immunol 59:511–516

158. Olynych TJ, Jakeman DL, Marshall JS (2006) Fungal zymosan induces leukotriene production by human mast cells through a dectin-1-dependent mechanism. J Allergy Clin Immunol 118:837–843
159. Henz BM, Maurer M, Lippert U et al (2001) Mast cells as initiators of immunity and host defense. Exp Dermatol 10:1–10
160. Selander C, Engblom C, Nilsson G et al (2009) TLR2/MyD88-depedent and –independent activation of mast cell IgE responses by the skin commensal yeast *Malassezia sympodialis*. J Immunol 182:4208–4216
161. Schmid-Grendelmeier P, Flückiger S, Disch R et al (2005) IgE-mediated and T-cell-mediated autoimmunity against manganese superoxide dismutase in atopic dermatitis. J Allergy Clin Immunol 115:1068–1075
162. Crameri R, Faith A, Hemmann S et al (1996) Humoral and cell-mediated autoimmunity in allergy to *Aspergillus fumigatus*. J Exp Med 184:265–270
163. Mayer C, Appenzeller U, Seelbach H et al (1999) Humoral and cell-mediated autoimmune reaction to human acidic ribosomal P_2 protein in individuals sensitized to *Aspergillus fumigatus* P_2 protein. J Exp Med 189:1507–1512
164. Stadler JF, Taieb A, Atherton DJ (1993) Severity scoring of atopic dermatitis: the SCORAD index. Consensus report of the European task force on atopic dermatitis. Dermatology 186:23–31
165. Andersson A, Scheynius A, Rasool O (2003) Detection of Mala f and Mala s allergen sequences within the genus *Malassezia*. Med Mycol 41:479–485
166. Zargari A, Midgley G, Bäck O et al (2003) IgE-reactivity to seven *Malassezia* species. Allergy 58:306–311
167. Zargari A, Doekes G, van Ieperen-van Dijk AG et al (1995) Influence of culture period on the allergenic composition of *Pityrosporum orbiculare* extracts. Clin Exp Allergy 25: 1235–1245
168. Johansson C, Sandström MH, Bartosik J et al (2003) Atopy patch test reactions to *Malassezia* allergens differentiate subgroups of atopic dermatitis patients. Br J Dermatol 148:479–488
169. Tengvall Linder M, Johansson C, Scheynius A et al (2000) Positive atopy patch test reactions to *Pityrosporum orbiculare* in atopic dermatitis patients. Clin Exp Allergy 30:122–131
170. Johansson C, Ahlborg N, Andersson A et al (2009) Elevated peripheral allergen-specific T cell response is crucial for a positive atopy patch test reaction. Int Arch Allergy Immunol 150:51–58
171. Elias PM, Steinhoff M (2008) "Outside-to-inside" (and now back to "outside") pathogenic mechanisms in atopic dermatitis. J Invest Dermatol 128:1067–1070
172. Müller UR (2003) Recent developments and future strategies for immunotherapy of insect venom allergy. Curr Opin Allergy Clin Immunol 3:299–303
173. Wallner M, Briza P, Thalhamer J et al (2007) Specific immunotherapy in pollen allergy. Curr Opin Mol Ther 9:160–167
174. Helbling A, Reimes A (2003) Immunotherapy in fungal allergy. Curr Allergy Asthma Rep 3:447–453

Malassezia Fungemia, Antifungal Susceptibility Testing and Epidemiology of Nosocomial Infections

8

Athanasios Tragiannidis, Andreas Groll, Aristea Velegraki and Teun Boekhout

Core Messages

> *Malassezia* catheter-related fungemia, and invasive infections in critically ill, premature neonates and immunocompromised individuals of all ages are issues addressed in relation to risk factors, host–pathogen interactions, pathogenesis, management, and outcome. The epidemiology of drug resistance of isolates from deep tissues, biological fluids, and skin is highlighted by assessing *Malassezia* susceptibility to conventional antifungals, synthetic, and naturally occurring compounds. Yet, lack of standardization for *Malassezia* susceptibility testing restricts associations of *in vitro* with *in vivo* responses to antifungals. The chapter also describes the capacity of PCR fingerprinting, pulsed field gel electrophoresis (PFGE), and amplified fragment length polymorphism (AFLP) analysis to dependably identify nosocomial outbreaks, but concludes that robust analytical methods such as multilocus sequence typing (MLST) would be required to resolve global *Malassezia* epidemiological issues.

8.1 *Malassezia* Fungemia and Invasive Infections

Athanasios Tragiannidis and Andreas Groll

Malassezia species are rare causes of invasive infections in critically ill low-birth-weight, infants and in immunocompromised children and adults. The clinical spectrum ranges from asymptomatic infection to life-threatening sepsis and disseminated disease, with

A. Velegraki (✉)
Mycology Laboratory, Medical School, National and Kapodistrian University of Athens, Athens 115 27, Greece
e-mail: aveleg@med.uoa.gr

T. Boekhout et al. (eds.), *Malassezia and the Skin*
DOI: 10.1007/978-3-642-03616-3_8, © Springer Verlag Berlin Heidelberg 2010

intravascular catheters and administration of lipid supplemented parenteral nutrition acting as the main risk factors [1–5]. Here, we review the epidemiology, risk factors, pathogenesis, clinical manifestations, diagnosis, treatment, and outcome, of invasive *Malassezia* infections in humans.

8.1.1
Epidemiology and Risk Factors

Malassezia species are well recognized as a cause of skin disorders in otherwise healthy individuals. However, information on *Malassezia* fungemia and invasive disease is limited. The overwhelming majority of invasive infections reported in the literature have been associated with *M. furfur sensu lato* and *M. pachydermatis*. *M. furfur*, which is an obligatory lipophilic yeast and a commensal in humans, has been described predominantly in conjunction with nosocomial outbreaks in neonatal intensive care units (NICU) and sporadically in severely immunocompromised patients. *M. pachydermatis*, in contrast, is a zoophilic yeast mainly associated with otitis externa and seborrhoic dermatitis in dogs. *M. pachydermatis* is only occasionally isolated from human skin, but has been implicated in nosocomial infections in hospitalized severely ill neonates [2, 3].

The first case of *Malassezia* sp. as an etiologic agent of bloodstream infection and sepsis was reported in 1981 by Redline et al. [6]; they reported a case of *Malassezia* pulmonary vasculitis in an infant receiving total parenteral nutrition via an indwelling central venous catheter (CVC). Until 1987, only few further reports of *Malassezia* fungemia in infants and adults emerged [7–11]. During the past two decades, however, there have been numerous reports on invasive *Malassezia* infections outbreaks in NICUs, particularly in neonates and infants receiving intravenous lipids [2, 12–14]. Cases have also been described in immunocompromised children and adults with CVCs, and more rarely, preceding abdominal surgery and other significant underlying conditions [15–19].

There are limited comprehensive data on the frequency of invasive *Malassezia* infections in immunocompromised patients that provide information on the overall clinical relevance of this opportunistic infection. However, studies investigating the colonization of central venous lines specifically by *Malassezia* spp. have demonstrated colonization rates of 2.4–32% in critically ill neonates [8, 20, 21] and 0.7% in unselected hospitalized adults [22]. Generally, *Malassezia* infections in bone marrow transplant (BMT) patients appear to be uncommon; as of 3,044 BMT patients, only six (0.2%) developed *Malassezia* infections, two of whom were with an involvement of the blood stream [15]. In a study of critically ill neonates, eight of 25 consecutively explanted CVCs grew *M. furfur*, and in one of the infants (4%) was found evidence of systemic infection [8]. While only routine blood cultures were utilized in the transplant patients, the study in neonates used media supplemented with olive oil, emphasizing the importance of methodological aspects in culture-based systematic epidemiological investigations.

Malassezia spp. may be isolated from the skin of 3% of healthy newborn infants, but 30–64% of hospitalized premature infants become colonized by the second week of life [5, 8, 14]. Bell et al. reported isolation of *Malassezia* yeasts identified then as *M. furfur* in 41% of critically ill newborns in the NICU, while less than 10% of hospitalized newborns in a nonintensive care setting were colonized [23]. Likewise, Aschner et al. [8] reported that 28% of infants in an NICU were colonized in the first week of life, whereas 84% of older infants in the NICU were skin culture positive for *M. furfur*. These and other data [24, 25] indicate that colonization in neonates and infants is associated with low gestational age, admission to the NICU, and length of hospitalization [26].

Although no systematic data exist on risk factors for invasive *Malassezia* infections in immunocompromised patients beyond the neonatal age, colonization and the presence of a central line appear to be obligatory prerequisites for fungemia [1, 3, 15]. The risk factors for invasive *Malassezia* infections in adults, neonates, and infants that are summarized in Table 8.1 indicate that while invasive infections may occur sporadically, in the last decade, nosocomial outbreaks regarding neonatal *M. furfur* and *M. pachydermatis* infections have been widely reported. As shown by molecular typing methods, infants become colonized by skin contact with parents or healthcare workers, which may further transmit the organism from an infected or colonized infant to others via their hands [2, 12, 14]. In that case,

Table 8.1 Summary of predisposing factors reported in the scientific literature, and putative clinical and laboratory criteria for the diagnosis of *Malassezia* systemic infections

	Predisposing factors	Certainty of diagnosis	
		Low	High
Adults and children	Crohn's disease and various other chronic illnesses[a], CAPD, intensive care and high APACHE score, immunosuppresion, CVC, TPN with LS, long term administration of antibiotics, invasive surgical procedures	Persistent fever unresponsive to antibiotics; eucocytosis/ leucocytopenia, thrombocytopenia; identification of *Malassezia* point source nosocomial epidemic; negative microscopy and culture	Positive microscopy for yeasts with *Malassezia* morphology in Gram/ Giemsa-stained smear of buffy coat or blood drawn through the CVC or involved tissue and/or positive culture
Infants	Comorbidities of prematurity, admission to neonatal intensive care unit, use of broad-spectrum antibacterial treatment, prolonged TPN with LS, CVC, use of soybean oil-containing creams	Persistent fever unresponsive to antibiotics; leucocytosis/ leucocytopenia; thrombocytopenia; cluster of invasive *Malassezia* cases in neonatal ICU; negative microscopy and culture	Positive microscopy for yeasts with *Malassezia* morphology of buffy coat Gram/Giemsa-stained smear of blood drawn through the CVC and/or positive culture

[a]Require case controlled studies for further evaluation

CAPD chronic ambulatory peritoneal dialysis; *APACHE* acute physiology and chronic health evaluation; *CVC* central venous catheter; *TPN* total parenteral nutrition; *LS* lipid supplementation; *ICU* intensive care unit patient-to-patient transmission may be effectively interrupted by improving hand-washing practices [12]. Other outbreak investigations have shown that *Malassezia* yeasts can also persist for prolonged time on incubator surfaces (Chap. 8.3), providing an additional source for continued transmission [27].

8.1.2 Host–Pathogen Interactions

Little is known about virulence factors and host immune responses in invasive *Malassezia* infections. *Malassezia* yeasts are able to exist in both yeast and mycelial forms; can grow under microaerophilic and anaerobic conditions; and can adhere and form biofilms on the surfaces of different materials [9, 28, 29]. They have an exceptionally thick cell wall in comparison with other yeasts with an additional lipid layer on the outside, which appears to be important for the organism's ability to suppress cytokine release, and down-regulate phagocytic uptake and killing. They also produce a range of enzymes and metabolites, including azelaic acid, which has been shown to decrease the production of reactive oxygen species (ROS) in neutrophils [28]. While these factors are in support of the general ability of the organism to cause invasive disease, their biological relevance *in vivo* remains to be elucidated.

At present, it is unclear which components of the immune system are most important in the host's defense against invasive infections. Studies examining cellular and humoral immune responses specific to *Malassezia* species in patients with superficial *Malassezia*-associated diseases and healthy controls have generally been unable to define significant differences in their immune response. *Malassezia* may stimulate the reticuloendothelial system and activate the complement cascade, but may also suppress cytokine release and down-regulate phagocytic uptake and killing. It appears that the lipid-rich external layer of the organism is pivotal in this alteration of phenotype. Thus, elucidating the nonspecific immune response to *Malassezia* species may be key to a better understanding of how these organisms live as commensals and yet can rarely cause invasive diseases [28].

8.1.3 Pathogenesis of Invasive Infections

Likely due to the sporadic nature of invasive infections, no clinical studies have addressed the immunological predisposition and responses to *Malassezia* in critically ill neonates or in immunocompromised children and adults [28]. Most patients had a serious underlying disease affecting the overlapping components of host responses; had an indwelling vascular access; and had received parenteral nutrition containing lipid emulsions (Table 8.1). As evidenced by outbreak investigations, the cutaneous commensal flora of patients or health care workers is the usual source of infection [2, 12, 14]. Apart from contamination during insertion, or following administration of a contaminated parenteral solution, catheters may become infected by migration of organism from the exit site along the outer catheter wall or from the hub through the lumen of the catheter. Adherence of the organism to the catheter lumen with consequent biofilm production results in local replication and shedding of the organism in the blood [9, 26, 28–31]. Microbial and host factors may play a role in localizing the organisms to the catheter or in progression to fungemia and clinical sepsis [18, 20]. However, even if host defenses are able to clear the organism from the blood, the infection may not be resolved until the catheter is removed.

Similar to catheter-related candidemia [32], catheter-related *Malassezia* fungemia has been associated with administration of parenteral lipid emulsions. While the exact mechanisms of this association are unclear, it is conceivable that lipids administered through the catheter may provide a growth advantage for *Malassezia* [12, 14, 33]. On the other hand, parenterally administered lipids may negatively affect host immunity by blocking the reticuloendothelial system, reducing the generation of reactive oxygen species (ROS), which are shown to decrease phagocytosis by neutrophils *in vitro*, thus contributing to clinical disease [28].

8.1.4
Clinical Manifestations

The clinical signs and symptoms of *Malassezia* fungemia and sepsis are generally non specific. Depending on the severity of the infection, the most commonly reported symptoms in critically ill, premature infants have been fever and respiratory distress; other, less frequent symptoms include lethargy, bradycardia, hepatomegaly, splenomegaly, seizures, and cyanosis [14, 31, 34]. Respiratory distress may result from pneumonia or bronchopneumonia, with an interstitial appearance on radiography. The main laboratory findings in this setting are leukocytosis or leukopenia, and thrombocytopenia. Affected patients usually are premature, low birth weight infants with multiple comorbidities, extended hospitalization, CVCs, and parenteral nutrition including lipid emulsions [2, 7, 10, 12, 35, 36].

Catheter-associated *Malassezia* fungemia is sporadic in immunocompromised children and in adults, and consequently, clinical manifestations are not as well described as in infants. Fever appears to be universal [31]; other symptoms and findings may include chills and rigors, myalgia, nausea and vomiting, respiratory distress with or without apnea, pneumonia, leukopenia, thrombcytosis, and less frequently, leukocytosis [1, 3, 15, 30]; signs of exit site inflammation are uncommon [1]. Similar to the neonatal setting, the most common patient profile includes prolonged hospitalization, the presence of CVCs, and the use of intravenous fat emulsions [1, 3, 15, 17]. Patients reported in literature presented with a variety of underlying diseases such as bowel surgery, Crohn's disease, hemorrhagic pancreatitis, cancer and bone marrow transplantation [1, 15, 16, 18, 37, 38]. However, individual cases without the typical risk factors have been reported [39, 40].

Catheter-associated *Malassezia* fungemia may result in embolic-metastatic infection of the heart and the lungs, and less frequently, dissemination to other organs such as the skin, the kidneys, the pancreas, the liver, the spleen and the brain [26, 34, 41]. Histopathologic changes include mycotic thrombi around the tips of catheters, vegetations on the endocardium, septic inflammatory lesions in the heart and the lungs [6, 34, 42]. Reported invasive *Malassezia* infections other than fungemia include individual cases of *Malassezia* mastitis, thrombophlebitis, sinusitis, malignant otitis externa, meningitis, septic arthritis, soft tissue abscesses, and catheter-associated peritonitis, in continuous ambulatory peritoneal dialysis patients [43–46].

8.1.5 Laboratory Diagnosis

As *Malassezia* yeasts represent an uncommon cause of fungemia and sepsis, a high index of suspicion is needed to diagnose the infection. However, while *Malassezia* fungemia has been increasingly recognized over the past two decades, its frequency may in fact be higher as the current clinical data suggest. Detection is complicated by the organism's lipid-dependent nature as most routinely used media do not support its growth [33, 43]. Use of lipid supplemented media may be warranted in certain specimens, especially if cultures appear sterile on routine media, and yeasts have been observed on microscopy. The patient group where this may be most appropriate are critically ill premature neonates receiving parenteral lipid emulsions through central venous lines. Supplementation of blood culture bottles with palmitic acid has been shown to improve recovery of *Malassezia* in this patient group [43].

Malassezia species can be detected in blood and other specimens by direct microscopic examination, by culture, and by molecular methods [33]. Examining Giemsa- or Gram-stained smears of blood or buffy coat of blood specimens (Table 8.4) obtained through the catheter is helpful and may provide the clue to culture the specimen on lipid-enriched fungal medium that supports growth of *Malazzesia* yeasts [33, 43, 47] (Chap. 2.1). Sabouraud agar supplemented with olive oil has been frequently used for this purpose. However, because only *M. furfur* and *M. pachydermatis* grow easily on this medium, other species such as *M. sympodialis* may be missed, as suspected in two cases of heart infection in Martinique with positive smear examination but negative cultures (Fig. 8.1) (N. Desbois, unpublished data). Furthermore, the time it takes to culture *Malassezia* (5 days and longer, dependent on the species) hinders rapid diagnosis, thus molecular methods can be an alternative diagnostic approach [48, 49]. A proposed PCR assay [49] used to detect blood culture-proven *M. furfur* fungemia in only four patients was of low sensitivity (25%). Therefore, improvement in sensitivity, standardization of the PCR assay, and thorough evaluation are issues that have to be resolved before employing it in the clinical setting.

8.1.6 Clinical Management

Because invasive *Malassezia* infections are rare and larger patient series are lacking, evidence-based treatment recommendations cannot be made. However, based on the association of the disease with CVCs and the ability of the organism to produce biofilms [9, 29], there is general consensus that the management of *Malassezia*-related fungemia and sepsis requires the prompt removal of any indwelling catheter, in addition to intravenous antifungal therapy, and less clear, the temporary discontinuation of parenteral nutritional lipid solutions

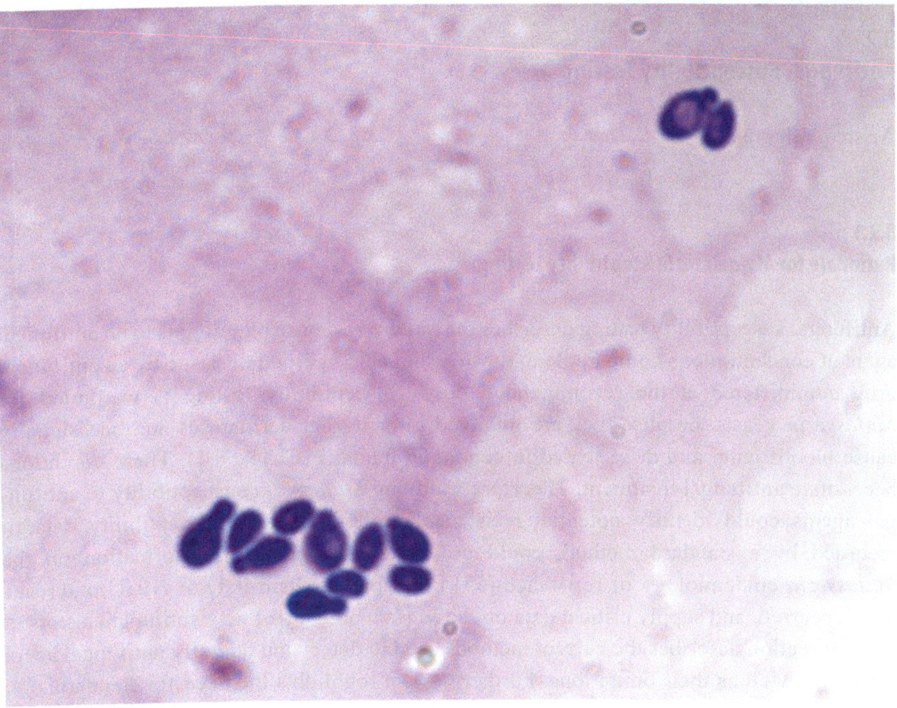

Fig. 8.1 Direct microscopy of a mitral valve apposition, showing typical ovoid *Malassezia* yeasts May-Grunwald-Giemsa stained ×1,000 (courtesy of Nicole Desbois, Fort de France, Martinique)

[33]. *In vitro* studies of superficial and invasive clinical *Malassezia* isolates consistently demonstrate susceptibility to amphotericin B and the triazoles, whereas flucytosine and the echinocandins appear to be inactive [30, 43, 50–53]. Thus, in the absence of experimental and comparative clinical data and based on the clinical experience with invasive *Candida* infections, fluconazole or voriconazole may be rational first-line options for antifungal chemotherapy with an amphotericin B product as back-up for refractory or life-threatening infections. While the duration of treatment has not been defined, we would advocate a course of 14 days of effective antifungal therapy after the last positive blood culture and catheter removal as recommended for invasive *Candida* infections [31, 32], and optional switch from initial intravenous to oral therapy depending on the individual patient's clinical response.

8.1.7 Outcome

Very little is known about the detailed morbidity and mortality of invasive *Malassezia* infections. While clearly, *Malassezia* can cause severe disease [38] and fatal cases have been reported in untreated patients [34, 54], available series of catheter-associated fungemia in premature neonates [2, 12] and in immunocompromised non-neonatal patients [1, 15] suggest low attributable mortality with appropriate management.

8.2 Antifungal Susceptibility Testing

Aristea Velegraki

8.2.1 Rationale for *Malassezia* Susceptibility Testing

Antifungal susceptibility testing describes the *in vitro* response of a fungus to an antifungal agent at concentrations comparable to those achievable in blood or tissue by an antifungal drug administered at the recommended dose. Susceptibility testing is warranted for *Malassezia* yeasts, because they are involved in or trigger dermatoses and occasionally cause bloodstream and deep seated infections in humans (Chapt. 8.1). These conditions necessitate antifungal treatment. Therefore, studying *Malassezia* susceptibility to antifungal agents could identify potential resistance profiles. Overall, susceptibility data, if recorded by a standard method, could provide clinically useful information on the *Malassezia* epidemiology of resistance [55] in hospitals where ICU or NICU outbreaks have occurred, and supply critical data on the *in vitro* potency of new antifungal agents.

This section describes the current methods used to detect and quantify antifungal agent activity as well as their limitations. Furthermore, it highlights the need for standardizing susceptibility testing for *Malassezia* species. Finally, it outlines conventional and novel methods for *in vitro* investigations, bioassay testing of new antifungal agents, as well as testing new formulations of established antifungals against *Malassezia* yeasts.

8.2.2 Standard Methods and Selection of *Malassezia* Colonies for Susceptibility Testing from Mixed Blood Cultures and Skin Specimens

Standardized microdilution susceptibility testing for yeasts is a recent, yet a rapidly evolving, *modus operandi*. In addition to the current Clinical and Laboratory Standards Institute (CLSI) susceptibility testing standard, [56] the Subcommittee on Antifungal Susceptibility Testing (AFST) of the European Committee for Antimicrobial Susceptibility Testing (EUCAST) has also recently proposed a standard testing scheme for yeasts [57]. Nevertheless, these methods cannot be directly applied on *Malassezia*, as media and test conditions are still not standardized and appropriately evaluated. This is due to (a) the low rating of these yeasts as systemic pathogens, which impaired intense and methodical interlaboratory assessment of test parameters; and (b) the lipophilic nature of the majority of *Malassezia* species that has encumbered assessment of suitable growth substrates for reliable and reproducible susceptibility testing.

In addition to standardization of media and test parameters, critical in obtaining consistent susceptibility results for any tested pathogen, is selection of single species colonies grown in cultures of clinical specimens. This is of key significance in testing *Malassezia*, as mixed species may be isolated from pathological specimens together with bacteria and *Candida* species that inhabit normal and diseased, human and animal skin. Besides mixed cultures that are obtained from skin specimens, mixed *Malassezia* sp. with *Candida* sp. or bacteria can also grow in blood cultures of high-risk (Sect. 8.1) pediatric and adult patients. In these cases, *Malassezia* colonies are frequently missed, as growth, even of *M. furfur,* may be very weak and colonies rare due to the lack of supplements required for growth of lipophilic *Malassezia* species. *Malassezia* colonies may be also missed due to the presence of antagonistic *Candida* yeasts or bacteria, the growth of which is better supported by standard blood culture media. In suspected cases of a yeast infection in high-risk patients, the choice of specialized mycological vials is critical for potential *Malassezia* co-isolation. In our experience, the BACTEC™ Myco/F lytic culture vials, with brain heart infusion broth, glycerol 0.10% w/v and Tween 80 0.0025% w/v, provide better growth support for *M. furfur* and *M. sympodialis* (A. Velegraki, unpublished observation). Therefore, differentiation of *Malassezia* cells from the prolific cells of *Candida* sp., which may co-isolate, is facilitated in the Gram-stained smears that are prepared directly from the vials. Consequently, this guides the laboratory to include modified Dixon's or Leeming-Notman agar in the subsequent subcultures to promote *Malassezia* growth.

Generally, in skin specimen cultures, it is imperative to confirm that there are single isolates of each *Malassezia* species before susceptibility testing. However, complete conventional identification (Chap. 2.1), comprising physiological and biochemical tests, as well as molecular identification are lengthy procedures (see Chap. 3.1). It would therefore be clinically realistic to pursue susceptibility testing after initial differentiation of each isolate from a mixed culture by microscopy, and to ascertain that cultures are free of bacteria by the rapid Bacto urea test as described in Chap. 2.1. Then, following subculture in *Malassezia* selective media, pure cultures of each isolate can be obtained, and susceptibility testing using single colonies can be performed in parallel with comprehensive molecular identification procedures.

8.2.3
In vitro Response of *Malassezia* Yeasts to Antifungal Agents

Hitherto, the susceptibility testing methodologies employed for the majority of *Malassezia* species have varied significantly, ranging from agar diffusion methods to modified CLSI broth microdilution methods using fatty acid RPMI (faRPMI) broth, modified Christensen's urea broth (mCUB) and modified Leeming-Notman (mLN) broth (Table 8.2). Yet, there are still no published trials on the *in vitro* response to established and newer antifungals of the recently recognized species *M. nana, M. caprae* and *M. equina.* Also, limited data exist on *Malassezia* species susceptibility against the expanding array of topical imidazole derivatives [58–61] and the oral triazole derivatives [50, 62, 63]. Even fewer reports cover the *in vitro* performance of broad-spectrum hydroxypyridone-derived synthetic topical antifungals [64, 65], the natural sesquiterpenoids [66], and essential oils, such as tea tree oil [67].

Despite scarcity of susceptibility data extracted thus far from nonstandardized microdilution and agar diffusion methods, there is an emerging pattern indicating possible species-dependent azole and allylamine (terbinafine) responses (Table 8.3). Even earlier [51],

Table 8.2 Antifungal drugs tested against *Malassezia* yeasts by agar dilution and broth microdilution methods and by corneofungimetry bioassays

Description of methods and media	*Malassezia* species studied	Drugs tested MIC, MFC, time-kill [66, 87] and synergy assays [88]	References
Solid-phase susceptibility *in vitro* assays			
Agar dilution method in mDA, mLNA	*M. furfur, M. sympodialis, M. slooffiae, M. restricta, M. globosa, M. obtusa, M. dermatis, M. japonica, M. pachydermatis*	Amphotericin B, fluconazole, itraconazole, pramiconazole (research code: R126638), ketoconazole, luliconazole (research code: NN-502), terbinafine, rilopirox, Tea tree oil	[5, 16–18, 21, 34, 51, 59, 64, 71, 80–82, 87–90]
Liquid-phase susceptibility *in vitro* assays			
CLSI M27 A; M27 A2; M27 A3 [56] broth microdilution in faRPMI 1640, mCUB [68, 69, 91] and mLN, [52, 77] mCUB supplemented with phenol red [82] and LNB supplemented with Alamar blue [50]	*M. furfur, M. slooffiae, M. sympodialis, M. globosa, M. dermatis, M. obtusa, M. restricta, M. yamatoensis, M. japonica, M. pachydermatis*	Amphotericin B, fluconazole, itraconazole, terbinafine, voriconazole, posaconazole, albaconazole, bifonazole, miconazole, econazole, ketoconazole, luliconazole (research code: NN-502), ciclopirox olamine, xanthorrhizole, Tea tree oil	[50, 52, 61, 66, 67, 69, 70, 72, 76–80, 87, 88, 91, 99]
Bioassays			
Corneofungimetry in cyanoacrylate skin surface strippings (CSSS) [92]	*M. furfur, M. globosa, M. restricta*	2% Ketoconazole gel and ream with or without topical corticosteroid (desonide) adjunct, 1% terbinafine, pramiconazole (research code: R126638)	[84, 86, 92]

MIC minimum inhibitory concentration; *MFC* minimum fungicidal concentration; *mDA* modified Dixon's agar; *mLNA* modified Leeming-Notman; *faRPMI* 1640 fatty acid RPMI 1640; *mCUB* modified Christensen's urea broth; *LNB* Leeming-Notman broth

in vitro data generated from agar dilution methods have identified terbinafine tolerance among strains of *M. furfur, M. globosa* and *M. obtusa,* and good *M. sympodialis* terbinafine response. Similar *in vitro* species-dependent responses have been noted with the polyene, amphotericin B [63]. Specifically, the mCUB (Table 8.2) microdilution method [68, 69] identified strains of *M. globosa* and *M. restricta* with poor response to voriconazole, itraconazole and ketoconazole (MIC ≥ 8 μg/mL). The faRPMI microdilution method [63] recognized strains of *M. globosa*, *M. restricta* and *M. furfur* with poor response to fluconazole (MIC ≥ 64 μg/mL), terbinafine (MIC ≥ 1 μg/mL) and amphotericin B (MIC ≥ 8 μg/mL), whereas increased fluconazole MIC (16 μg/mL) was noted for *M. pachydermatis* and increased terbinafine MIC (4 μg/mL) for *M. dermatis*. Interestingly, *M. globosa* strains have also been reported with MICs of itraconazole (16 μg/mL), voriconazole (≥16 μg/mL) and fluconazole (≥64 μg/mL) when tested by microdilution in mLN broth [52]. These high MIC values are indicative of tolerance to these drugs.

It is important to emphasize that the broth microdilution azole and terbinafine MIC data, compiled in Table 8.3, represent values reported by independent research groups using different test media, inocula ranging from 2.5–4 × 10^3 [52, 63, 70] to 10^6 [68, 69] and incubation times varying from 72–96 h [63, 70] to 168 h, [52, 68] depending on the growth rate of each *Malassezia* species. Furthermore, incubation times also depended on the ability of the test medium used in each protocol to support optimum growth for each *Malassezia* species. Likewise, lower inoculum size, in conjunction with using RPMI 1640 as test medium, could explain the lower MIC values compared to those recorded for *Malassezia* species tested in the mCUB substrate. Variable intraspecies *in vitro* responses of *Malassezia* strains to antifungals is indeed to be expected; nonetheless, definitive conclusions on potential species-specific response to antifungals can be reached once test media and conditions are evaluated using a panel of reference strains in a quality control interlaboratory exercise.

It is noteworthy that faRPMI and mCUB, used so far in broth microdilution tests, assay different properties. First, microdilution susceptibility testing in RPMI denotes growth inhibition with no color reaction designating growth. Therefore, reading MIC endpoints can be difficult due to medium turbidity caused by the lipid supplements and the button-like growth in the bottom of the microtitre plate wells [53, 63, 69]. Because of this, *M. pachydermatis* minimum fungicidal concentration (MFC) readings are following MIC readings through subcultures from each well into potato dextrose agar (PDA) plates, after incubation for 2–5 days at 32°C [69]. Application of this procedure to lipophilic *Malassezia* species, using selective media such as modified Dixon's agar may prove useful. Second, susceptibility testing using mCUB measures metabolic activity. The reported *M. pachydermatis* negative broth urease test (34.4%) among 32 veterinary isolates, [71] requires further consideration regarding the impact of mCUB in reading *M. pachydermatis* MICs. Therefore, besides screening isolates for urease activity to ensure pure *Malassezia* cultures before performing susceptibility studies, the test would also account for each pure isolate's urease activity.

Overall, the primary objective of susceptibility testing is the recognition of resistance to help selection of targeted therapy. The contribution of drug resistance in high relapse rate reported for pityriasis versicolor (PV) [62, 72, 73] is not yet evaluated. This high relapse rate is often attributed to lack of patient compliance [74] rather to treatment failure, but in the absence of a standardized susceptibility assay for *Malassezia* yeasts, correlation of *in vitro* susceptibility data with treatment outcome cannot be supported.

8.2.4 New Formulations and Newer Antifungal Agents

In the last decade, a number of new formulations of established broad-spectrum antifungals with prominent anti-*Malassezia* activity have been introduced and clinically evaluated [75]. In addition, new azoles and other compounds have so far been evaluated for either

Table 8.3 Effect of test media and inoculum size on the broth microdilution (BMD)-generated MICs for 402 [52, 63, 68–70] clinical isolates

Species/antifungal drugs	No. of strains [63]	MIC range (μg/mL)/faRPMI with low inoculum[a] [63]	No of strains [69]	MIC range (μg/mL)/mCUB with high inoculum[b] [69]	No of strains [52]	MIC range (μg/mL)/mLN with low inoculum[a] [52]	No of strains [68]	MIC range (μg/mL)/mCUB with high inoculum[b] [68]	Σ
M. furfur	12		52		74		13		151
Fluconazole		0.5–32		–		<0.125–16		–	
Itraconazole		0.03–0.06		0.03–0.5		<0.03–0.25		3.2–25	
Ketoconazole		0.03–1		0.03–1		<0.03–4		–	
Voriconazole		0.03–16		0.03–1		<0.03–0.5		–	
Posaconazole		0.03–32		–		–		–	
Terbinafine		0.03–0.5		–		–		3.2–50	
M. globosa	22		74		2		1		99
Fluconazole		0.03–0.125		–		<0.125–>64		–	
Itraconazole		0.03–0.125		0.015–≥8		0.03–16		0.8–6.3[c]	
Ketoconazole		0.03–0.125		0.0015–≥8		<0.03–05		–	
Voriconazole		0.03–0.125		0.03–≥8		<0.03–>16		–	
Posaconazole		0.03–0.06		–		–		–	
Terbinafine		0.03–0.125		–		–		0.8–6.3[c]	
M. obtusa	2		6		8		6		22
Fluconazole		2–4[d]		–		<0.125–16		–	
Itraconazole		0.03–0.25		0.03–0.125		<0.03–0.25		0.1–1.6	
Ketoconazole		0.03–0.6		0.06–0.25		<0.03–0.5		–	
Voriconazole		0.03–0.6		0.125–0.25		<0.03–1		–	
Posaconazole		0.03[d]		–		–		–	
Terbinafine		0.03–0.125		–		–		0.8–6.3	
M. restricta	2		16		–		1		19
Fluconazole		0.5–1		–		–		–	

Table 8.3 (continued)

Itraconazole		0.03[d]		0.015–>8		–		1.6–6.3[c]	
Ketoconazole		0.03[d]		0.015–≥8		–		–	
Voriconazole		0.03[d]		0.06–8		–		–	
Posaconazole		0.03[d]		–		–		–	
Terbinafine		0.125–1		–		–		6.3–25[c]	
M. sympodialis	8		50		11		7		76
Fluconazole		0.5–16		–		<0.125–>64		–	
Itraconazole		0.03–0.06		0.015–2		<0.03–16		0.025–0.1	
Ketoconazole		0.03–0.06		0.015–4		<0.03–0.5		–	
Voriconazole		0.03–0.06		0.015–1		<0.03–>16		–	
Posaconazole		0.03–0.06		–		–		–	
Terbinafine		0.03–0.06		–		–		0.05–0.8	
M. pachydermatis	21 [63, 70]		2		–		12		35
Fluconazole		8–16		–		–		–	
Itraconazole		≤0.03–0.25		0.03–0.125		–		0.8–6.3	
Ketoconazole		≤0.03–0.06		0.06–0.25		–		–	
Voriconazole		0.03^2		0.125–0.25		–		–	
Posaconazole		0.125[d]		–		–		–	
Terbinafine		0.125[d]		–		–		3.2–25	
Total number of strains	67		200		95		40		402

faRPMI fatty acid RPMI broth; *mCUB* modified Christensen's urea broth; *mLN* modified Leeming–Notman broth

[a] $2.5–4 \times 10^3$

[b] Approximately 1×10^6

[c] Range of a single isolate tested multiple times

[d] Only one value. Either one strain tested, or MICs of two strains tested identical on multiple experiments

Table 8.4 New imidazole formulations, synthetic and naturally occurring compounds evaluated for anti-*Malassezia* activity

Families of antifungal agent or compound	Antifungals	References
Topical imidazole derivatives	Anhydrous ketoconazole gel Eberconazole Lanoconazole Luliconazole	[75, 100, 93] [58] [60] [59, 61]
Oral triazole derivatives	Albaconazole Posaconazole Pramiconazole	[50] [63] [85, 97]
Topical synthetic hydroxypyridones	Ciclopirox Rilopirox	[64] [65]
Natural sesquiterpenoids	Xanthorrhizol	[66]
Essential oils	Australian tea tree oil	[67]
Cationic antimicrobial polypeptides	Cathelicidin	[83]

in vitro or *in vivo* anti-*Malassezia* activity (Table 8.4) as well as for topical preparations used in the veterinary field [76].

8.2.5 *Malassezia* Susceptibility to Synthetic and Naturally Occurring Compounds

8.2.5.1 Ciclopirox olamine

Ciclopirox is a hydrophobic broad-spectrum antifungal agent with anti-inflammatory properties that has been tested by broth microdilution against *M. furfur*. The compound is reported [77] with low MICs (range 0.001–0.25 μg/mL). Independent clinical trials have shown that ciclopirox 1% shampoo is a safe agent that improves scalp SD [78, 79]. However, there are no studies correlating the *in vivo Malassezia* yeast responses to ciclopirox shampoo with MIC values.

8.2.5.2 Rilopirox

The *in vitro* activity of rilopirox has been evaluated against *M. furfur* clinical strains from dandruff, PV, and SD, by microdilution and the agar dilution method [64]. Reported agar dilution method and microdilution MICs range from 12 to 50 μg/mL and from 16 to 128 μg/mL, respectively. The clinical relevance of the *in vitro* rilopirox performance against *Malassezia* yeasts remains to be clarified.

8.2.5.3
Xanthorrhizol

Xanthorrhizol, is an active component methanol extracted from the rhizome of *Curcuma xanthorrhiza* (Zingiberaceae) that is traditionally used in Indonesia for medicinal purposes. Its *in vitro* anti-*M. furfur* and *M. pachydermatis* activity was recently evaluated [66] by a modified CLSI microdilution method. MICs, MFCs, as well as time-kill curves have been determined. The MFCs were 5 μg/mL for *M. furfur* and 2.7 μg/mL *for M. pachydermatis,* whereas time kill readings showed that 25 μg/mL and 20 μg/mL kill 100% of the *M. furfur* and *M. pachydermatis* inoculated cells, in 5 h and in 20 min respectively. Again, the clinical relevance of these *in vitro* findings has yet to be evaluated.

8.2.5.4
Australian Tea Tree Oil

Susceptibility data on the essential oil of *Melaleuca alternifolia* are limited. Yet, there are data on the *in vitro* activity of tea tree oil, generated by agar dilution and modified CLSI microdilution methods [80] showing good *in vitro* activity against *M. furfur, M. sympodialis, M. slooffiae, M. globosa, M. obtusa* and *M. pachydermatis* [81] (MIC range for all *Malassezia* species tested by both methods: 0.03–0.25 μg/mL). It has been estimated that the recorded MICs are below the therapeutic concentrations (5–10% solutions or concentrated Tea tree oil [82]).

8.2.5.5
Cathelicidin

Indeed, the activity of cathelicidin is not tested by traditional susceptibility methods, yet, it is included herein as its quantifiable activity provides new insight into therapeutic approaches of *Malassezia*-induced skin diseases. The cathelicidin antimicrobial peptides have been isolated from many different mammalian species and play an important part in innate immune responses against bacterial infections. *M. furfur* susceptibility to synthetic and human (LL-37) antimicrobial peptides has been recently studied [83]. It has been shown that the normally expressed cathelicidins are processed to active peptides in skin biopsies of patients with PV.

8.2.6
Corneofungimetry: A Bioassay Predicting Drug Efficacy

Ketoconazole is an established, potent broad-spectrum oral antifungal with known auxiliary anti-inflamatory, antiseborrheic, and antiproliferative properties that significantly contribute in relieving the symptoms of seborrheic dermatitis (SD) (also referred to as seborrhoic eczema) (Chap. 7.1). However, its hepatotoxicity has led to the development of topical formulations (creams shampoos and powders) containing different ketoconazole concentrations. The *in vitro* activity of other topical imidazoles and oral azoles that have been developed in the recent years (Table 8.4) has been evaluated by non-standard agar dilution or microdilution susceptibility testing (Table 8.3). Nevertheless, predicting clinical activity, even after testing susceptibility with standard methods, may be unsound due to the variable host responses to treatment and to the variable fungal load in the sites of infection. Correlating *in vitro* with *in vivo* and *ex vivo* results would therefore give a more pragmatic view on a drug's efficacy. In that respect, bioassay methods, such as corneofungimetry, can complement *in vitro* susceptibility testing to offer clinically realistic data on the topical or oral efficacy of antifungals.

Corneofungimetry [84, 85] utilizes the finding [62] that *Malassezia* yeasts can grow *ex vivo* on sebum-coated human corneocytes. By testing human stratum corneum specimens from patients undergoing topical or oral treatment, the individualized therapeutic response and the drug levels achieved in the stratum corneum can be determined.

Interestingly, corneofungimetry bioassays against *M. furfur, M. globosa*, and *M. restricta* have shown that the efficacy of the new topical 2% ketoconazole gel alone, and in combination with low-potency topical corticosteroids such as 0.05% desonide [86], which is used for the treatment of *Malassezia*-aggravated atopic dermatitis (AD also referred to as atopic eczema), instigates good clinical outcome. Similarly, good therapeutic response has been observed in patients with *Malassezia*-associated SD by a single 200 mg pramiconazole dose [84, 85]. This entailed reduction of the pre-treatment *Malassezia* yeast population at the sites of infection by approximately one-third on day seven post-treatment [85]. Assessment of the published data so far indicates that corneofungimetry bioassays can generate clinically useful data by essentially evaluating the levels achieved in the stratum corneum of patients under treatment. However, these data may not safely predict the duration of treatment for chronic or well-established infections.

8.3 Molecular Epidemiology of Nosocomial Outbreaks

Teun Boekhout

The epidemiology of nosocomial outbreaks caused by *M. furfur* and/or *M. pachydermatis* has been studied only in a few cases by molecular biology methods. An outbreak at a neonatal intensive care unit (NCIU) ward in a Dutch hospital has been studied by means of pulsed field gel electrophoresis (PFGE) and RAPD analysis [27, 94]. It is important to note that this outbreak was caused by both *M. pachydermatis* and *M. furfur,* which, as far as we know, has not been reported elsewhere. Outbreak isolates of both species were studied by RAPD using prokaryotic repeat consensus primers based on enterobacterial repetitive intergenic consensus (ERIC) or repetitive extragenic palindrome (REP) motifs and PFGE, and the resulting banding patterns were compared with a set of reference isolates from the reference collection of CBS Fungal Biodiversity Centre (www.cbs.knaw.nl). Primers ERIC IR, ERIC2, REP1R-I and REP2 generated identical banding patterns for the outbreak isolates, when compared to those of the reference strains that showed some variation [27]. Karyotyping of intact chromosomes with PFGE, however, identified minor chromosomal length variation in outbreak and reference isolates alike, suggesting involvement of different strains in the outbreak. This supposition, however, was challenged by the RAPD-generated data that showed no genetic diversity in the outbreak isolates, whereas some diversity was present in the reference strains.

It was therefore concluded that the outbreak was, most likely, caused by a single strain. As also discussed in Chap. 3.1., it seems that PCR typing is a more sensitive epidemiological tool than PFGE. The outbreak was controlled by improving hygiene measures in the NICU. In this case, no attempt was made to identify the source of the outbreak.

In another *M. pachydermatis* outbreak in a NICU in the USA, pulsotyping of HaeIII digested chromosomes indicated that the banding patterns of all 15 isolates obtained from the neonates were identical, and these also matched those from isolates obtained from a health care worker and those from his pet dog. The authors concluded, although with some uncertainty, that *M. pachydermatis* was most likely introduced into the ward through the colonized hands of dog-owner health care workers who subsequently transmitted it from patient to patient. Similar to the outbreak in the Dutch hospital, improvement of hygiene measures, such as meticulous hand washing practices and use of a new pair of disposable gloves while caring for each neonate, eliminated the outbreak [12]. Contrary to the evidence for the aforementioned outbreaks that have been attributed to a single strain, two other outbreak reports that occurred in two distinct geographic locations indicate involvement of multiple *M. pachydermatis* strains.

RAPD typing of 14 catheter-related outbreak isolates from different patients in a Sweedish NICU, demonstrated genetic heterogeneity, whereas *M. pachydermatis* isolates from different body sites of the same neonate yielded the same banding pattern. The investigators also assumed that the neonates were either initially colonized by *M. pachydermatis* from the skin of their parents, or from the hands of nursing staff, and that, subsequently, infection occurred via permanent catheters [2]. Using amplified fragment length polymorphism (AFLP) analysis (Fig. 8.2) of nine isolates of *M. pachydermatis,* originating from a

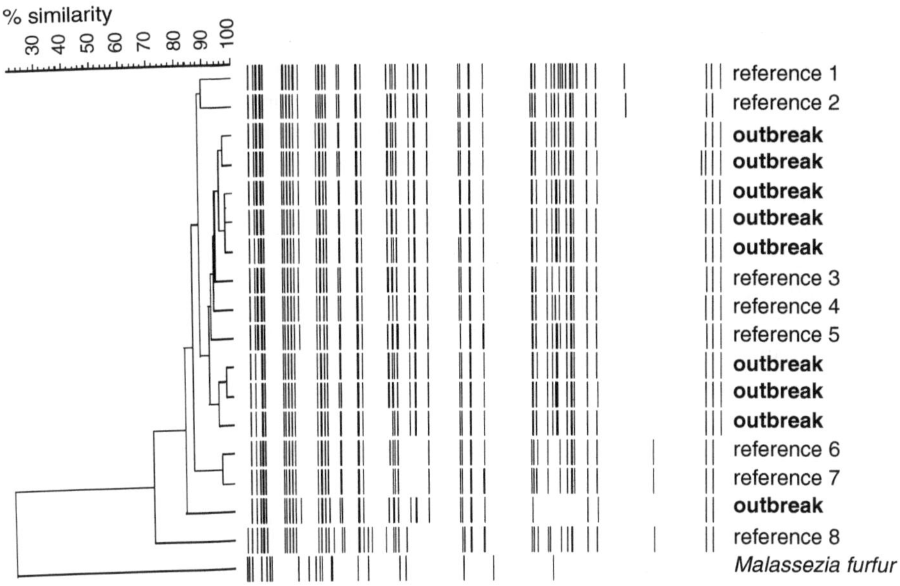

Fig. 8.2 Genetic variation, as determined with amplified fragment length polymorphism (AFLP), of a series of isolates from a nosocomial outbreak and reference strains from a culture collection. Note that the amount of genetic variation is similar to that of the genetically nonrelated reference strains, thus indicating that multiple isolates were involved in the outbreak

nosocomial outbreak in a USA-based hospital, some genetic diversity was observed, similar to the level present in a number of reference isolates obtained from dogs, but from different geographical regions [95].

8.4 Conclusions

Over the past two decades, *Malassezia* has been increasingly recognized as cause of catheter-associated fungemia in critically ill premature neonates and immunocompromised individuals of all ages. However, since many routine blood culture systems do not effectively support its growth *in vitro*, the incidence will most likely be even higher as current clinical data suggest. A high index of suspicion is required to recognize *Malassezia* species as cause of invasive infections. Microbiological diagnosis still relies on the growth of the organism in appropriate culture media, but molecular methods offer some hope for improved detection, thus contributing in collecting consistent epidemiological data in the near future. While no evidence-based treatment recommendations can be made, systemic therapy with antifungal triazoles or amphotericin B and removal of indwelling devices are considered the cornerstones of clinical management.

As *Malassezia*-induced or exacerbated infections require systemic and/or topical antifungal chemotherapy, susceptibility testing represents a *sine qua non* clinical laboratory procedure. Thus far, testing protocols by agar dilution and broth microdilution formats imply possible *M. restricta, M. globosa* and *M. sympodialis* resistance to certain azole drugs. Systematic studies correlating clinical outcome with susceptibility results would elucidate this point, as the existing sporadic data do not sponsor definitive conclusions. However, standardization is needed, for it is obvious that appropriate validation is required before adopting a susceptibility assay intended for patient care and welfare. Susceptibility testing is a highly controlled procedure. As with any assay, its stochastic nature subjects it to random strain and procedures-related events, which may lead to erroneous results. Clinical relevance of susceptibility results for topical and oral agents targeting the skin can be assessed with standardized susceptibility testing procedures, co-evaluated with pharmacodynamic and pharmacokinetic studies in natural systems such as the human stratum corneum. Besides the impact of *Malassezia* susceptibility testing assays in the clinical domain, routine performance of standardized testing would also provide useful information on the epidemiology of resistance to antifungal agents in the hospital and in the community.

Evidence thus far supports the hypothesis that *Malassezia*-related nosocomial outbreaks in neonatal wards can be caused by either a single or multiple *M. pachydermatis* strains. The only *M. furfur* outbreak that was studied by molecular means was, most likely, caused by a single strain.

From the foregoing discussion, it is clear that different methods have different resolution capacities, and, consequently, the generated epidemiological data are interpreted accordingly. Based on our experiences, PCR-fingerprinting methods using ERIC or REP primers, as well as AFLP may be preferred over PFGE. The main disadvantage of these fingerprint methods is the lack of time-spatial reproducibility thus impeding objective comparison of the results. A promising method, which, unfortunately, has not yet been used in epidemiological studies of *Malassezia*-related outbreaks, is the multilocus sequence typing (MLST) in which several housekeeping genes (usually up to 7–10) are being sequenced across all available isolates [96–101]. We foresee the future development of a MLST-typing scheme of *Malassezia* species that may be coordinated by the ISHAM working group on *Malassezia*. The main advantage of MLST is that the results are reliable and reproducible. Another advantage is that storing of the MLST data in a central depository, such as the http://www.mlst.net, will make it possible for other investigators, epidemiologists, and clinicians to compare isolates from their studies. Finally, if a sufficient body of epidemiological data is generated, our global *Malassezia* epidemiological perspective will improve.

References

1. Barber GR, Brown AE, Kiehn TE et al (1993) Catheter-related *Malassezia furfur* fungemia in immunocompromised patients. Am J Med 95:365–370
2. Chryssanthou E, Broberger U, Petrini B (2001) *Malassezia pachydermatis* fungaemia in a neonatal intensive care unit. Acta Paediatr 90:323–327
3. Dankner WM, Spector SA, Fierer J et al (1987) *Malassezia* fungemia in neonates and adults: complication of hyperalimentation. Rev Infect Dis 9:743–753
4. Groll AH, Walsh TJ (2001) Uncommon opportunistic fungi: new nosocomial threats. Clin Microbiol Infect 7(suppl 2):8–24
5. Shattuck KE, Cochran CK, Zabransky RJ et al (1996) Colonization and infection associated with *Malassezia* and *Candida* species in a neonatal unit. J Hosp Infect 34:123–129
6. Redline RW, Dahms BB (1981) *Malassezia* pulmonary vasculitis in an infant on long-term intralipid therapy. N Engl J Med 305:1395–1398
7. Alpert G, Bell LM, Campos JM (1987) *Malassezia furfur* fungemia in infancy. Clin Pediatr 26:528–531
8. Aschner JL, Punsalang A, Maniscalo W et al (1987) Percutaneous central venous catheter colonization with *Malassezia furfur*: incidence and clinical significance. Pediatrics 80:535–539
9. Cannizzo FT, Eraso E, Ezkurra PA et al (2007) Biofilm development by clinical isolates of *Malassezia pachydermatis*. Med Mycol 45:357–361
10. Powell DA, Aungst J, Snedden S et al (1984) Broviac catheter-related *Malassezia furfur* sepsis in five infants receiving intravenous fat emulsions. J Pediatr 105:987–990
11. Redline RW, Redline SS, Boxerbaum B et al (1985) Systemic *Malassezia furfur* infections in patients receiving intralipid therapy. Hum Pathol 16:815–822
12. Chang HJ, Miller H, Watkins N et al (1998) An epidemic of *Malassezia pachydermatis* in an intensive care nursery associated with colonization of health care workers' pet dogs. New Engl J Med 338:706–711
13. Surmont I, Gavilanes A, Vandepitte J, et al (1989) *Malassezia furfur* fungaemia in infants receiving intravenous lipid emulsions. A rarity or just underestimated? Eur J Pediatr 148: 435–438
14. Welbel SF, McNeil MM, Pramanik A et al (1994) Nosocomial *Malassezia pachydermatis* bloodstream infections in a neonatal intensive care unit. Pediatr Infect Dis J 13:104–108
15. Morrison VA, Weisdorf DJ (2000) The spectrum of *Malassezia* infections in the bone marrow transplant population. Bone Marrow Transplant 26:645–648

16. Schleman KA, Tullis G, Blum R (2000) Intracardiac mass complicating *Malassezia furfur* fungemia. Chest 118:1828–1829
17. Schoepfer C, Carla H, Bezou MJ et al (1995) *Malassezia furfur* septicemia after bone marrow graft. Arch Pediatr 2:245–248
18. Shparago NI, Bruno PP, Bennett J (1995) Systemic *Malassezia furfur* infection in an adult receiving total parenteral nutrition. J Am Osteopath Assoc 6:375–377
19. Weiss SJ, Schoch PE, Cuhna BA (1991) *Malassezia furfur* fungemia associated with central venous catheter lipid emulsion infusion. Heart Lung 20:87–90
20. Masure O, Minoui A, Legall P et al (1997) Etude prospective de la colonisation des catheters vasculaires par les levures *Malassezia*. J Mycol Méd 7:33–36
21. Sizun J, Karangwa A, Giroux JD et al (1994) *Malassezia furfur*-related colonization and infection of central venous catheters. A prospective study in a pediatric intensive care unit. Intensive Care Med 20:496–499
22. Curvale-Fauchet N, Botterel F, Legrand P et al (2004) Frequency of intravascular catheter colonization by *Malassezia* spp. in adult patients. Mycoses 47:491–494
23. Bell LM, Alpert G, Slight PH et al (1985) Skin colonization of hospitalized and non- hospitalized infants with lipophilic yeasts. Abstr ICAAC 519 25th ICAAC, 29 September – 2 October, Minneapolis, Minnesota, USA
24. Ahtonen P, Lehtonen OP, Kero P et al (1990) *Malassezia furfur* colonization of neonates in an intensive care unit. Mycoses 33:543–547
25. Ashbee HR, Leck AK, Puntis JW et al (2002) Skin colonization by *Malassezia* in neonates and infants. Infect Control Hosp Epidemiol 23:212–216
26. Maki DG (1982) Infections associated with intravascular lines. In: Remington JS, Swartz MN (eds) Current clinical topics in infectious diseases, 3rd edn. McGraw-Hill, New York, pp 309–363
27. Van Belkum A, Boekhout T, Bosboom R (1994) Monitoring spread of *Malassezia* infections in a neonatal intensive care unit by PCR-mediated genetic typing. J Clin Microbiol 32: 2528–2532
28. Ashbee HR, Evans EG (2002) Immunology of diseases associated with *Malassezia* species. Clin Microbiol Rev 15:21–57
29. Powell DA, Marcon MJ, Durrell DE et al (1987) Scanning electron microscopy of *Malassezia furfur* attachment to Broviac catheters. Hum Pathol 18:740–745
30. Marcon MJ, Powell DA (1987) Epidemiology, diagnosis, and management of *Malassezia furfur* systemic infection. Diagn Microbiol Infect Dis 7:161–175
31. Marcon MJ, Durrell DE, Powell DA et al (1987) *In vitro* activity of systemic antifungal agents against *Malassezia furfur*. Antimicrob Agents Chemother 31:951–953
32. Pappas PG, Rex JH, Sobel JD et al (2004) Guidelines for treatment of candidiasis. Clin Infect Dis 38:161–189
33. Marcon MJ, Powell DA (1992) Human infections due to *Malassezia* spp. Clin Microbiol Rev 2:101–119
34. Shek YH, Tucker MC, Viciana AL et al (1989) *Malassezia furfur* disseminated infection in premature infants. Am J Clin Pathol 92:595–603
35. Devlin RK (2006) Invasive fungal infections caused by *Candida* and *Malassezia* species in them neonatal intensive care unit. Adv Neonatal Care 6:68–77
36. Zomorodian K, Mirhendi H, Tarazooie B et al (2008) Molecular analysis of *Malassezia* species isolated from hospitalized neonates. Pediatr Dermatol 25:312–316
37. Masure O, Leostic C, Abalain ML et al (1991) *Malassezia furfur* septicaemia in a child with leukaemia. J Infect 23:335–336
38. Middleton C, Lowenthal RM (1987) *Malassezia furfur* fungemia as a treatable cause of obscure fever in a leukemia patient receiving parenteral nutrition. Aust N Z J Med 17:603–604
39. Myers JW, Smith R, Youngberg G et al (1992) Fungemia due to *Malassezia furfur* in patients without the usual risk factors. Clin Infect Dis 14:620–621

40. Wurst RM, Knospe WN (1988) *Malassezia furfur* fungemia in a patient without the usual risk factors. Ann Intern Med 109:432–433
41. Kessler AT, Kourtis AP, Simon N (2002) Peripheral thromboembolism associated with *Malassezia furfur* sepsis. Pediatr Infect Dis J 21:356–357
42. Hassall E, Ulich T, Ament M (1983) Pulmonary embolus and *Malassezia* pulmonary infection related to urokinase therapy. J Pediatr 102:722–725
43. Ashbee HR (2007) Update on the genus *Malassezia*. Med Mycol 45:287–303
44. Kim E, Cohen RS, Ramachandran P et al (1993) Adhesion of percutaneously inserted Silastic central venous lines to the vein wall associated with *Malassezia furfur* infection. JPEN J Parenter Enteral Nutr 17:458–460
45. Nguyen ST, Lund CH, Durand DJ (2001) Thrombolytic therapy for adhesion of percutaneous central venous catheters to vein intima associated with *Malassezia furfur* Infection. J Perinatol 21:331–333
46. Rosales CM, Jackson MA, Zwick D (2004) *Malassezia furfur* meningitis associated with total parenteral nutrition subdural effusion. Pediatr Dev Pathol 7:86–90
47. Nelson SC, Yau YC, Richardson SE et al (1995) Improved detection of *Malassezia* species in lipid-supplemented Peds Plus blood culture bottles. J Clin Microbiol 33:1005–1007
48. Gaitanis G, Velegraki A, Frangoulis E et al (2002) Identification of *Malassezia* species from patient skin scales by PCR-RFLP. Clin Microbiol Infect 8:162–173
49. Tirodker U, Nataro J, Smith S et al (2003) Detection of fungemia by polymerase chain reaction in critically ill neonates and children. J Perinatol 23:117–122
50. Garau M, Pereiro M Jr, del Palacio A (2003) *In vitro* susceptibilities of *Malassezia* species to a new triazole, albaconazole (UR-9825), and other antifungal compounds. Antimicrob Agents Chemother 47:2342–2344
51. Gupta AK, Kohli Y, Li A et al (2000) *In vitro* susceptibility of the seven *Malassezia* species to ketoconazole, voriconazole, itraconazole and terbinafine. Br J Dermatol 142:758–765
52. Miranda KC, de Araujo CR, Costa CR et al (2007) Antifungal activities of azole agents against the *Malassezia* species. Int J Antimicrob Agents 29:281–284
53. Prado MR, Brito EH, Brilhante RS et al (2008) Subculture on potato dextrose agar as a complement to the broth microdilution assay for *Malassezia pachydermatis*. J Microbiol Methods 75:341–343
54. Richet HM, McNeil MM, Edwards MC et al (1989) Cluster of *Malassezia furfur* pulmonary infections in infants in a neonatal intensive-care unit. J Clin Microbiol 27:1197–1200
55. Dimopoulos G, Velegraki A, Falagas ME (2009) A 10-year survey on antifungal susceptibility of candidemia isolates from intensive care unit patients in Greece. Antimicrob Agents Chemother 53:1242–1244
56. CLSI (2008) Reference method for broth dilution antifungal susceptibility testing of yeasts; Approved standard M27-A3. In: Wayne PA (ed) Clinical and laboratory standards institute, 3rd edn, vol 28, No 14
57. EUCAST Definitive Document ED ef 7.1 (2008) Method for the determination of broth dilution MICs of antifungal agents for fermentative yeasts Subcommittee on Antifungal Susceptibility Testing (AFST) of the ESCMID European Committee for Antimicrobial Susceptibility Testing (EUCAST). Clin Microbiol Infect 14:398–405
58. Barbanoj MJ, Antonijoan R, Gea G et al (2005) Eberconazole cream: topical and general tolerability, sensitization potential and systemic availability. Methods Find Exp Clin Pharmacol 27:227–234
59. Koga H, Nanjoh Y, Makimura K et al (2008) *In vitro* antifungal activities of luliconazole, a new topical imidazole. Med Mycol 29:1–8
60. Rubin AI, Bagheri B, Scher RK (2002) Six novel antimycotics. Am J Clin Dermatol 3:71–81
61. Uchida K, Nishiyama Y, Yamaguchi H (2004) *In vitro* antifungal activity of luliconazole (NND-502), a novel imidazole antifungal agent. J Infect Chemother 10:216–219

62. Faergemann J (1989) A new model for growth and filament production of *Pityrosporum ovale (orbiculare)* on human stratum corneum in vitro. J Invest Dermatol 92:117–119
63. Velegraki A, Alexopoulos EC, Kritikou S et al (2004) Use of fatty acid RPMI 1640 media for testing susceptibilities of eight *Malassezia* species to the new triazole posaconazole and to six established antifungal agents by a modified NCCLS M27–A2 microdilution method and Etest. J Clin Microbiol 42:3589–3593
64. Nenoff P, Haustein UF (1997) *In vitro* susceptibility testing of *Malassezia furfur* against rilopirox. Skin Pharmacol 10:275–280
65. Nenoff P, Reinl P, Haustein UF (2001) The yeast fungus *Malassezia*: pathogen, pathogenesis and therapy. Hautarzt 52:73–86
66. Rukayadi Y, Hwang JK (2007) *In vitro* anti-*Malassezia* activity of xanthorrhizol isolated from *Curcuma xanthorrhiza* Roxb. Lett Appl Microbiol 44:126–130
67. Weseler A, Geiss HK, Saller R et al (2002) Antifungal effect of Australian tea tree oil on *Malassezia pachydermatis* isolated from canines suffering from cutaneous skin disease. Schweiz Arch Tierheilkd 144:215–221
68. Nakamura A, Kano A, Murai T et al (2000) Susceptibility testing of *Malassezia* species using the urea broth microdilution method. Antimicrob Agents Chemother 44:2185–2186
69. Rincón S, Cepero de García MC, Espinel-Ingrof A (2006) A modified Christensen's urea and CLSI broth microdilution method for testing susceptibilities of six *Malassezia* species to voriconazole, itraconazole, and ketoconazole. J Clin Microbiol 44:3429–3431
70. Brito EHS, Fontenelle ROS, Brilhante RSN et al (2007) *In vitro* activity of an ear rinse containing tromethamine, EDTA, benzyl alcohol and 0.1% ketoconazole on *Malassezia* organisms from dogs with otitis externa. Vet Dermatol 18:115–119
71. Brito EHS, Fontenelle ROS, Brilhante RSN et al (2007) Phenotypic characterization and *in vitro* antifungal sensitivity of *Candida* spp. and *Malassezia pachydermatis* strains from dogs. Vet J 174:147–153
72. Gupta AK, Plott T (2004) Ciclopirox: a broad-spectrum antifungal with antibacterial and anti-inflammatory properties. Int J Dermatol 43(Suppl 1):3–8
73. He SM, Du WD, Yang S et al (2008) The genetic epidemiology of tinea versicolor in China. Mycoses 51:55–62
74. Plewig G, Jansen T(1999) Seborrheic dermatitis. In: Freedberg IM, Eisen AZ, Wolff K et al (eds) Fitzpatrick's dermatology in general medicine. McGraw-Hill, New York, pp 1482–1489
75. Borgers M, Degreef H (2007) The role of ketoconazole in seborrheic dermatitis. Cutis 80: 359–363
76. Swinney A, Fazakerley J, McEwan N et al (2008) Comparative *in vitro* antimicrobial efficacy of commercial ear cleaners. Vet Dermatol 19:373–379
77. Kokjohn K, Bradley M, Griffiths B et al (2003) Evaluation of *in vitro* activity of ciclopirox olamine, butenafine HCl and econazole nitrate against dermatophytes, yeasts and bacteria. Int J Dermatol 42(suppl 1):11–17
78. Abeck D (2004) Rationale of frequency of use of ciclopirox 1% shampoo in the treatment of seborrheic dermatitis: Results of a double-blind, placebo-controlled study comparing the efficacy of once, twice, and three times weekly usage. Int J Dermatol 43(suppl 1):13–16
79. Lebwohl M, Plott T (2004) Safety and efficacy of ciclopirox 1% shampoo for the treatment of seborrheic dermatitis of the scalp in the US population: Results of a double-blind, vehicle-controlled trial. Int J Dermatol 43(suppl 1):17–20
80. Hammer KA, Carson CF, Riley TV (2000) *In vitro* activities of ketoconazole, econazole, miconazole, and *Melaleuca alternifolia* (tea tree) oil against *Malassezia* species. Antimicrob Agents Chemother 44:467–469
81. Sidrim JJC, Rocha MFG (1007) Phenotypic characterization and *in vitro* antifungal sensitivity of *Candida* spp. and *Malassezia pachydermatis* strains from dogs. Vet J 174:147–153

82. Nenoff P, Haustein UF, Brandt W (1996) Antifungal activity of the essential oil of *Melaleuca alternifolia* (tea tree oil) against pathogenic fungi *in vitro*. Skin Pharmacol 9:388–394
83. López-García B, Lee PHA, Gallo RL (2006) Expression and potential function of cathelicidin antimicrobial peptides in dermatophytosis and tinea versicolor. J Antimicrob Chemother 57:877–882
84. Piérard-Franchimont C, Ausma J, Wouters L et al (2005) Activity of the triazole antifungal R126638 as assessed by corneofungimetry. Skin Pharmacol Physiol 19:50–56
85. Piérard GE, Ausma J, Henry F et al (2007) A pilot study on seborrheic dermatitis using pramiconazole as a potent oral anti-*Malassezia* agent. Dermatology 214:162–169
86. Piérard-Franchimont C, Vroome V, Cauwenbergh G et al (2005) Corneofungimetry bioassay on *Malassezia* spp. under ketoconazole and desonide influences. Skin Pharmacol Physiol 18:98–102
87. Sugita T, Tsubuku H, Tajima M et al (2006) Time-kill assay of itraconazole against *Malassezia* species. Microbiol Immun 50:625–627
88. Pietschmann S, Hoffmann K, Voget M et al (2009) Synergistic effects of miconazole and polymyxin B on microbial pathogens. Vet Res Commun 33:489-505
89. Faergemann J, Ausma J, Borgers M (2006) *In vitro* activity of R126638 and ketoconazole against *Malassezia* species. Acta Derm Venereol 86:312–315
90. Sancak B, Ayhan M, Karaduman A et al (2005) *In vitro* activity of ketoconazole, itraconazole and terbinafine against *Malassezia* strains isolated from neonates. Mikrobiyol Bul 39: 301–308
91. Murai T, Nakamura Y, Kano R et al (2002) Susceptibility testing of *Malassezia pachydermatis* using the urea broth microdilution method. Mycoses 28:84–87
92. Arrese JE, Fogouang L, Piérard-Franchimont C et al (2002) Euclidean and fractal computer-assisted corneofungimetry: A comparison of 2% ketoconazole and 1% terbinafine topical formulations. Dermatol 204:222–227
93. Brito EHS, Fontenelle ROS, Brilhante RSN, et al (2009) The anatomical distribution and antimicrobial susceptibility of yeast species isolated from healthy dogs. Vet J 182:320-326
94. Boekhout T, Bosboom R (1994) Karyotyping of *Malassezia* yeasts: taxonomic and epidemiological implications. Syst Appl Microbiol 17:146–153
95. Theelen B, Silvestri M, Guého E et al (2001) identification and typing of *Malassezia* yeasts using amplified fragment length polymorphism (AFLP), random amplified polymorphic DNA (RAPD) and denaturing gradient gel electrophoresis (DGGE). FEMS Yeast Res 1:79–86
96. Bougnoux ME, Tavanti A, Bouchier C et al (2003) Collaborative consensus for optimized multilocus sequence typing of *Candida albicans*. J Clin Microbiol 41:5265–5266
97. Cliff PR, Sandoe JA, Heritage J et al (1008) Use of multilocus sequence typing for the investigation of colonisation by *Candida albicans* in intensive care unit patients. J Hosp Infect 69:24–32
98. Jacobsen MD, Gow NA, Maiden MC et al (2007) Strain typing and determination of population structure of *Candida krusei* by multilocus sequence typing. J Clin Microbiol 45:317–323
99. Jacobsen MD, Boekhout T, Odds FC (2008) Multilocus sequence typing confirms synonymy but highlights differences between *Candida albicans* and *Candida stellatoidea*. FEMS Yeast Res 8:764–770
100. Maiden MC, Bygraves JA, Feil E et al (1998) Multilocus sequence typing: a portable approach to the identification of clones within populations of pathogenic microorganisms. Proc Natl Acad Sci USA 95:3140–3145
101. Tavanti A, Davidson AD, Johnson EM et al (2005) Multilocus sequence typing for differentiation of strains of *Candida tropicalis*. J Clin Microbiol 43:5593–5600

Genomics and Pathophysiology: Dandruff as a Paradigm

9

Jun Xu, Teun Boekhout, Yvonne DeAngelis, Tom Dawson
and Charles W. Saunders

Core Messages

- The recent sequencing of the genomes of dandruff-associated basidiomycetous yeasts, *Malassezia globosa* and *Malassezia restricta,* disclosed that the *M. globosa* genome is among the smallest for a free-living fungus. *M. globosa* produces a similar set of secreted hydrolases as the human pathogen *Candida albicans.* Although phylogenetically more closely related to the plant pathogen *Ustilago maydis*, *M. globosa* produces a different set of secreted hydrolases, which is a likely adaptation to the host niche and may be involved in pathogenicity. *M. globosa* is apparently missing several enzymes in fatty acid metabolism, including fatty acid synthase, Δ9 desaturase, and $\Delta^{2,3}$ enoyl CoA isomerase. The two former enzymes are apparently missing also in another skin microbe, *Corynebacterium jeikeium. M. globosa* has six lipase genes in each of two lipase families, which, compared with the lipases from a related fungus *U. maydis,* had undergone duplications since divergence from the *Ustilago*-containing lineage. There is also evidence for duplication of other *M. globosa* genes for secreted enzymes such as aspartyl proteases, phospholipases C, and acid sphingomyelinases. The *M. globosa* genome encodes proteins similar to all *Malassezia* allergens, the coding sequences of which have been isolated, and genes associated with mating, although mating has not yet been observed in *Malassezia.*

C. W. Saunders (✉)
Miami Valley Innovation Center, The Procter & Gamble Co, Cincinnati, OH 45253-8707, USA
e-mail: saunders.cw@pg.com

T. Boekhout et al. (eds.), *Malassezia and the Skin*
DOI: 10.1007/978-3-642-03616-3_9, © Springer Verlag Berlin Heidelberg 2010

9.1 Introduction

Malassezia and their role in dandruff and seborrheic dermatitis (SD) are of particular interest, as elucidation of pathogenetic mechanisms would instigate research into more effective treatments (Fig. 9.1).

Until recently, very little was known about the *Malassezia* genome. A recent report [1] on the genome sequences of *M. globosa* and *M. restricta* has brought new insights into the pathogenic capacity of these organisms. In this chapter, it is not our goal to repeat the results described in that paper and its 52-page supplementary material. Instead, we will provide additional perspective to the conclusions of that paper as well as some additional insights gleaned from the genome sequence.

Malassezia spp. are considered to have an essential role in dandruff and SD [2], although Koch's postulates have not been satisfied. The role of the fungus is complicated in that it is present essentially on all human scalps and yet many people lack symptoms, which is not unusual in diseases where microorganisms play a role. Disease-implicated microorganisms are present on many people without causing symptoms. Such microorganisms include *Staphylococcus aureus* [3], *Mycobacterium tuberculosis* [4], dermatophytes [5], and *C. albicans* [6]. In all of these cases, there are host attributes that determine whether the

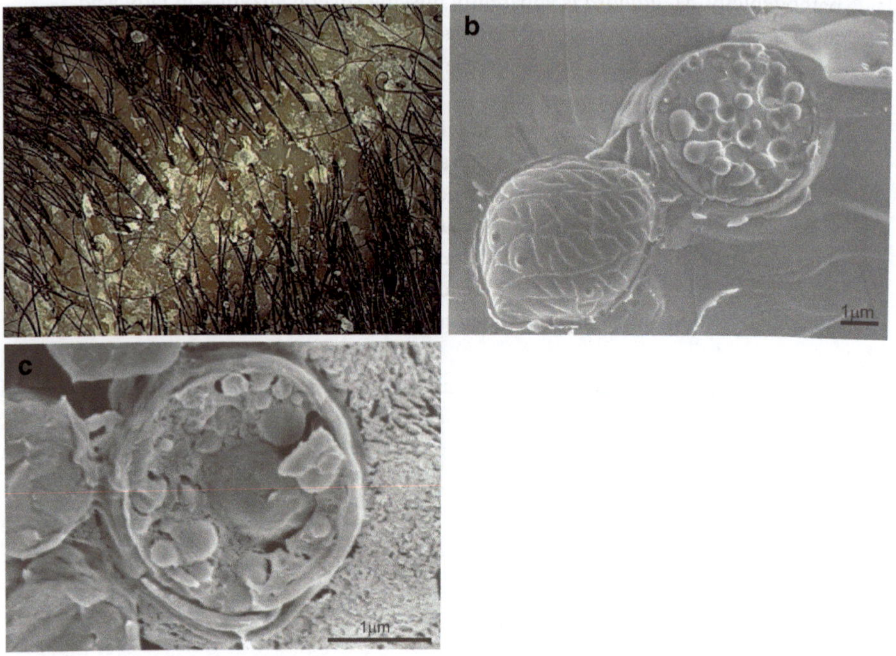

Fig. 9.1 Dandruff and *Malassezia*. (**a**) Scalp flaking known as dandruff. Note excessive flaking without overt irritation. Photo was taken with a Sony High-scope at 20× magnification. (**b**) Scanning electron micrograph of *Malassezia globosa* CBS strain 7966 after 14 days growth on LNA agar. Note the spiral structure in the cell wall (see also Chap. 2.3). (**c**) *Malassezia restricta* CBS strain 7877 grown under the same conditions as (**b**)

microorganism can cause disease. With dandruff, one host susceptibility component that has been described is sensitivity to oleic acid, which is a product of *Malassezia* lipase activity. Topically applied oleic acid can induce dandruff in dandruff-susceptible people only [7].

The role of *Malassezia* in dandruff and seborrheic dermatitis is supported by the efficacy of a variety of antifungal agents (i.e., pyrithione zinc, selenium sulfide, ciclopirox olamine, climbazole, and ketoconazole) [8] with no other obvious related biological activity. Upon treatment, the fungi are largely removed, followed by alleviation of symptoms [9]. With termination of treatment, the fungal numbers increase, and the symptoms often return [9]. While a different fungus may possibly contribute to these symptoms, no such fungus has been reported from the scalp. There are multiple hypotheses for the molecular mechanisms by which *Malassezia* cause dandruff and SD. One possibility is that lipase-mediated hydrolysis of sebum triglycerides leads to an excess of oleic acid that aggravates the scalp [10, 11]. Another possibility is that *Malassezia* produce tryptophan metabolites that are active on the aryl hydrocarbon receptor, thus producing inflammation [12]. An altered immune response has also been implicated (reviewed in Chap. 5).

Among the *Malassezia* species, *M. restricta* and *M. globosa* are the most commonly found on the scalp [13–16]. Genome-sequencing efforts have been focused on the *M. globosa* type strain CBS 7966, as the tRFLP pattern of its rDNA region matches that of the *M. globosa* strain(s) most commonly found on human scalps (Gemmer and Dawson unpublished observations). A sevenfold (7×) sequence coverage of the *M. globosa* genome and a one-fold (1×) *M. restricta* type strain CBS 7877 genome sequence coverage were generated. In this chapter, we focus on the *M. globosa* genome, with some discussion on what was elucidated through the incomplete sequencing of the *M. restricta* genome.

9.2 Genome Size

At 8.9 Mb, the *M. globosa* genome is among the smallest for free-living fungi (Table 9.1).

The microsporidian, *Encephalitozoon cuniculi*, now classified as a fungus, contains a much smaller genome (2.9 Mb), but is incapable of growth outside animal cells [17]. Another obligate intracellular pathogen, *Pneumocystis carinii,* has a genome of only 7.0 Mb, not including the telomeric ends and centromeres (the *Pneumocystis* genome proj-

Table 9.1 Some properties of the *M. globosa* nuclear genome

Size (bp)	8.9 M
G+C content (%)	52
Protein-coding gene number	4,285
Percent coding	69
Percent genes with introns	27
Average gene length (bp)	1,484
Average coding length (bp)	1,447
tRNA genes	82

ect. http://pgp.cchmc.org). The free-living yeast *Eremothecium (Ashbya) gossypii* has a 9.2-Mb genome [18], a size similar to that of the *M. globosa* genome. Other sequenced yeast and fungal genomes are larger, with the largest at 80 Mb [1]. While most characterized bacterial genomes are smaller, some are larger, such as the 13 Mb *Sorangium cellulosum* [19] genome and the 9.7 Mb genome of *Rhodococcus* RHA1 [20]. The relatively small size of the *M. globosa* genome can be attributed to the small number of protein-coding genes (n=4,285), the few introns (present in 27% of genes), and the few repetitive elements (comprising 0.78% of the genome). The number of *M. globosa* protein-coding genes is smaller than that of all other free-living fungi the genome of which has been sequenced. Among those bacteria the genome of which has been sequenced, most encode fewer genes than *M. globosa*. However, a larger number of proteins are encoded by some bacteria, such as *Streptomyces coelicolor* A3 (2) which encodes 7,825 proteins [21].

Small genomes are characteristic of organisms with specialized niches, with obligate intracellular parasites at the extreme. The genome size of *M. globosa*, relative to the genome size of its yeast and fungal relatives, suggests that *M. globosa* may have limited capacity to survive in diverse environments although this capacity is not as limited as in the intracellular microsporidia. The notion of a limited niche for *M. globosa* is consistent with the observation that *Malassezia* yeasts are found on the skin of warm-blooded animals but rarely in any other environment. This limited niche is reflected in the requirements for suitable culture media for growth *in vitro* (as described in Chap. 2.1).

9.3 Phylogenetic Relationship with Other Fungi

In a previous work using the sequence similarities of ribosomal RNA-encoding genes, *Malassezia* had been placed in a phylogenetic tree adjacent to plant pathogens such as *Ustilago maydis* [22, 23]. Further analysis, using a set of protein-coding genes [1], confirmed this placement. Most (75%) of the *M. globosa* genes show *U. maydis* genes as their best matches in the NCBI NR database (excluding the *M. restricta* sequences). Some of these plant pathogens, namely the smut fungi, have had a large impact through the ages on human food. Further study of these fungi should yield novel insights into the mechanisms by which these microorganisms interact with their hosts.

9.4 Lipid and Amino Acid Metabolism

Malassezia globosa requires lipid for growth in culture [24], an observation explained by the lack of a fatty acid synthase gene within the genome [1]. Most other characterized *Malassezia* spp., including *M. restricta*, also require lipid supplementation for growth. Genome sequencing suggests that *M. restricta* too, lacks a fatty acid synthase gene, as fragments of such a large gene would have been detected even with the 1× sequence coverage of that genome. Among free-living fungi with sequenced genomes, only these *Malassezia* species lack a fatty acid synthase gene. By contrast, fatty acid synthase genes are lacking in *Entamoeba histolytica* [25], *Encephalitozoon cuniculi* [17], and *Corynebacterium jeikeium* K411 [26].

In addition to its inability to synthesize fatty acids, *M. globosa* apparently lacks a Δ9 desaturase (EC 1.14.19.2) gene, the product of which places a double bond into common fatty acids, such as oleic acid. A Δ9 desaturase gene was not detected in the *M. restricta* genome either [1]. As Δ9 fatty acids, such as oleic acid, are commonly found on human skin [27], it is likely that the human host is satisfying *M. globosa* needs for such unsaturated fatty acids. In contrast, the genome of the close relative, *U. maydis* contains a Δ9 desaturase gene (UM00955.1).

The *M. globosa* genome encodes the enzymes capable of degrading saturated fatty acids via β-oxidation. However, there is an intriguing limitation. With fatty acids containing a *cis*-double bond, this must be modified to a *trans*-double bond before oxidizing that region of the fatty acid. Furthermore, if the double bond occurs at an odd-numbered carbon, then the double bond must be shifted to an even-numbered carbon. Both of these modifications must happen for β-oxidation of the common skin lipid, oleic acid. The *Saccharomyces cerevisiae ECI1*product, a $\Delta^{3,2}$-enoyl CoA isomerase (EC 5.3.3.8), catalyzes both reactions, and *ECI1*mutant *S. cerevisiae* strains are limited in growth on unsaturated fatty acids [28]. The *M. globosa* genome shows no indication of an *ECI1*-like gene, suggesting that *M. globosa* either (1) has limited capacity for β-oxidation of unsaturated fatty acids, (2) contains a $\Delta^{3,2}$ enoyl CoA isomerase that is different from *ECI1* and not found by similarity searches, or (3) uses a different biochemical pathway for unsaturated fatty acid oxidation. In addition, no evidence for an *ECI1*homolog was found in the *M. restricta* genome sequences. In contrast, a *S. cerevisiae ECI1*homolog (UM01599.1) was found in *U. maydis*. The most likely explanation is that, *M. globosa* and *M. restricta* have lost the *ECI1* and Δ9 desaturase gene homologs since *Malassezia* spp. last shared a common ancestor with *U. maydis,* although further biochemical experiments and genomic analysis will be needed to test this hypothesis. In that respect, it would be interesting to know which of these lipid metabolic enzymes are encoded by *M. pachydermatis*, the one known *Malassezia* species that, while lipophilic, does not require lipid for growth. Also, a comparison of the lipid metabolic capacity of *Cor. jeikeium* and *M. globosa* would be particularly interesting, as both organisms live on skin, lack a fatty acid synthase gene, and depend on exogenous lipids for growth in the laboratory. However, unlike *M. globosa*, *Cor. jeikeium* is probably capable of complete oxidation of oleic acid. The $\Delta^{3,2}$ enoyl CoA isomerase, apparently missing from *M. globosa,* is likely part of a multi-enzyme complex encoded by a homolog of *E. coli fadB* [26, 29].

Corynebacterium jeikeium may share with *M. globosa* a defect in oleic acid synthesis. There are two bacterial pathways for generation of the double bond in oleic acid. One is based on a Δ9 desaturase [30], and the second is based on 3-hydroxydecanoyl-ACP-dehydrogenase [31] that acts during fatty acid elongation. The latter process would not be expected in *Co. jeikeium* because of the absence of fatty acid synthesis in this organism. It is noteworthy that there was no evidence for a *Cor. jeikeium* homolog of a Δ9 desaturase gene, when the *S. cerevisiae OLE1* or the *Mycobacterium tuberculosis desA3* was used in a similarity search [30], although a *desA3* homolog in the genome of the nonlipophilic *Cor. diptheriae* was identified. In addition, no evidence was found for a *Cor. jeikeium* homolog of 3- hydroxydecanoyl-ACP dehydrogenases, as in *E. coli fabA* or *fabZ*.

In contrast to lipid metabolism, where *M. globosa* takes advantage of the host provision of fatty acid sources, *M. globosa* is apparently self reliant for amino acids. The *M. globosa* genome appears to encode all of the enzymes necessary for synthesis of all twenty canonical amino acids. It was expected that amino acids should be available to the yeast through the skin, because the stratum corneum contains amino acids as a result of proteolysis of filaggrin [32]. Besides the lipid metabolic pathways, *Cor. jeikeium* also resembles *M. globosa* in its amino acid metabolic pathways, in that both organisms can apparently synthesize the twenty standard amino acids [26]. It would be interesting to know if the amino acid synthesis genes are transcriptionally active on the skin and whether there is a selective advantage for these microorganisms to synthesize amino acids.

9.5 Lipases

Lipases and phospholipase C may be used by *Malassezia* to generate fatty acids from sebum triglycerides, therefore compensating for the lack of fatty acid synthase. Beyond this nutritional role, lipases may contribute to dandruff symptoms (see above). A large number of gene copies are often interpreted as a sign that the corresponding proteins are of particular significance. Among enzymes encoded by the *M. globosa* genome, lipases and aspartyl proteases are represented by the highest gene copy number. The only characterized *M. globosa* lipase, the LIP1 product, hydrolyzes diglycerides and monoglycerides [10]. However, *M. globosa* cells contain a triglyceride-hydrolyzing activity, so other lipases must be active against triglycerides.

Malassezia globosa encodes twelve lipases that can be partitioned equally into either of two PFAM families and are depicted on a phylogenetic tree of lipases (Fig. 9.2). Our research group constructed the tree following a BLAST search in the NCBI NR database, using the amino acid sequence of each of these twelve lipases. Every lipase, that was one of the top fifteen matches to any of the *M. globosa* lipases, as well as any *U. maydis* and *M. restricta* lipases were included in the tree. Of course, many lipases were among the top matches to multiple *M. globosa* lipases. Although partial sequencing of the *M. restricta* genome [1] resulted in many gene fragments, it was, in four cases, possible to assemble sufficient lipase sequences to place these in the phylogenetic tree. However, some *M. restricta* lipases may be missing.

Among the lipases, one family is PFAM category Lipase 3 (PF01764). *U. maydis* encodes one lipase within this family, whereas *M. globosa* encodes six lipases in the family. As described previously [1], the *M. globosa* genome contains clusters of genes for secreted enzymes. One such cluster contains four consecutive lipase genes (MGL_0797 through MGL_0800), each potentially transcribed in the same direction. These lipases are more similar to each other than to any other protein. It is likely that this cluster arose from lipase gene duplication, after the separation of *Malassezia* and *Ustilago* from a common

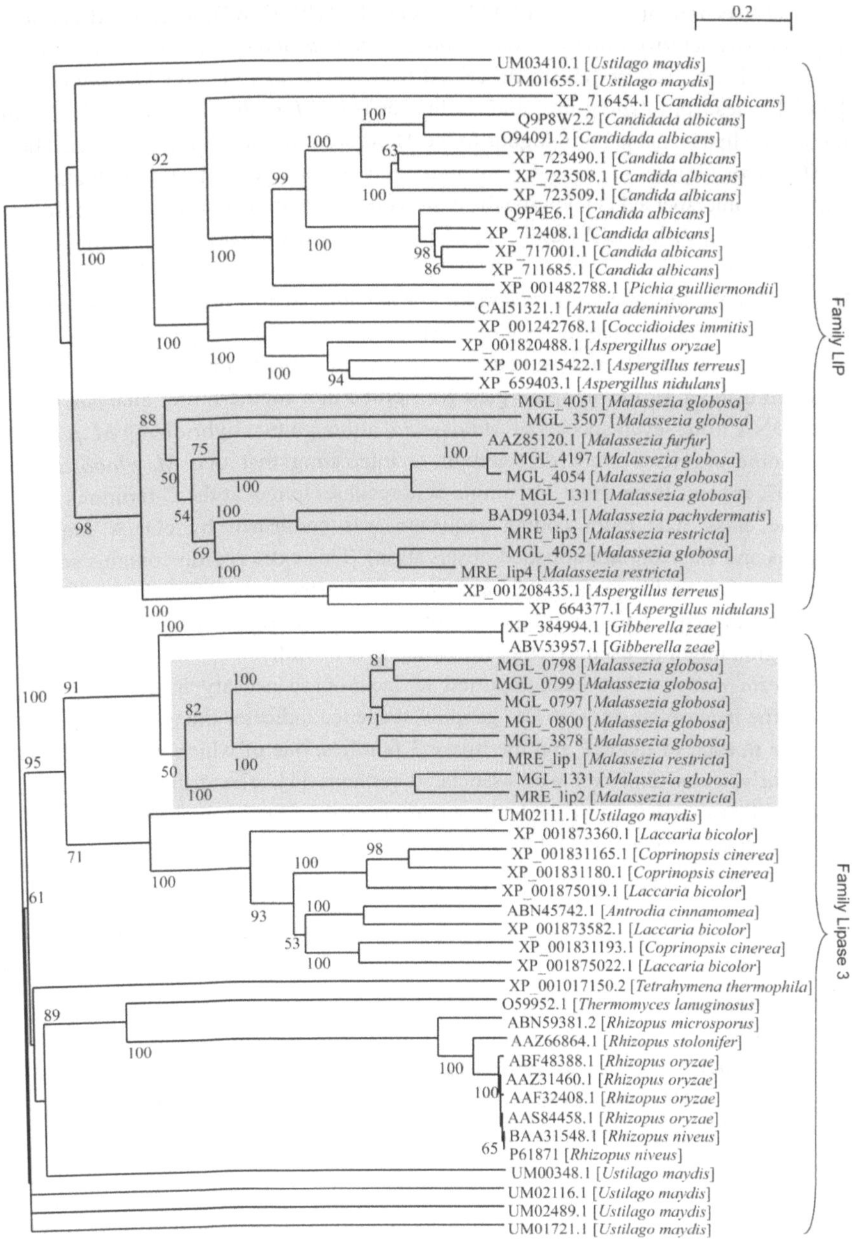

Fig. 9.2 Phylogenetic tree of lipases. To construct the tree, the protein sequences of twelve *M. globosa* lipases that belong to PFAM families LIP and Lipase_3 were searched against the NCBI NR database. The top fifteen matches to any of the *M. globosa* lipases were combined and duplicates removed. Partial open reading frames of putative *M. restricta* lipases were also included. A multiple sequence alignment was generated using MAFFT [33] L-INS-I model. The tree was constructed using PHYML as described previously [1]. The numbers shown on the tree are bootstrap values derived from 100 bootstrap datasets (only those that have a bootstrap value greater or equal than 50 were shown)

ancestor. The second family is the PFAM category LIP (PF03583), with one division containing well-characterized lipases from *Candida albicans* and one lipase from *U. maydis*. A second division contains lipases described from *M. furfur* [34] and *M. pachydermatis* [35], two lipases from *Aspergillus*, and six lipases from *M. globosa*. It is not clear if there is an *Ustilago* lipase that is homologous to the *Malassezia* lipases in this division. Three of these *M. globosa* lipases are encoded by a secreted enzyme gene cluster, with an aspartyl protease gene interspersed [1]. Unlike the case with the set of PFAM Lipase 3 lipase genes, the gene cluster of PFAM LIP members is not a set of most similar genes, making interpretation of its history more speculative. Although members of the LIP family clearly belong to the same family, they are not very similar to each other, with a protein sequence identity of about 30–45%. The nucleotide sequence has no apparent similarity to each other, which may be the reason why Brunke and Hube [34] were not able to identify additional lipase genes when using a *M. furfur* lipase gene as a probe in a southern blot analysis with the genomic DNA from seven species of *Malassezia*, although two hybridizing *M. pachydermatis* genome segments were detected. It is interesting that one *M. globosa* lipase, MGL_3507, appears to be about 110 amino acid residues longer at the C-terminus than the rest of the class. The extra coding sequence was confirmed by cDNA sequencing (DeAngelis and Dawson, unpublished observation). This extra region contains some low complexity amino acid sequence and is rich in serine, lysine and glycine. In the absence of experimental evidence and homology with other known protein sequences, we speculate that this region might be used to tether the lipase to cell wall.

Malassezia restricta has been reported to lack lipase activity in culture [10, 13]. However, the incomplete *M. restricta* genome sequence indicates the presence of at least two lipases in each of the LIP and the lipase 3 families, one of which was found among *M. restricta* secreted and cell wall-associated proteins [1]. The difficulty in detecting *M. restricta* lipase activity may be due to a combination of factors including culture optimization for lipase production and the choice of the appropriate lipase assay.

9.6 Aspartyl Proteases

Malassezia globosa encodes seventeen aspartyl protein genes, an uncommonly large number [1] (Fig. 9.3). Proteases could hydrolyze host proteins to supply nutrients, degrade host tissues, modify host cells to facilitate adhesion, or alter the immune response.

A phylogenetic tree of aspartyl proteases was created, using the methodology described for lipases; the only difference being that we have included in the tree, the top ten matches for each of the *M. globosa* aspartyl proteases. In one case, neighboring genes encoded aspartyl proteases with the highest similarity (MGL 3330 and 3331). In other cases, two *M. globosa* protease genes, despite being highly similar to each other, were not adjacent in the genome. As was the case with lipase genes, this pattern suggests gene duplication events more recent than the split from the common ancestor shared with *Ustilago*.

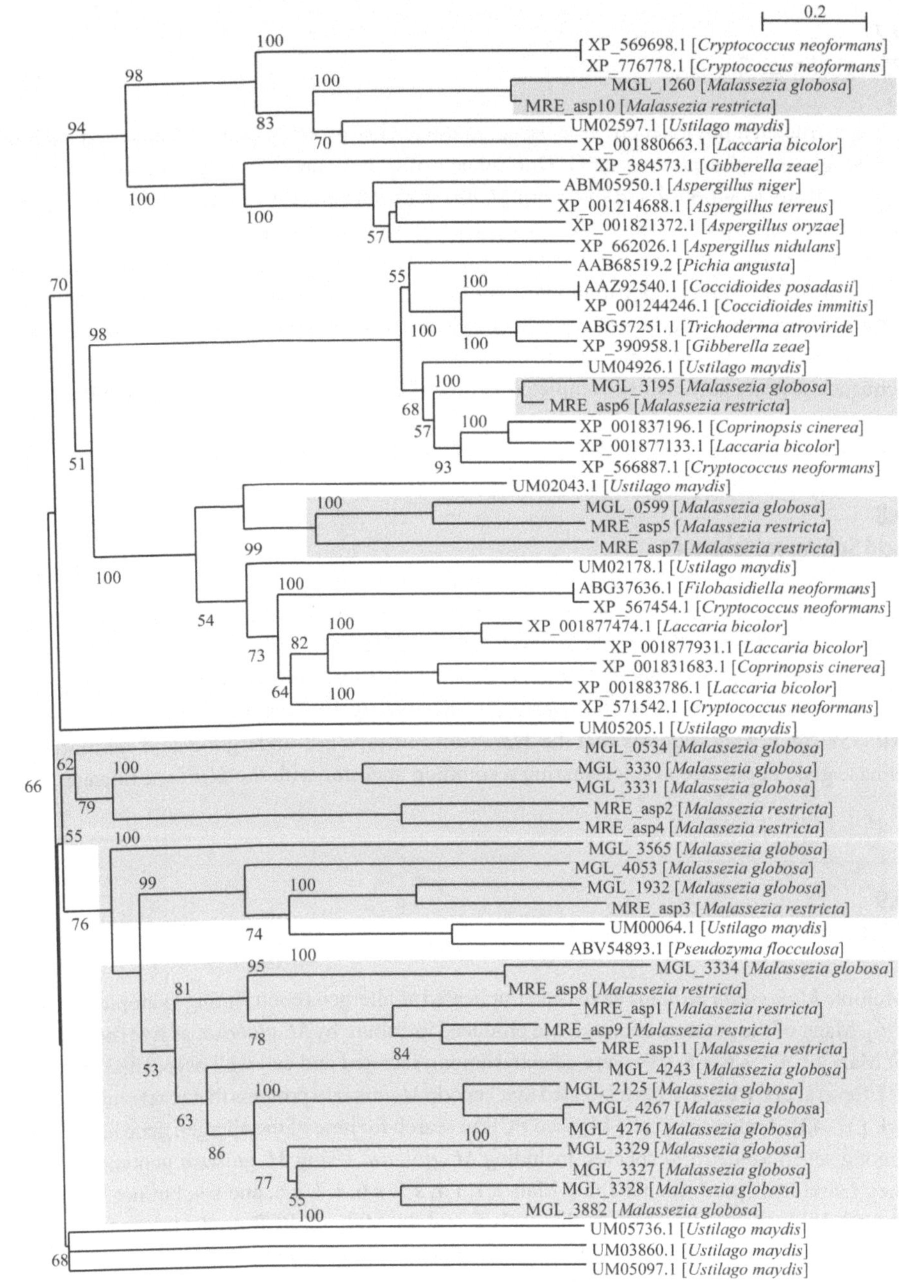

Fig. 9.3 Phylogenetic tree of aspartyl proteases. The tree was constructed using the same method as described in Fig. 9.2 except that only the top ten hits from the NCBI NR databases were included

9.7 Phospholipase C

We constructed a phylogenetic tree based on the twenty proteins that best matched each of the *M. globosa* phospholipases C. One branch of a tree contains only *Malassezia* genes, with six from *M. globosa* and five from *M. restricta.* The closest relatives are a set of ascomycete phospholipases, with a more distant set of bacterial phospholipases.

There are no basidiomycete phospholipases related closely enough to be shown (Fig. 9.4). Three of the *M. globosa* phospholipase genes are found adjacent to each other, and these three are each other's closest relatives within *M. globosa.* Each of these phospholipases has for its most similar protein a *M. restricta* phospholipase C. Our *M. restricta* genome sequence was not sufficiently complete to know if these genes are adjacent in the genome.

9.8 Acid Sphingomyelinases

A phylogenetic tree (not shown) indicates that one branch contains three adjacent *M. globosa* acid sphingomyelinase genes (MGL_1573, 1574, and 1575), as well as a fourth acid sphingomyelinase gene (MGL_568) and three *M. restricta* genes (MRE asm1, asm2, and asm3). Most similar to these *Malassezia* acid sphingomyelinase genes is a *U. maydis* gene, XP_758084.1. This suggests that the *Malassezia* lineage has undergone acid sphingomyelinase gene duplications since sharing a common ancestor with the *Ustilago* lineage.

9.9 Allergens

Multiple *Malassezia* proteins have been implicated as allergens contributing to atopic eczema [36]. Many of the putative allergens are produced in culture by *M. globosa*, as five (homologs of Mala s 1, f 2, s 6, s 8, s 9) were detected among secreted and cell wall-associated proteins [1]. Several cDNAs have been isolated that encode *Malassezia* proteins that react with human IgE [37–43]. Andersson et al. [44] used PCR to search for nine of the allergen gene homologs among seven *Malassezia* species, including *M. globosa*. Using *M. globosa* genomic DNA, they found PCR products matching Mala s 1, f 4, s 5, s 6, s 7, s 8, and s 9, but not Mala f 2 and f 3. With *M. globosa* mRNA, they only found the Mala s 6 PCR product.

Based on genome sequence, *M. globosa* encodes proteins similar to all of the putative *Malassezia* allergens. However, there is no simple one-to-one relationship among all of these potential allergens (Table 9.2). For example, Mala s 7 shows a weak match to two *M. globosa* proteins, MGL_0968 and MGL_2673. A single protein from *M. globosa,* MGL_4042, shows similarity to three previously described allergens, Mala f 2, f 3 and s 5, with the highest similarity to Mala f 2 (Table 9.2).

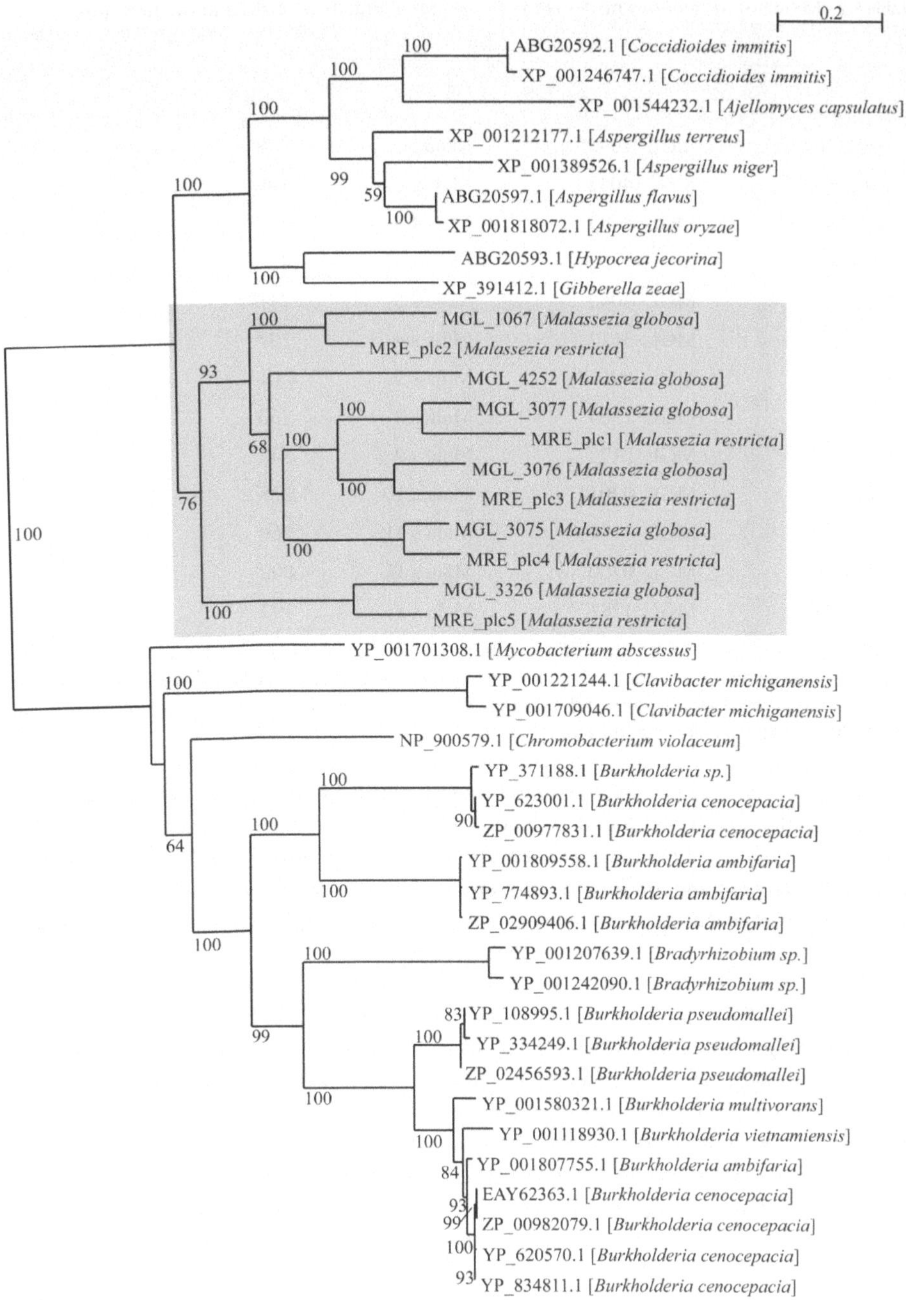

Fig. 9.4 Phylogenetic tree of phospholipases C. The tree was constructed using the same method as described in Fig. 9.2 except that only the top twenty hits from the NCBI NR databases were included

Table 9.2 Match of *M. globosa* proteins to *Malassezia* allergens described in the literature

Accession No.	Best matched *M. globosa* gene	Name	Blast bit-score	In *M. globosa* EST sequence database
AB011804	MGL_4042	Mala f 2	288	–
AB011805	MGL_4042	Mala f 3	199	–
AF084828	MGL_2703	Mala f 4	517	X
X96486	MGL_1303	Mala s 1	497	X
AJ011955	MGL_4042	Mala s 5	211	–
AJ011956	MGL_3612	Mala s 6	297	–
AJ011957	MGL_0968	Mala s 7	48.1	–
AJ011958	MGL_1304	Mala s 8	250	–
AJ011959	MGL_2179	Mala s 9	342	X
AJ428052	MGL_0201	Mala s 10	1,328	X
AJ548421	MGL_3190	Mala s 11	269	X
AJ871960	MGL_0750	Mala s 12	802	X
AJ937746	MGL_1781	Mala s 13	192	–

The accession number is the nucleotide sequence accession number. The bit-score is using the translated amino acid sequence of the allergens against the proteome of *M. globosa*. The match of Mala s 7 has a very low bit-score and only matches partially, but it is the best match to Mala s 7 in the NCBI nr database other than Mala s 7 itself

These three allergens were known to be similar [38] and thought to encode a peroxiredoxin. Similar genes are found in many other fungal species, including *U. maydis*, *Coprinopsis cinerea*, *Yarrowia lipolytica*, *Paracoccidioides brasiliensis*, *Laccaria bicolor*, *Aspergillus terreus*, and *Ajellomyces capsulatus*. The absence of the Mala f 2 and f 3 products in the PCR reactions of Andersson et al. [44] may be explained by the lack of a good match to MGL_4042 by the primers used for PCR.

There is some controversy over the presence of a Mala s 1 homolog in *Malassezia* other than *M. sympodialis*. Andersson et al. [44] found this homolog in four *Malassezia* species, including *M. globosa* 7966, whereas Gaitanis *et al.* [45] did not find a Mal s 1 homolog in several *Malassezia* species, including *M. globosa* 7966. The two groups collaborated on a letter [46] suggesting that choice of primers may have contributed to the discrepancy and that the Mala s 1 homologs may not closely match each other. Our alignment of Mala s 1 with its *M. globosa* homolog, MGL_1303, indicates that the two genes share about 70% identity in amino acid sequence. However, the nucleotide sequences are not highly conserved, and less than 1/5 of the nucleotide sequence can be aligned. We were not able to find a convincing match using the two sets of primers used by Gaitanis et al. [45] and Andersson et al. [44]. The lack of a good match to the PCR primers likely explains the difficulty in obtaining a PCR product for Mala s 1 in *M. globosa*. It is interesting that the *M. globosa* MGL_1303 gene (Mala s 1 homolog) and the MGL_1304 gene (Mala s 8 homolog) are located near each other in the chromosome, separated by only 605 base pairs.

In summary, genomic and proteomic analysis indicates that *M. globosa* encodes proteins similar to the allergens that have been identified in *M. furfur* and *M. sympodialis*.

Zargari et al. [47] have used Western blots to show that human IgE recognizes many proteins from *M. globosa*, and recognition of only some of these proteins is inhibited by competition with *M. sympodialis* protein. This implies that *M. globosa* produces many IgE-reactive epitopes that are not produced by *M. sympodialis*. Therefore, there may still be other *M. globosa* proteins that react with IgE and have not yet been characterized. More research is required to understand the role of these potential allergens in human disease.

9.10 Tryptophan Metabolites

In addition to the secreted enzymes that are likely to interact with the host, *Malassezia* also produce small molecules that may interact with the host. *Malassezia* are reported to produce tryptophan metabolites with a role in SD [12] and pityriasis versiolor [48]. Zuther et al. [48] used *U. maydis* as a model organism and demonstrated that this plant pathogen produced a similar set of tryptophan metabolites as does *M. furfur*. This analysis led to the discovery of two genes with a role in pigment production. The tryptophan aminotransferase Tam1p catalyzes the formation of indole pyruvate, which reacts to form pigments. We observed that *M. globosa* encodes a *TAM1* homolog, MGL_2601. Zuther et al. [48] also found that mutations in the sulphite reductase gene, *SIR1*, led to defective pigment production, *M. globosa* contains a *SIR1* homolog, MGL_0080.

9.11 Mating

The *M. globosa* genome contains a mating locus, with pheromone receptor and bW and bE homeodomain genes, with some similarity to other basidiomycete mating-type loci [49]. This suggests that *M. globosa* may be capable of undergoing a sexual cycle. There are precedents for genome analysis providing the first indications of sexuality among fungi. With *C. albicans,* the discovery of mating- type genes was followed by the discovery of mating [50]. *Aspergillus oryzae* and *A. fumigatus* were thought to be asexual until the genome sequence indicated the presence of many genes associated with mating [51]. Perhaps, the existence of mating genes within the *M. globosa* genome will prompt the discovery of mating in these fungi. Mating can be important in the spread of virulence traits [52] and may be important in the distribution of pathogenic strains of *Malassezia.*

9.12 Adaptation to Animal Skin

As pointed out by Xu et al. [1], *M. globosa* and *C. albicans* are capable of living on animal skin and secrete a similar set of enzymes; lipases, aspartyl proteases, phospholipases, and acid sphingomyelinases. In contrast, *U. maydis,* more closely related phylogenetically to *M. globosa,* secretes a different set of enzymes, including many glycosyl hydrolases. We propose that the sets of secreted enzymes by *M. globosa* and *C. albicans* represent an adaptation to life on animal skin. In this chapter, we have extended the previous observations by generating phylogenetic trees of the lipases, aspartyl proteases, phospholipases C, and acid sphingomyelinases, providing for the first time, evidence that *Malassezia* secreted enzyme genes have undergone duplications since sharing a common ancestor with *Ustilago.* As additional genomes of related organisms become available, it will be interesting to analyze the apparent losses and duplications of genes for secreted enzymes.

9.13 Conclusion

The *M. globosa* genome reveals duplications of genes for secreted enzymes, including lipases, aspartyl proteases, phospholipases C, and acid sphingomyelinases. These duplications likely arose after the divergence of *Malassezia* from its ancestors, shared with the plant pathogen *Ustilago.* These enzymes are also encoded by multiple genes in *C. albicans,* a human opportunistic pathogen found on skin, and it is therefore likely that these multicopy genes are a result of adaptation to animal skin.

Analysis of the *M. globosa* and *M. restricta* genomes discloses multiple pathways with potential roles in the pathogenesis of dandruff and seborrheic dermatitis, providing new intervention targets for the development of novel, more effective treatments.

Acknowledgments We thank the coauthors of our genome and proteome analysis publication [1] as well as all the individuals acknowledged in that publication. We especially highlight our appreciation to Pamela Trotter for her insights into lipid metabolism. We thank Julie Baker for the electron micrographs and Holly Krigbaum for the dandruff photograph.

References

1. Xu J, Saunders CW, Hu P et al (2007) Dandruff-associated *Malassezia* genomes reveal convergent and divergent virulence traits shared with plant and human fungal pathogens. Proc Natl Acad Sci USA 104:18730–18735
2. Ashbee HR, Evans EG (2002) Immunology of diseases associated with *Malassezia* species. Clin Microbiol Rev 15:21–57
3. Feng Y, Chen CJ, Su LH et al (2008) Evolution and pathogenesis of *Staphylococcus aureus*: lessons learned from genotyping and comparative genomics. FEMS Microbiol Rev 32:23–37
4. Zahrt TC (2003) Molecular mechanisms regulating persistent *Mycobacterium tuberculosis* infection. Microbes Infect 5:159–167
5. White TC, Oliver BG, Gräser Y et al (2008) Generating and testing molecular hypotheses in the dermatophytes. Eukaryot Cell 7:1238–1245
6. Calderone RA, Fonzi WA (2001) Virulence factors of *Candida albicans*. Trends Microbiol 9:327–335

7. Ro BI, Dawson TL (2005) The role of sebaceous gland activity and scalp microfloral metabolism in the etiology of seborrheic dermatitis and dandruff. J Investig Dermatol Symp Proc 10:194–197
8. Schwartz JR, Cardin CW, Dawson TL (2004) Dandruff and seborrheic dermatitis. In: Baran R, Maibach HI (eds) Textbook of cosmetic dermatology. Martin Dunitz, London, UK
9. Crespo Erchiga V, Delgado Florencio V (2002) *Malassezia* species in skin diseases. Curr Opin Infect Dis 15:133–142
10. DeAngelis YM, Saunders CW, Johnstone KR et al (2007) Isolation and expression of a *Malassezia globosa* Lipase Gene, LIP1. J Invest Dermatol 127:2138–2146
11. Troller JA (1971) Model system for the investigation of dandruff. J Soc Cosmet Chem 22:187–198
12. Gaitanis G, Magiatis P, Stathopoulou K et al (2008) AhR ligands, malassezin, and indolo[3, 2-b]carbazole are selectively produced by *Malassezia furfur* strains isolated from seborrheic dermatitis. J Invest Dermatol 128:1620–1625
13. Aspiroz C, Moreno LA, Rezusta A et al (1999) Differentiation of three biotypes of *Malassezia* species on human normal skin. correspondence with *M. globosa, M. sympodialis* and *M. restricta*. Mycopathologia 145:69–74
14. Gemmer CM, DeAngelis YM, Theelen B et al (2002) Fast, noninvasive method for molecular detection and differentiation of *Malassezia* yeast species on human skin and application of the method to dandruff microbiology. J Clin Microbiol 40:3350–3357
15. Midgley G (2000) The lipophilic yeasts: state of the art and prospects. Med Mycol 38(Suppl 1): 9–16
16. Paulino LC, Tseng CH, Blaser MJ (2008) Analysis of *Malassezia* microbiota in healthy superficial human skin and in psoriatic lesions by multiplex real-time PCR. FEMS Yeast Res 8: 460–471
17. Katinka MD, Duprat S, Cornillot E et al (2001) Genome sequence and gene compaction of the eukaryote parasite *Encephalitozoon cuniculi*. Nature 414:450–453
18. Dietrich FS, Voegeli S, Brachat S et al (2004) The *Ashbya gossypii* genome as a tool for mapping the ancient *Saccharomyces cerevisiae* genome. Science 304:304–307
19. Schneiker S, Perlova O, Kaiser O et al (2007) Complete genome sequence of the myxobacterium *Sorangium cellulosum*. Nat Biotechnol 25:1281–1289
20. McLeod MP, Warren RL, Hsiao WW et al (2006) The complete genome of *Rhodococcus* sp. RHA1 provides insights into a catabolic powerhouse. Proc Natl Acad Sci USA 103: 15582–15587
21. Bentley SD, Chater KF, Cerdeno-Tarraga AM et al (2002) Complete genome sequence of the model actinomycete *Streptomyces coelicolor* A3(2). Nature 417:141–147
22. Begerow D, Bauer R, Boekhout T (2000) Phylogenetic placements of ustilaginomycetous anamorphs as deduced from nuclear LSU rDNA sequences. Mycol Res 104:53–60
23. Begerow D, Stoll M, Bauer R (2006) A phylogenetic hypothesis of Ustilaginomycotina based on multiple gene analyses and morphological data. Mycologia 98:906–916
24. Chen TA, Hill PB (2005) The biology of *Malassezia* organisms and their ability to induce immune responses and skin disease. Vet Dermatol 16:4–26
25. Loftus B, Anderson I, Davies R et al (2005) The genome of the protist parasite *Entamoeba histolytica*. Nature 433:865–868
26. Tauch A, Kaiser O, Hain T et al (2005) Complete genome sequence and analysis of the multiresistant nosocomial pathogen *Corynebacterium jeikeium* K411, a lipid-requiring bacterium of the human skin flora. J Bacteriol 187:4671–4682
27. Stewart ME, Steele WA, Downing DT (1989) Changes in the relative amounts of endogenous and exogenous fatty acids in sebaceous lipids during early adolescence. J Invest Dermatol 92:371–378
28. Gurvitz A, Mursula AM, Firzinger A et al (1998) Peroxisomal Delta3-cis-Delta2-trans-enoyl-CoA isomerase encoded by *ECI1* is required for growth of the yeast *Saccharomyces cerevisiae* on unsaturated fatty acids. J Biol Chem 273:31366–31374

29. Yang SY, Li JM, He XY et al (1988) Evidence that the fadB gene of the fadAB operon of *Escherichia coli* encodes 3-hydroxyacyl-coenzyme A (CoA) epimerase, delta 3-cis-delta 2-trans-enoyl-CoA isomerase, and enoyl-CoA hydratase in addition to 3-hydroxyacyl-CoA dehydrogenase. J Bacteriol 170:2543–2548
30. Phetsuksiri B, Jackson M, Scherman H et al (2003) Unique mechanism of action of the thiourea drug isoxyl on *Mycobacterium tuberculosis*. J Biol Chem 278:53123–53130
31. Wang H, Cronan JE (2004) Functional replacement of the FabA and FabB proteins of *Escherichia coli* fatty acid synthesis by *Enterococcus faecalis* FabZ and FabF homologues. J Biol Chem 279:34489–34495
32. Rawlings AV, Harding CR (2004) Moisturization and skin barrier function. Dermatol Ther 17(Suppl 1):43–48
33. Katoh K, Kuma K, Toh H et al (2005) MAFFT version 5: improvement in accuracy of multiple sequence alignment. Nucleic Acids Res 33:511–518
34. Brunke S, Hube B (2006) MfLIP1, a gene encoding an extracellular lipase of the lipid-dependent fungus *Malassezia furfur*. Microbiology 152:547–554
35. Shibata N, Okanuma N, Hirai K et al (2006) Isolation, characterization and molecular cloning of a lipolytic enzyme secreted from *Malassezia pachydermatis*. FEMS Microbiol Lett 256: 137–144
36. Casagrande BF, Fluckiger S, Linder MT et al (2006) Sensitization to the yeast *Malassezia sympodialis* is specific for extrinsic and intrinsic atopic eczema. J Invest Dermatol 126: 2414–2421
37. Andersson A, Rasool O, Schmidt M et al (2004) Cloning, expression and characterization of two new IgE-binding proteins from the yeast *Malassezia sympodialis* with sequence similarities to heat shock proteins and manganese superoxide dismutase. Eur J Biochem 271: 1885–1894
38. Limacher A, Glaser AG, Meier C et al (2007) Cross-reactivity and 1.4-A crystal structure of *Malassezia sympodialis* thioredoxin (Mala s 13), a member of a new pan-allergen family. J Immunol 178:389–396
39. Lindborg M, Magnusson CG, Zargari A et al (1999) Selective cloning of allergens from the skin colonizing yeast *Malassezia furfur* by phage surface display technology. J Invest Dermatol 113: 156–161
40. Onishi Y, Kuroda M, Yasueda H et al (1999) Two-dimensional electrophoresis of *Malassezia* allergens for atopic dermatitis and isolation of Mal f 4 homologs with mitochondrial malate dehydrogenase. Eur J Biochem 261:148–154
41. Schmidt M, Zargari A, Holt P et al (1997) The complete cDNA sequence and expression of the first major allergenic protein of *Malassezia furfur*, Mal f 1. Eur J Biochem 246:181–185
42. Yasueda H, Hashida-Okado T, Saito A et al (1998) Identification and cloning of two novel allergens from the lipophilic yeast, *Malassezia furfur*. Biochem Biophys Res Commun 248: 240–244
43. Zargari A, Selander C, Rasool O et al (2007) Mala s 12 is a major allergen in patients with atopic eczema and has sequence similarities to the GMC oxidoreductase family. Allergy 62: 695–703
44. Andersson A, Scheynius A, Rasool O (2003) Detection of Mala f and Mala s allergen sequences within the genus *Malassezia*. Med Mycol 41:479–485
45. Gaitanis G, Menounos P, Katsambas A et al (2003) Detection and mutation screening of *Malassezia sympodialis* sequences coding for the Mal s 1 allergen implicated in atopic dermatitis. J Invest Dermatol 121:1559–1560
46. Gaitanis G, Velegraki A, Rasool O et al (2005) Clarifications on conflicting results published on the amplification of the Mala s 1 allergen gene sequences. J Invest Dermatol 124:479
47. Zargari A, Midgley G, Back O et al (2003) IgE-reactivity to seven *Malassezia* species. Allergy 58: 306–311

48. Zuther K, Mayser P, Hettwer U et al (2008) The tryptophan aminotransferase Tam1 catalyses the single biosynthetic step for tryptophan-dependent pigment synthesis in *Ustilago maydis*. Mol Microbiol 68:152–172
49. Bakkeren G, Kämper J, Schirawski J (2008) Sex in smut fungi: structure, function and evolution of mating-type complexes. Fungal Genet Biol 45(Suppl 1):S15–21
50. Miller MG, Johnson AD (2002) White-opaque switching in *Candida albicans* is controlled by mating-type locus homeodomain proteins and allows efficient mating. Cell 110:293–302
51. Galagan JE, Calvo SE, Cuomo C et al (2005) Sequencing of *Aspergillus nidulans* and comparative analysis with *A. fumigatus* and *A. oryzae*. Nature 438:1105–1115
52. Fraser JA, Giles SS, Wenink EC et al (2005) Same-sex mating and the origin of the Vancouver Island *Cryptococcus gattii* outbreak. Nature 437:1360–1364

Malassezia Yeasts in Animal Disease 10

Ross Bond, Jacques Guillot and F. Javier Cabañes

Core Messages

› During the last 20 years, interest in the genus *Malassezia* has increased dramatically among veterinarians. *M. pachydermatis* is now recognised as an important cause of dermatitis and otitis externa in dogs, but appears to be a relatively uncommon pathogen in other animal hosts. Cases of dermatitis or otitis externa in dogs and cats are normally managed by topical or systemic antifungal therapy and by correcting factors predisposing to infection. Lipid-dependent *Malassezia* are more frequently isolated in horses and domestic ruminants. This chapter reviews current knowledge on colonisation and infection of dogs, cats and other animals by *Malassezia* spp., including pathogenesis, clinical signs, diagnosis and treatment. The potential for zoonotic transfer of *Malassezia* spp. from animals to humans is also discussed.

10.1 *Malassezia pachydermatis* as a Pathogen in the Dog

10.1.1 Historical Aspects of Infection in Dogs

Malassezia pachydermatis was first described by Weidman [1], who observed numerous bottle- or 'bowling pin'-shaped yeasts in the skin scales from an Indian rhinoceros (*Rhinoceros unicornis*) with a generalised exfoliative dermatitis. The organism was cultured on an unspecified medium, forming smooth, entire, domed, doughy, cream-yellow colonies; the morphology of the cells in culture was the same as described in the skin scales.

R. Bond (✉)
Senior Lecturer in Veterinary Dermatology, Department of Veterinary Clinical Studies, Royal Veterinary College, Hawkshead Lane, North Mymms, Hatfield, Herts, AL9 7TA, UK
e-mail: rbond@rvc.ac.uk

T. Boekhout et al. (eds.), *Malassezia and the Skin*
DOI: 10.1007/978-3-642-03616-3_10, © Springer Verlag Berlin Heidelberg 2010

Weidman was unable to match the organism to any existing descriptions, but commented on the similarities to *Pityrosporum ovale*. However, the organism from the rhinoceros was smaller and less irregular in shape than *P. ovale*, and he proposed the name *Pityrosporum pachydermatis*.

Gustafson [2] was the first to suggest that *M. pachydermatis* was involved in the pathogenesis of canine otitis externa (OE), having isolated more than 200 strains from affected dogs. The yeasts were round or oval and showed mono-polar budding with a broad base, without mycelium or pseudomycelium formation. None of the 20 strains tested fermented or assimilated sugars, and he concluded that these yeasts belonged to the genus *Pityrosporum*. Unlike the previously recognised *Pityrosporum* yeasts, *P. orbiculare* and *P. ovale*, the strains grew well without lipid-supplementation. Gustafson was able to induce OE by applying *M. pachydermatis* to the external ear canal, whereas a '*P. ovale*' strain of human origin had no effect. Although Slooff [3] suggested that Gustafson's strains were probably identical to *P. pachydermatis*, Gustafson proposed the name *Pityrosporum canis* for his strains, the cultural requirements of the former species being not known. Fraser [4] isolated non lipid-dependent *Pityrosporum* yeasts from the ear canals of both healthy dogs and dogs with OE. He considered that his strains were consistent with the original descriptions of Weidman [1] and termed these *P. pachydermatis*. The yeast was isolated in comparable frequencies from healthy dogs and dogs with OE, but he found that affected dogs tended to have larger numbers of the yeast on smears of cerumen. Subsequent studies either confirmed Fraser's findings [5] or reported an increased prevalence of the yeast in OE [6, 7]. Studies in which cases of OE in dogs responded to antifungal therapy were described [8–10], supporting the current opinion that the yeast acts as an opportunistic secondary pathogen within the canine external ear canal.

In 1970, Slooff [3] assigned all strains of *Pityrosporum* capable of growing without lipid-supplementation to the single species *P. pachydermatis*. The International Commission on the Taxonomy of Fungi [11] recommended that the generic name *Malassezia* should take priority over *Pityrosporum*, and Yarrow and Ahearn [12] adopted the name *M. pachydermatis* in the third edition of 'The yeasts, a taxonomic study.' Subsequent molecular studies have confirmed that *M. pachydermatis* may be regarded as a species separate from the lipid-dependent species, although there is quite marked genetic heterogeneity within the species [13, 14]. These aspects are discussed in more detail in Chaps. 2 and 3.

Dufait [15] made the first reports of *M. pachydermatis* as a cause of more widespread dermatitis in the dog. He described a series of 50 dogs with pruritic dermatitis from which the yeast could be readily recovered by cytology or culture and which responded to antifungal therapy. Skin lesions consisted of erythema and hyper-pigmentation that most often affected the ventral abdomen, although the face, feet and perineal regions were also commonly affected. Similar findings were described in 23 cases from Brazil [16] and a severe pruritic seborrhoeic skin disorder was described, from which *M. pachydermatis* could be isolated in eight West Highland white terriers [17]. Mason and Evans [18] described 11 cases in detail and demonstrated the presence of the yeast in the stratum corneum in all nine cases from which skin biopsies were obtained. Concurrent allergic diseases were recognised in four dogs, and one dog had hepatic disease. In each case, antibacterial therapy was without benefit, but all dogs responded to either systemic ketoconazole or topical antifungal therapy. This seminal report forced the wider veterinary dermatology community to consider the potential role of *M. pachydermatis* as a cause of canine skin disease beyond the external ear canal, and during the 1990s, there was a gradual acceptance that the yeast was an opportunistic secondary pathogen.

10.1.2
Factors Predisposing to Diseases

The ability of fungi to colonise or infect mammalian skin is determined at the molecular level by complex interactions between the fungal cell and its virulence factors and the host tissues and components of the immune system. Processes such as fungal cell adherence to corneocytes, and innate and specific immune responses that are mounted by the host, are important in determining the outcome of colonisation. At the clinical level, concurrent skin diseases are often recognised in dogs with skin infections caused by commensal microbes of the skin, including *Staphylococcus pseudintermedius* and *M. pachydermatis*, and identification and correction of these disorders can reduce the recurrence of infection.

10.1.2.1
Breed Predilections

In a survey of dogs in the United States using a cytological technique [19], it was reported that basset hounds and dachshunds were predisposed to the development of abnormally high populations of *M. pachydermatis*. In another study of 40 dogs, conducted in the South-East of England, elevated cutaneous *Malassezia* sp. populations were identified [20]. Various breeds and crossbreeds were represented; basset hounds, cocker spaniels, and West Highland white terriers were significantly over-represented compared with samples of both the general and dermatological hospital population, and boxers were over-represented compared with the general population. In a series of cases examined in France, West Highland white terriers, Shar Pei and basset hounds were most frequently affected [21]. By contrast, both Mason [22] and Dufait [15] reported the poodles to be the most frequently affected breed in their case series, in Australia and Belgium, respectively, whereas poodles were not present in the group of dogs seen in England. These differences may reflect regional variations in breed popularity and/or differences in the regional gene pool.

10.1.2.2
Concurrent Diseases

Canine atopic dermatitis (AD) is a genetically predisposed inflammatory and pruritic allergic skin disease with characteristic clinical features that is associated most commonly with IgE antibodies to environmental allergens [23]. In one of the original cases series of canine *Malassezia* dermatitis, concurrent atopic dermatitis was diagnosed in 2 out of 11 cases [18].

The relationship between AD and *M. pachydermatis* carriage was investigated by comparing yeast population sizes in the axilla and groin regions of 12 normal and 12 atopic dogs using tape-strips and contact plates [24]. *Malassezia* yeasts were identified using the tape-strip method in all normal and atopic dogs studied although counts were low in the healthy dogs and in six out of 12 atopic dogs, whereas the remaining six atopic dogs had higher counts. Both the numbers of sites from which the yeast was observed and the tape-strip counts in the atopic dogs significantly ($P<0.05$) exceeded those of healthy dogs at the groin, but not the axilla. Using the contact plate method, *M. pachydermatis* was isolated on contact plates from the groin of one healthy dog and from the axilla of another, whereas *M. pachydermatis* was cultured from eight of the atopic dogs. The number of dogs and number of sites in both the axilla and groin from which *M. pachydermatis* was cultured was significantly greater in the atopic group ($P<0.05$, $P<0.001$ and $P<0.001$, respectively).

In a study of 25 atopic and 25 healthy dogs conducted in France, yeast counts were obtained from four body areas (inter-digital skin of the feet, lip folds, ears and beneath the tail) by cytological assessment of slides prepared using swabs or superficial scrapings [25]. Atopic dogs were shown to have significantly higher yeast scores from the inter-digital skin and from beneath the tail. Within the atopic dogs, yeast counts were higher from inter-digital webs that were erythematous. In a recent study of a group of 41 atopic dogs, *M. pachydermatis* was isolated from the inter-digital skin of 71% and ears of 63% of the dogs [26]. Taken together, these studies indicate that elevated *M. pachydermatis* counts may be found in some dogs with AD.

Although AD tends to be the most frequently diagnosed concurrent disease in dogs with *Malassezia* dermatitis [20], it is important to appreciate that not all dogs with AD have *Malassezia* dermatitis, and that *Malassezia* dermatitis occurs in association with disorders other than AD [27]. Intertrigo, endocrinopathies and primary cornification defects are also recognised in some affected dogs [20, 21]. It is also important to consider that a concurrent disease might occur co-incidentally without necessarily favouring the yeast infection.

10.1.3 Pathogenesis

Malassezia pachydermatis is a normal inhabitant of healthy canine skin and mucosae; the yeast is most commonly isolated from the lip region, inter-digital skin, anal mucosa and external ear canal [2, 28–30]. In a temporal study of *M. pachydermatis* carriage in eight healthy beagles, 'resident' carriage, defined in this study as yeast isolation on at least 5 out of 6 occasions at intervals of 2 weeks, was identified in the external ear canal and anus of 7 and 3 dogs, respectively [31]. Skin populations in healthy basset hounds, a breed predisposed to *Malassezia* dermatitis, were found to exceed those of healthy dogs of other breeds, which may reflect the susceptibility of this breed to the disease [32, 33].

Dogs with skin disease associated with *M. pachydermatis* may have 100–10,000-fold increases in skin population densities in comparison with healthy dogs [32, 33]. *M. pachydermatis* populations at the nose, prepuce and vulva in affected basset hounds also exceeded those of healthy dogs, indicating that the factors which enable the development of abnormally high cutaneous populations also permit proliferation of the yeast at certain mucosal sites [32]. The presence of high population densities of the yeast in lesional skin in dogs with dermatitis refractory to antibacterial and anti-inflammatory therapy, and the clinical response following antifungal treatments and the associated reduction in yeast numbers provides good evidence for a pathogenic role [15, 18, 34], but it is not known whether there is a threshold populations density needed for infection. However, the density of organisms on the skin required to induce disease may vary in different circumstances. Atopic dogs sensitised to *M. pachydermatis* allergens [35], in a manner analogous to the apparent hypersensitivity to lipid-dependent *Malassezia* yeasts observed in human patients with the 'head-neck' form of AD [36], could show marked cutaneous inflammation with relatively low yeast populations. In addition, strain differences in the 'virulence' of the yeast may also be of importance, and quorum sensing events described in bacteria [37] may alter the expression of virulence characteristics dependent upon the population density.

Adherence, the specific attachment of micro-organisms to cells and tissues, is thought to play an important role in the colonisation and infection of mammals by pathogenic fungi [38]. An assay of *M. pachydermatis* adherence to canine corneocytes *in vitro* was used to examine whether the adhesive capacities of the epithelial cells could be related to population sizes of the yeast in the skin of seborrhoeic and healthy basset hounds, and in healthy Irish setter control dogs [31]. *M. pachydermatis* population sizes in the healthy basset hounds ($n=5$) were significantly greater ($P<0.001$) than those of the Irish setters ($n=6$), but were significantly lower ($P<0.001$) than those of the seborrhoeic basset hounds ($n=6$). The numbers of adherent yeast cells of the two *M. pachydermatis* strains tested were significantly greater ($P<0.001$) on corneocytes taken from the healthy basset hounds, compared with corneocytes derived from healthy Irish setters and seborrhoeic basset hounds. The inverse relationship between population densities *in vivo* and corneocyte adherence in vitro suggests that increased corneocyte receptivity for the yeast is not important in the pathogenesis of seborrhoeic dermatitis (SD) associated with *M. pachydermatis* in basset hounds. These findings are similar to those reported in a previous investigation of lipid-dependent *Malassezia* adherence in humans with SD; the yeasts were better able to adhere to stratum corneum cells from healthy individuals than to cells from patients with SD [39]. Studies of *M. pachydermatis* adherence to canine stratum corneum cells are reviewed in Chap. 5.

Like other cutaneous microbes, *M. pachydermatis* produces a range of enzymatic substances that are involved in its nutrition. Liberation of enzymes by the fungus ensures that a local supply of appropriate nutrients can enter the cell in a form suitable for further metabolism. The *M. globosa* genome has an abundance of genes encoding secreted hydrolyases (including lipases, phospholipases, aspartyl proteases and acid sphingomyelinases) with similarities to those expressed by *Candida albicans* [40]. These substances are likely to be involved in inducing host immune and inflammatory responses; however, the relative importance of the various enzymatic products from *M. pachydermatis* in inducing cutaneous inflammation has not been fully established. The enzymatic activities of *Malassezia* spp. are discussed more fully in Chap. 4.

Immediate intra-dermal test reactivity has been observed in atopic dogs following the injection of *M. pachydermatis* extracts at concentrations that cause no reaction in healthy dogs, suggesting that hypersensitivity responses to yeast allergens may be involved in the pathogenesis in some cases of atopic disease [35, 41]. These findings are supported by serological evidence of IgE antibody responses to the yeast in dogs based on ELISA [42], Western immunoblotting [43] and passive cutaneous anaphylaxis [44]. These observations are analogous to the suggestion that IgE-mediated hypersensitivity to *Malassezia*-derived allergens may be important in the pathogenesis of the 'head-neck' form of AD in humans [36, 45]. Some dermatologists now include *Malassezia* allergens in allergen-specific immunotherapy courses for atopic dogs, but the therapeutic benefit of this approach requires further assessment since most immunotherapy courses comprise multiple allergens and not just yeast allergens alone [46]. Since *Malassezia* dermatitis can exist in the absence of immediate hypersensitivity to the yeast, [47, 48] immunotherapy alone is unlikely to replace the need for antifungal therapy in dogs with high yeast counts in lesional skin.

In a study of basset hounds with *Malassezia* dermatitis tested intra-dermally using extracts prepared from two *M. pachydermatis* isolates, 2 out of 18 hounds showed strong immediate intra-dermal test reactivity [47]. In contrast, examination of test sites at 24 h post-injection showed frequent lesions characterised by erythematous plaques in the healthy and affected basset hounds. Skin biopsy specimens from these delayed reactions showed either moderate (2 cases) or severe (4 cases) inflammatory changes principally characterised by peri-vascular, interstitial or peri-adnexal infiltrates of neutrophils and lymphocytes. Patch testing with *M. pachydermatis* extracts were evaluated in seven healthy basset hounds and in seven basset hounds with *Malassezia* dermatitis, of which one healthy basset hound and five affected hounds showed positive reactivity characterised histologically by mild epidermal hyperplasia and mild to moderate peri-vascular, peri-adnexal and interstitial infiltrates of neutrophils and $CD3^+$ lymphocytes [48]. These studies indicate that immediate intra-dermal test reactivity is infrequent in affected basset hounds whereas delayed intra-dermal and patch test reactivity occurs more frequently. Immunological responses of the skin to *Malassezia* spp. is discussed in more detail in Chap. 5, but it is clear that a spectrum of cutaneous reactivity to *M. pachydermatis* antigens can be seen in dogs with *Malassezia* dermatitis. The variable immune responsiveness seen among dogs suggests that any future immunomodulatory therapeutic or prophylactic approach may have to target different components of the skin immune system.

10.1.4 Clinical Signs

Skin lesions associated with *M. pachydermatis* in dogs are usually erythematous, with varying degrees of traumatic alopecia [18, 21, 34]. Scaling is often prominent and a greasy exudate is a feature of lesions in inter-triginous and inter-digital areas in some dogs; significant exudation is normally accompanied by prominent malodour. Inter-digital lesions are common (Fig. 10.1) and in more severe cases, erythema and alopecia extend to affect the accessory carpal areas and medial aspects of the limbs. Some dogs with severe limb lesions have marked skin thickening, resulting in the formation of erythematous, scaling, alopecic ridges (Fig. 10.2). Pedal skin disease may also progress to involve the claw folds with red–brown staining of the claw and exudation in the claw fold [49]. Abdominal lesions often initially consist of symmetrical, well-demarcated, circular or elliptical areas of erythema that develop into scaly plaques. Alternatively, more diffuse erythema with scattered papules may be seen. Lesions in skin folds are normally characterised by erythema and a sticky exudate. Hyperpigmentation and lichenification are frequently observed in animals with

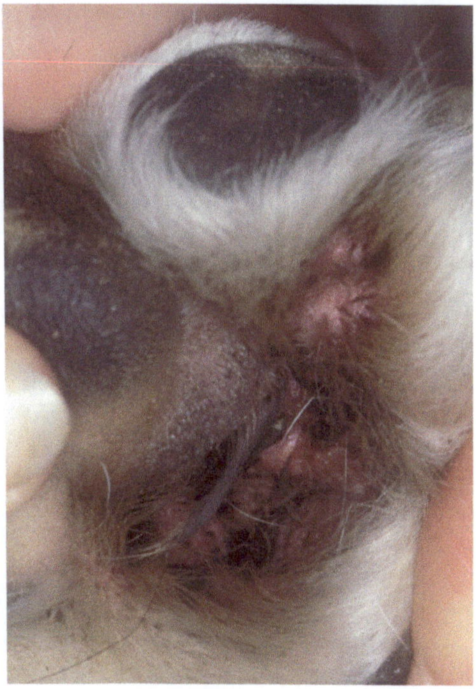

Fig. 10.1 Severe inter-digital erythema and greasy exudation in a dog with *Malassezia* dermatitis. The exudation should raise clinical suspicion of a microbial infection whereas uncomplicated allergic diseases involving this site generally show only erythema

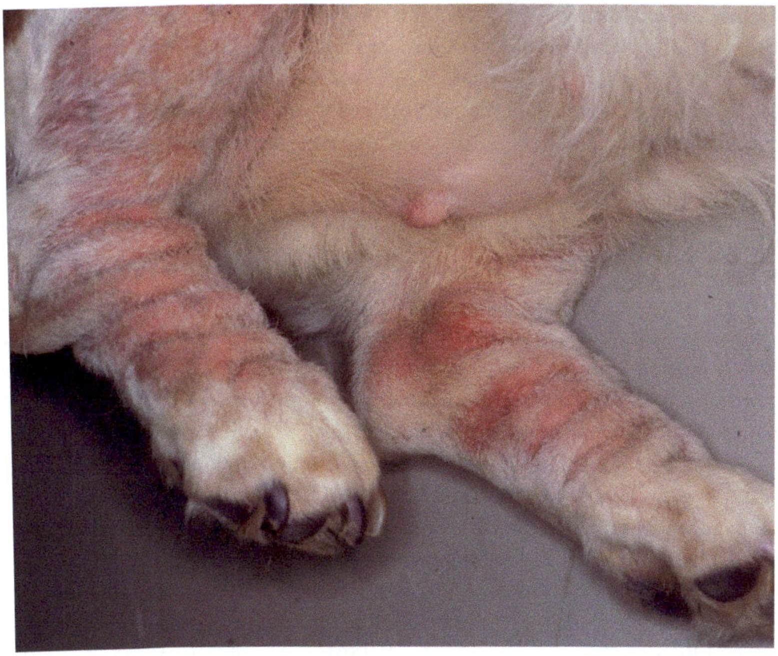

Fig. 10.2 Severe lesions of *Malassezia* dermatitis in this basset hound are characterised by intense erythema and traumatic alopecia over the hind limbs. The folded skin at this site is a predisposing factor for infection, although high yeast counts can also be demonstrated at mucosal sites in this breed

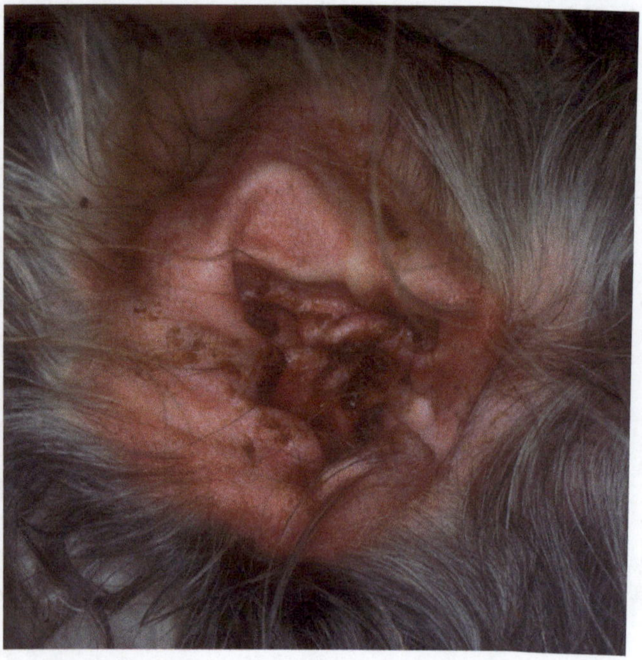

Fig. 10.3 *Malassezia*-associated otitis externa in dogs most often presents with an erythemato-ceruminous rather than a purulent discharge

chronic disease and is particularly common in West Highland white terriers. Dogs with concurrent OE show erythematous vertical ear canals and pinnae with varying degrees of lichenification and scaling, accompanied by a yellow or brownish ceruminous discharge (Fig. 10.3). Although skin lesions may be confined to one area, multiple regions are usually affected, especially the limbs (Figs. 10.1 and 10.2), ventrum, ears (Fig. 10.3) and face.

10.1.5 Diagnosis

The clinical signs of *Malassezia* dermatitis are not pathognomonic, and the disease should be routinely suspected in dogs with inflammatory skin diseases, especially those with erythema and or greasy exudation as a dominant presenting sign. *Malassezia* dermatitis may mimic or complicate atopic disease and dietary sensitivity. Seborrhoeic presentations and fold-associated diseases should also prompt an evaluation of the presence and numbers of the yeast.

The diagnosis is based on clinical signs, presence of elevated numbers of the yeast in lesional skin, and a clinical and mycological response to antifungal therapy. In veterinary practice, yeast numbers are most usefully assessed by cytological examinations of direct impression smears or smears of exudate obtained by swabbing. However, the tape-strip method may be more convenient and reliable [50]; clear adhesive tape is pressed on the surface of the skin thus collecting the stratum corneum cells and any superficial microbes (Fig. 10.4). This method is suitable in most anatomical sites and in dry or greasy lesions

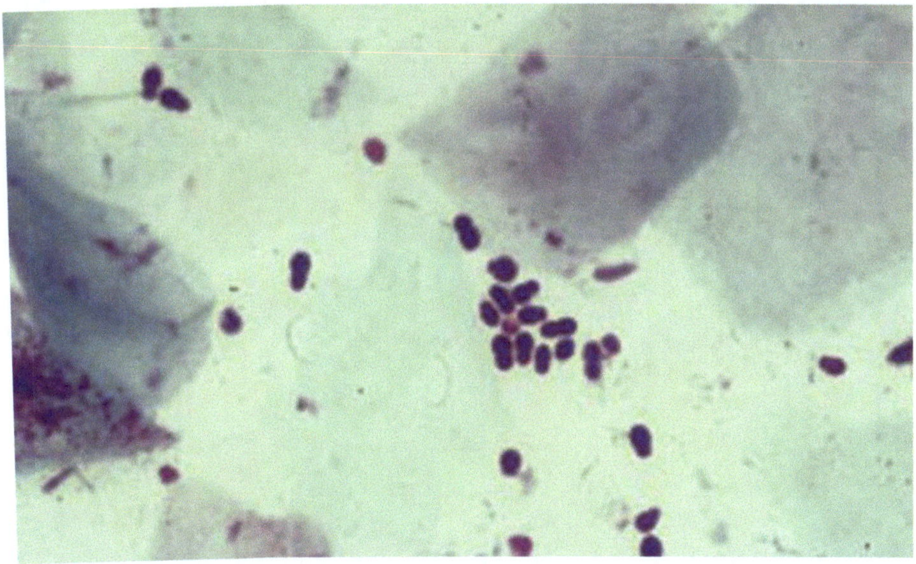

Fig. 10.4 Numerous yeast cells with morphologies typical of *M. pachydermatis* can be seen amongst squames in this tape-strip preparation taken from a dog with *Malassezia* dermatitis

although it is important to select a brand of tape that resists the staining process such as certain Scotch 3M tapes, or Sellotape Diamond Clear. The tape is stained, usually with a modified Wright's stain, and examined using the light microscope. The characteristic peanut-shaped *Malassezia* spp. yeast cells are normally rarely seen in healthy skin but are usually readily identified in specimens from affected individuals.

Some authors have proposed guidelines on the numbers of yeasts needed for significance (such as one or more yeast per high-powered field), but this approach does not accommodate the overlap seen in yeast population densities in skin samples from healthy and diseased dogs [32], nor does it consider the potential for relatively small numbers of organisms to generate skin disease in sensitised individuals. Furthermore, quantitative cultural investigations have demonstrated important breed and anatomical site differences in population sizes in healthy dogs. These data indicate that trial therapy is an important component in the diagnostic evaluation of *Malassezia* dermatitis; therapy should be given whenever the yeast is readily identified in cytological specimens obtained from consistent lesions.

Quantitative cultural techniques involving contact plates and detergent scrubs [51] are primarily used in research rather than routine clinical practise, although the contact plate method (Fig. 10.5) is a very useful method of isolating the yeast for the skin [24].

Skin biopsy specimens typically show marked irregular epidermal hyperplasia and spongiosis that extends to the follicular infundibula, frequently associated with a prominent scalloping of the lower epidermis, and ortho-keratosis or para-keratosis [21, 52, 53]. Dermal inflammation is characterised by a superficial peri-vascular or interstitial infiltrate of mononuclear cells, with focal accumulations of neutrophils, eosinophils and mast cells. Exocytosis of lymphocytes is frequent whereas eosinophilic or neutrophilic micro-abscesses are less

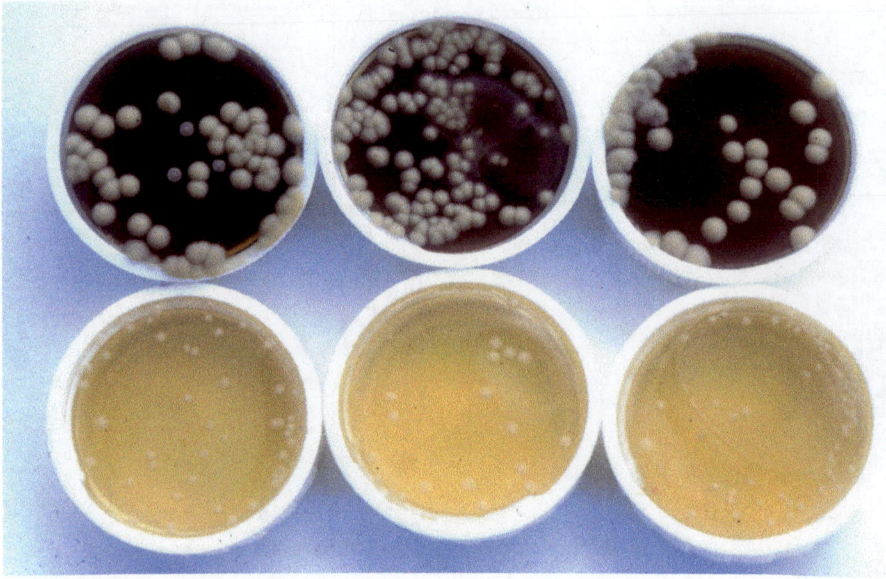

Fig. 10.5 Contact plate cultures of *M. pachydermatis* from canine skin. The *top plates* are composed of modified Dixon's agar which allows the development of larger colonies when compared with Sabouraud's dextrose agar (*below*). This technique permits a convenient semi-quantitative assessment of population sizes on the skin

often seen. Mast cells may be aligned in a linear pattern at the dermo-epidermal junction in some cases [53]. Yeast cells may be seen in the ortho-keratotic or para-keratotic stratum corneum or in crusts, and less often at the follicular ostia [21], but the disruption of this layer that occurs during routine tissue processing may lead to an absence of the yeast; biopsy has a low sensitivity for yeast detection when compared with cytology.

Clinical and cytological assessments should be repeated to determine the efficacy of antifungal therapy, and to establish whether there is evidence of concurrent diseases. Relapsing infection is common in cases where predisposing factors are not identified or corrected.

10.1.6 Treatment of *Malassezia* Otitis and Dermatitis in Dogs

In vitro studies have shown that *Malassezia* yeasts are normally susceptible to azole antifungal drugs such as clotrimazole, miconazole, ketoconazole and itraconazole [54–56]. Among the more recently developed antifungal agents, pozaconazole and voriconazole are often highly active against *Malassezia* spp. [54, 57, 58], but commercial products containing these agents are very expensive and are not currently licensed for veterinary use. The polyene macrolides, nystatin and natamycin (pimaricin), are also active against *M. pachydermatis*, but generally at higher concentrations *in vitro*, compared with azoles

[9, 10, 59, 60]. Silver sulphadiazine, an antibacterial sulphonamide compound used in the treatment of *Pseudomonas* OE as in dogs, was also been reported to have modest anti-*Malassezia* activity *in vitro* [55]. Plant derived substances, such as tea tree oil [61–63] and β-thujaplicin, a substance extracted from the acid oil of Aomori Hiba (*Thujopsis dolabrata*) [60, 64] have also been reported to have anti-*Malassezia* properties. The methods for conducting and interpreting susceptibility tests of antifungal drugs against *Malassezia* spp. *in vitro* are discussed in detail in Chap. 6.

10.1.6.1 Treatment of *Malassezia* Otitis Externa in Dogs

A detailed review of the investigation and management of canine OE is beyond the scope of this chapter, and this section will focus on the treatment of OE and media associated with *M. pachydermatis* in dogs.

It is commonly accepted that this commensal yeast is an opportunistic secondary pathogen whose proliferation often reflects the presence of underlying predisposing factors and primary diseases. Common predisposing factors include pendulous pinnae, narrow external canals, hairy ears and regular swimming. As discussed elsewhere in this section, AD is a commonly identified disease in dogs with *Malassezia* otitis, but other diseases, such as dietary sensitivity and keratinisation defects, might also favour *Malassezia* otitis. Dogs with *Malassezia* otitis should receive a complete dermatological evaluation since failure to identify and correct predisposing, primary and other perpetuating factors may result in persistent or recurrent disease.

Although nystatin mono-therapy was reportedly effective in an early study on the treatment of canine *Malassezia* OE [65], current products licensed for the treatment of canine OE associated with *M. pachydermatis* normally contain antifungal agents combined with an antibacterial drug and a glucocorticoid, reflecting the multi-factorial aetiology of most cases. The combination with glucocorticoids is often advantageous since inflammation may be important in creating a suitable micro-environment for yeast proliferation. In addition, prevention or correction of stenosis of the external ear canal by the use of anti-inflammatory drugs is important if irreversible pathological changes are to be avoided in chronic cases. The inclusion of an antibacterial agent may also be beneficial in case of a mixed infection with the common aural bacterial pathogens, such as staphylococci or pseudomonads. Combined antibacterial and antifungal drug administration may also prevent the switch from bacterial to yeast infection, or vice versa, that may be encountered when antibacterial or antifungal mono-therapy is utilised in dogs with otitis externa or media. Cytological evaluations of aural discharges are especially useful for the assessment of cases and monitoring the effects of treatment; ear wax smears enable the clinician to rapidly determine an estimation of the population density of the yeast before and during treatment.

Despite the very wide usage of the commercially available combination ear drop products, there is surprisingly little clinical trial data describing their efficacy. This may reflect in part the difficulty in standardising subject animals, where multiple different factors such as diverse underlying diseases and varied or mixed microbial infections may be present. In some cases of *Malassezia* OE, successful antimycotic therapy may not be associated with clinical remission due to ongoing concurrent disease, making interpretation difficult.

Miconazole is an imidazole with a spectrum of activity that includes *Malassezia* spp., dermatophytes and gram-positive bacteria [66]. The efficacy of a preparation containing miconazole, polymyxin B and prednisolone was evaluated in an open, uncontrolled study of 130 dogs in five veterinary practises in Germany [8]. *M. pachydermatis* was isolated from 57% of ears and it was the only pathogen isolated in half of these cases. In 86% of cases, the dogs were judged to be clinically cured or well-controlled 4 weeks after treatment was introduced. In another open study, the same product was used to treat 101 episodes of OE in a group of labrador retrievers [67]. *Malassezia* yeasts were isolated from 44 out of 49 ears cultured, either alone (71%) or in association with bacteria (18%). The product was judged to be satisfactory in 73% of cases, and unsatisfactory due to recurrence in 27% of cases. The miconazole, polymyxin B and prednisolone product was more frequently effective when compared with products containing nystatin and monosulfiram as the antifungal agent.

More recently, the efficacy of the miconazole, polymyxin B and prednisolone preparation (A) was compared with a clotrimazole-marbofloxacin-dexamethasone otic suspension (B) in a randomised single-blind study design involving 140 dogs in 15 European veterinary practises [68]. Yeasts, principally *M. pachydermatis*, were isolated from 45% of the dogs. The two products were judged to be equivalent based on cure rate in dogs with fungal otitis (4 out of 6 with B, 3 out of 8 with A) and mixed bacterial and fungal otitis (9 out of 19 with B, 8 out of 21 with A), although the authors reported a trend for the clotrimazole-containing product to be more effective. Bensignor and Grandemange [69] treated 20 dogs with bilateral *Malassezia* OE using the clotrimazole-marbofloxacin-dexamethasone otic suspension in one ear and a miconazole-containing product in the other. The reductions in yeast counts based on cytological assessments were comparable after either treatment, but ears treated with the combination product had lower post-treatment scores for erythema, pruritus and the quantity of cerumen. Although clinical scores also improved with miconazole-alone, these data provide justification for the use of combination products in the routine management of canine otitis externa.

Ciclopirox olamine, the 2-aminoethanol salt of 6-cyclohexyl-1-hydroxy-4methyl-2(1H)-pyridone, is a broad-spectrum antifungal drug of the family of 1-hydroxy-2-pyridones that has been shown to be of value in the treatment of scalp SD and onychomycosis in humans [70, 71]. Ciclopirox olamine has *in vitro* activity against various fungi including dermatophytes, *Candida* spp., *M. globosa*, *M. restricta*, and *M. pachydermatis*; synergistic fungicidal effects have also been observed with zinc pyrithione against *Malassezia* spp, and against other fungi with terbinafine or itraconazole [72–74]. The mode of action may relate to chelation of iron and other metallic enzyme cofactors [75]. A single application of piroctone olamine (1-hydroxy-4methyl-6-(2, 4, 4,-trimethylpentyl)-2(1H)-pyridone) was found to significantly reduce ear populations of *M. pachydermatis* for at least 4 days in a group of 10 healthy laboratory beagle dogs [76]. The value of hydroxy-pyridones in the management of fungal infections of dogs requires further assessment.

The concurrent use of ear cleaners is a common practise in the medical treatment of otitis externa, and some of these may have a degree of antifungal activity that may assist in treatment, and also in prevention of relapse in animals with excessive cerumen [77].

10.1.6.2
Malassezia Otitis Media

Malassezia otitis media is most likely to develop as a consequence of rupture of the tympanic membrane in dogs with chronic OE, but it is important to consider the presence of otitis media even when the membrane is intact since a previous rupture may heal, sealing the yeast infection within the middle ear [78]. In these circumstances, myringotomy is useful, both for diagnosis and flushing of the bulla.

The manufacturers of many of the combination products licensed for the treatment of canine OE state that their products should not be used in cases where the tympanic membrane is ruptured, due to the potential ototoxicity of the active ingredients or the vehicle. A 1% solution of clotrimazole licensed for the treatment of fungal otitis media in humans is favoured by some veterinary dermatologists. An alternative approach is to administer itraconazole orally either alone or in combination with topical clotrimazole, although clinical trial data on this approach is not available, and oral itraconazole may not be effective in the management of *Malassezia* OE [79].

10.1.6.3
Treatment of *Malassezia* Dermatitis in Dogs

Since *M. pachydermatis* is located within the stratum corneum of dogs, topical therapy alone may be sufficient to resolve the clinical signs of infection. Systemic therapy is often used when topical therapy is impractical due to compliance issues relating to the pet or owner, in severe cases of generalised dermatitis. As in cases of OE, identification and correction of predisposing diseases is often essential for successful management, although many dogs with *Malassezia* dermatitis require regular maintenance therapy to prevent relapse.

A randomised double-blind parallel study compared the clinical and anti-microbial efficacies of 2% miconazole/2% chlorhexidine shampoo with a selenium sulphide shampoo for the treatment of *Malassezia* dermatitis in 33 basset hounds in the U.K [34]. All 16 miconazole-chlorhexidine treated hounds and 11 of 17 selenium sulphide treated hounds improved when shampooed at 3-day intervals for 3 weeks, but the miconazole-chlorhexidine treated hounds showed significantly greater reductions in pruritus, erythema, exudation and overall severity, and in counts of *M. pachydermatis* and coagulase-positive staphylococci. Improvements in coat condition and scaling did not vary significantly between the two groups. The availability of this highly effective product might explain the trend for U.K.-based dermatologists to use topical rather than systemic treatment for canine *Malassezia* dermatitis [80]. Other topical agents used include enilconazole rinses, and ketoconazole, selenium sulphide or chlorhexidine shampoos.

Systemic therapy is indicated when topical therapy is ineffective or impractical due to compliance issues, or in severe cases where a rapid response is required. Ketoconazole, a dioxalane imidazole derivative with both antifungal and anti-inflammatory properties, has been widely used in the oral treatment of canine *Malassezia* dermatitis. Suggested doses range from 5 to 10 mg/kg, once or twice daily [18, 33, 81]. Side-effects include anorexia, vomiting, diarrhoea, thrombocytopenia, teratogenicity and hepatotoxicity. This drug should not be used in dogs known to have liver disease, and in atopic dogs receiving ciclosporin.

Itraconazole is a triazole antifungal agent that is used for the treatment of various fungal diseases in humans and for the treatment of dermatophytosis in cats. The keratinophilic and lipophilic properties of this drug enable intermittent administration for the treatment of fungal infections of the stratum corneum [82]. Itraconazole may be better tolerated in dogs when compared with ketoconazole, and side effects (principally anorexia) are uncommon at a dose of 5 mg/kg, although more frequent anorexia and skin ulceration/vasculitis was noted in dogs with blastomycosis that received 10 mg/kg [83]. The efficacy of once daily oral administration at 5 mg/kg for 21 days was compared with a pulse regimen (2 consecutive days per week) in 20 dogs with *Malassezia* dermatitis or otitis [79]. Both regimens were effective in the treatment of *Malassezia* dermatitis, and pulse therapy had the advantage of reducing costs and the potential for side-effects and potentially improving compliance. However, severe claw fold infections may require longer treatment or higher doses [79].

Terbinafine is an allylamine antifungal agent used primarily in the treatment of dermatophytosis in humans. *In vitro* studies have shown that terbinafine susceptibility varies amongst the different *Malassezia* spp., although *M. pachydermatis* appears to be susceptible [54, 58]. Administrating terbinafine at 30 mg/kg once daily for 21 days was apparently well-tolerated and reduced skin *M. pachydermatis* populations in seven healthy basset hounds [33], and in eight dogs with *Malassezia* dermatitis [81]. Pending further safety and efficacy studies, terbinafine may be an alternative treatment in dogs that react adversely to ketoconazole or itraconazole.

10.2 *Malassezia* spp. as Pathogens in Cats

10.2.1 Factors Predisposing to Disease

Malassezia yeasts appear to be relatively infrequent pathogens in cats, at least when compared with other carnivores like dogs. This may reflect the less frequent carriage rates detected in healthy cats [5, 84–87]. *M. pachydermatis* can be isolated from the external ear canal and mucosae of healthy animals as well as cats with otitis externa and dermatitis. In healthy cats, reported percentages of isolation of this species ranges from less than 10% [85, 87] to approximately 20% [86, 88–90] and up to 40% [91]. In parallel, *M. pachydermatis* has been identified in 41–71% of cats with OE [88–91]. Lipid-dependent species identified as *M. furfur* [90, 92, 93], *M. globosa* [85, 90, 91] and *M. sympodialis* [84, 88, 89, 92, 94] have been also reported from the skin or the external auditory canal of healthy cats or other felids (see also Chap. 3.2).

In 2004, a novel species, *M. nana,* was described from the otic discharges of a cat in Japan [95]. *Malassezia nana* seems to be the most common lipid-dependent species isolated from cats. This species may have been identified as *M. sympodialis* in the past. *Malassezia nana* and *M. slooffiae* have been isolated from cats with different disorders

[96], as well as *M. slooffiae* from the claw folds of healthy cats and animals with seborrhoeic dermatitis [97].

Malassezia yeasts may be recovered from various breeds but Devon Rex cats seem to be more frequently colonised [92, 97]. The presence of a high number of lipophilic yeasts on cytology correlated with the clinical observation of brown, greasy material in the nail folds of Devon Rex cats [92]. In another investigation, the frequencies of isolation and population sizes of *Malassezia* yeasts in the axillae and groin in healthy or seborrhoeic Devon Rex cats exceeded those of healthy domestic short-haired cats [97], whereas *Malassezia* counts in Cornish Rex cats, a breed with similar coat characteristics to Devon Rex cats, had counts comparable to those of domestic short-haired cats [98]. Further investigations are needed to explain why the skin of Devon Rex cats is so favourable to colonisation by *Malassezia* spp.

The isolation of *Malassezia* has also been associated with retroviral infections [99], para-neoplastic syndromes [96, 100, 101], thymoma [101, 102] and diabetes mellitus [103]. Based on these findings, *Malassezia* overgrowth may be considered as a marker of life-threatening, underlying diseases in some cats. *Malassezia* yeasts have also been associated with feline acne [104, 105] and idiopathic facial dermatitis [106].

Atopic dermatitis has been described as a common predisposing factor for *Malassezia* dermatitis in dogs whereas this association has been reported less frequently in cats [49, 107]. A series of 18 allergic cats with multifocal *Malassezia* spp. overgrowth was reported [108]. AD was diagnosed in 16 of these animals. All the cats were otherwise healthy and those tested (16 out of 18) were free from retroviral infections. The beneficial effects of azole antifungal therapy alone in 5 out of 7 of these cats led the authors to conclude that *Malassezia* spp. yeasts can exacerbate the clinical signs of allergy in cats as well as dogs.

10.2.2 Clinical Signs

Ceruminous OE responsive to antifungal ear drops remains the commonest clinical presentation of *Malassezia*-associated diseases in cats. Like in dogs, *Malassezia* otitis is characterised by a waxy, moist, brown or yellow exudate with variable erythema and pruritus. Yeasts are frequently associated with ear mites (*Otodectes cynotis*) (Fig. 10.6). Recently, secondary *Malassezia* infection was described in four cats with an unusual proliferating and necrotizing otitis externa [109].

Occasional cases of localised or generalised *Malassezia* dermatitis have also been described in cats. In atopic animals, cutaneous lesions related to *Malassezia* overgrowth commonly occur on the face, ventral neck, abdomen and ear canals. These lesions are characterised by some degree of alopecia, erythema, greasy adherent brownish scales, hyperpigmentation, easily plucked hair and follicular casts [110]. Hyperpigmentation and lichenification commonly observed in chronic canine *Malassezia* dermatitis are seldomly (hyperpigmentation) or not seen (lichenification) in cats. Exfoliative erythroderma, greasy exudation and varying degrees of pruritus may be seen in cases secondary to a severe, systemic diseases [101–103].

The presence of brown, greasy material in the nail folds may be associated with the presence of *Malassezia* yeasts. However, the presence of *Malassezia* yeasts in the nail

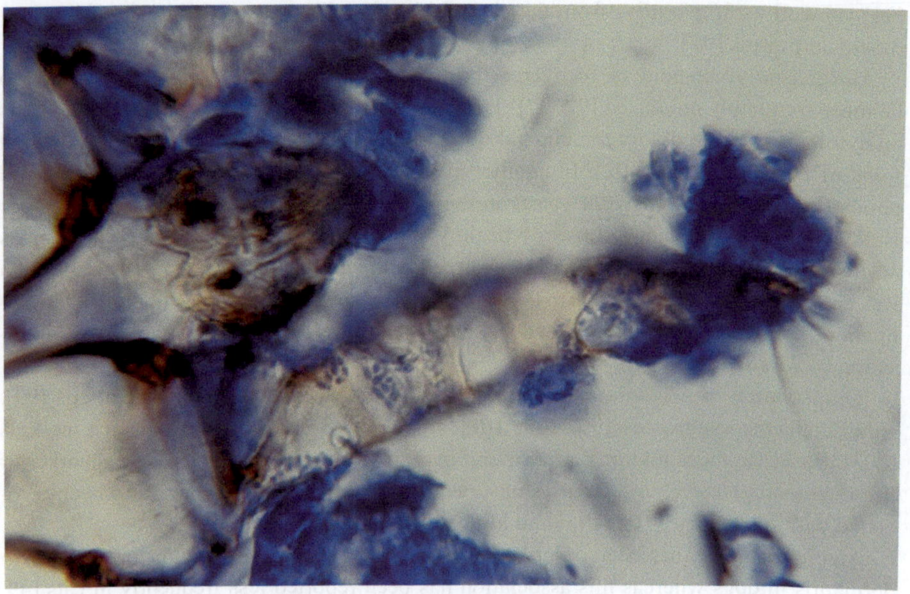

Fig. 10.6 Large number of *Malassezia* cells are present on the legs of this ear mite (*Otodectes cynotis*), a common aural parasite of pet carnivores (service de Parasitologie, ENVA)

folds and concurrent greasy material was not associated with pruritus, even in Devon Rex cats with high numbers and the presence of more than one *Malassezia* species [92].

10.2.3 Approach to Diagnosis

The criteria required for the diagnosis of *Malassezia* dermatitis have not been definitively established in pet carnivores. However, it may be proposed that such a diagnosis is appropriate when a cat with elevated numbers of *Malassezia* yeasts on lesional skin shows good clinical and mycological responses to antifungal therapy [80]. The situation in cats is more complex than in dogs because feline skin may be colonised by different *Malassezia* species and because the exact role of these species in the pathogenesis of the disease remains unknown.

The list of differential diagnoses may be lengthy because *Malassezia* dermatitis is usually associated with other diseases. In Devon Rex cats, *Malassezia* overgrowth should be systematically suspected in cases of SD (without evidence of systemic disease).

Cytological examination is considered as the most useful technique for assessment of *Malassezia* density on the skin surface in cats. Several cytological criteria have been proposed to diagnose canine *Malassezia* dermatitis including the observation of more than two yeasts per high power field in skin specimens. However, no report has been published that provided details of yeast density on the surface of normal feline skin and the use of threshold values is not recommended in cats.

Contact-plate cultural techniques provide a rapid and convenient method for isolation of the yeasts [24, 97]. This quantitative method is superior to standard swabbing methods because population sizes can be determined, and in most cases, population densities in affected areas greatly exceed those of healthy skin. Because of the presence of lipid-dependent yeasts on the skin of cats, the use of lipid-supplemented media, especially the modified Dixon's medium, is required.

In cats, histopathological examination of skin biopsy specimens typically shows hyperkeratosis and irregular epidermal hyperplasia [101]. Other features corresponding to the primary cutaneous disease may be seen. As in dogs, *Malassezia* yeasts are not always visible in the epidermal stratum corneum, even in cases where large numbers are seen cytologically; this probably reflects disruption of this layer during processing.

10.2.4
Treatment and Management

Therapeutic measures are similar to those used in dogs. For OE, ear drops containing an antifungal drug are very efficient as long as the potential predisposing factor (ear mite infestation for example) has been identified and correctly treated.

Feline *Malassezia* dermatitis is usually treated by systemic antifungal drugs like ketoconazole or itraconazole [110]. A pulse treatment regimen with oral itraconazole at 5 mg/kg once daily for 7 days on, 7 days off and 7 days on was effective in the treatment of five Devon Rex cats with *M. pachydermatis*-associated SD. A 2% miconazole-2% chlorhexidine product has been shown to have good efficacy in dogs, but topical treatments are usually not very convenient for cats. Identification and correction of underlying causes should be attempted in relapsing cases and when there is the potential for serious concurrent systemic disease.

The skin of cats may be colonised by different *Malassezia* species and significant differences in the *in vitro* susceptibility to antifungal drugs have been demonstrated among lipophilic yeasts [54, 58, 61]. As a consequence in future, the specific identification of the *Malassezia* species present on a cat with lesional skin might be useful for the selection of the most appropriate treatment.

10.3
Malassezia spp. as Pathogens in Other Animals

10.3.1
Malassezia pachydermatis

At the moment, few studies have been published on distribution and pathogenicity of the different *Malassezia* spp. on animals following the last taxonomic revision of this genus [111]. Traditionally, the lipid-dependent species were thought to occur only on human skin, while *M. pachydermatis* was restricted to animal skin and in particular to carnivores. In addition to dogs and cats, *M. pachydermatis* is found in the skin of a wide range of wild and domestic animal species, mainly in carnivores [86, 112]. This species is more frequently isolated from dogs than cats and appears to be a relatively infrequent pathogen in other animals [88, 112], although its prevalence in other animals is probably underestimated.

Both the first association of this yeast with skin disease and the first description of this species (as *Pityrosporum pachydermatis*) was published by Weidman in 1925 [1] in a case of an exfoliative dermatitis in an Indian rhinoceros (*Rhinocerus unicornis*). A similar case of exfoliative dermatitis was described more recently in a captive southern white rhinoceros (*Ceratotherium simum simum*) [113]. Co-infection of large numbers of *M. pachydermatis* and *Candida parapsilosis* were found in the lesions. Application of a 2.5% natamycin solution resulted in complete recovery. This yeast has been shown to be a component of the normal skin microbiota of different rhinoceros species [86, 114, 115]. Isolates from wide-mouthed rhinoceroses had the exclusive LSU rRNA sequence type Ic [13]. There is no genetic information about *M. pachydermatis* isolates from other rhinoceros species.

More recently, *M. pachydermatis*-associated dermatitis has been also reported in sea lions. In a Californian sea lion (*Zalophus californianus*) this species was found responsible for cutaneous lesions on both flanks and chest (Fig. 10.7), which were successfully treated with topical 0.2% enilconazole [116]. All isolates from this animal belonged to the LSU rRNA sequence type Id that is typically isolated from dogs. To date, only one additional report has been published affecting, in this case, a South American sea lion (*Otaria byronia*) with generalised dermatitis [117]. The same yeast was isolated from water in which the animal was kept.

Only a single case of SD in a goat associated with *M. pachydermatis* has been reported [118]. Cytological examination showed yeasts of variable morphology with long, cylindrical, ovoid or globose cells suggesting that several species of *Malassezia* were present on the skin of this animal. Colonies compatible with *M. pachydermatis* were recovered from the lesional skin on Sabouraud's glucose agar with gentamicin. However, no specific culture media were used for recovering lipid-dependent yeasts. This animal was successfully treated with chlorhexidine shampoo and a 0.2% solution of enilconazole.

In addition to dogs and cats, OE associated to *M. pachydermatis* has been also cited in fennecs and ferrets [86]. An outbreak of otitis and pinnal necrosis in ferrets associated with *Malassezia* yeast infection where mites were probably the primary pathogen has been also described [119]. Isolates from these animals have the special LSU rRNA sequences type Ib and Ig, respectively, different from the dog and cat strains that are distributed among the types Ia, Id and Ie [13]. In addition to these animals, OE associated to *M. pachydermatis* has been also reported in pigs and in one camel [120, 121].

Malasezia pachydermatis has been also involved in a skin disorder in a black bear (*Ursus americanus*) found dead in a pasture with patches of mild alopecia and scaling on the thorax and ears [122]. Although no ticks, mites, or apparent effects of trauma were observed in this case, these authors pointed out that it was difficult to determine whether *M. pachydermatis* was an infectious agent or merely a commensal of the skin. These same authors [123] reported the association of *M. pachydermatis* with sarcoptic mange in red foxes (*Vulpes fulva*), porcupines (*Erethizon dorsatum*) and coyotes (*Canis latrans*).

Although *M. pachydermatis* is not recovered from lagomorphs [86] a combined *Sarcoptes* and *Malassezia* spp. infection was diagnosed in a rabbitry [124]. High numbers of round

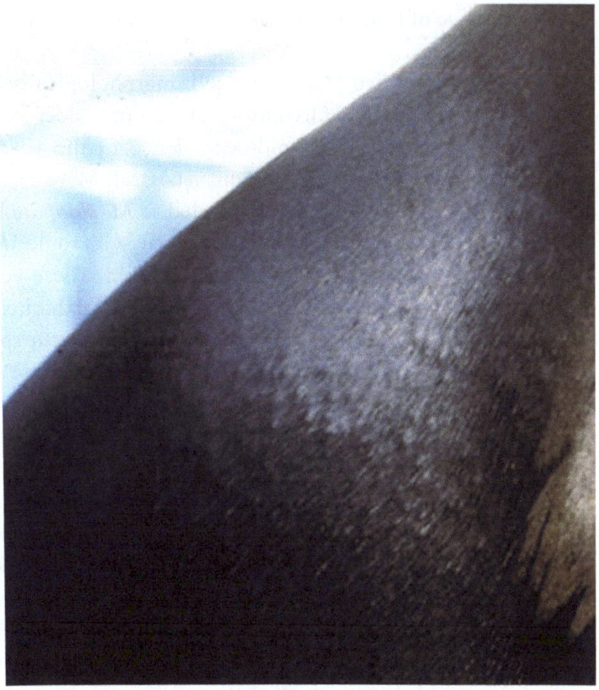

Fig. 10.7 *M. pachydermatis*-associated dermatitis in a Californian sea lion characterised by a papular dermatitis of the flank

to oval budding yeast cells were detected in different skin sections but a culture was not done, and cell morphology showed in this publication suggest a *Malassezia* spp. distinct from *M. pachydermatis*. It has been isolated also from acoustic meatus of bats (*Molossus molossus*) of the Amazon region, Brazil [125], but without associated infection.

Malassezia. pachydermatis was isolated repeatedly from a scarlet macaw; this is the first report of birds affected by this yeast [126]. *Malassezia* yeasts were not detected in a recent study of the distribution of these organisms on the skin of healthy psittacine birds and psittacine birds with feather-destructive behaviour [127]. No specific culture media was used in this study for recovering lipid-dependent yeasts. However, *Malassezia* yeasts have isolated from different species of birds, including parrots, in other previous studies [128, 129].

10.3.2 Lipid-Dependent *Malassezia* Species

The skin of animals can be also colonised by lipid-dependent species in addition to *M. pachydermatis*, but very little is known about their pathogenic role in animal skin. These lipid-dependent species constitute the major population of the lipophilic mycobiota in herbivores such as horses, goats, sheep and cows [130]. Different species such as *M. furfur, M. globosa, M. obtusa, M. slooffiae, M. sympodialis* and *M. restricta* were identified using phenotypic methods, although a large number of isolates could not be identified.

The occurrence of lipid-dependent species in these animals was more frequent than that of *M. pachydermatis*.

Bliss [131] identified *Malassezia* cells morphologically distinct from *M. pachydermatis* in milking goats that had circular lesions on the udder similar to those seen in pityriasis versicolor in humans. The lipid-dependence of the yeasts could not be demonstrated because attempts at isolation were unsuccessful. More recently, a case of generalised dermatitis in an adult Pygmy goat associated to *M. slooffiae* has been reported [132]. Large numbers of yeasts and hyphal forms consistent with some *Malassezia* species were detected in skin scrapings from affected areas of the skin. These yeasts were identified in the skin by DNA sequencing of the LSU rRNA gene that exhibited almost 100% homology with *M. slooffiae* AJ249956 sequence. However, *Malassezia* spp. were not isolated from skin scrapings inoculated onto modified Dixon's agar, Leeming's medium, or Sabouraud's dextrose agar supplemented with olive oil. Skin disease was not reported in the other animals in the herd. *M. caprae*, a new lipid-dependent species isolated mainly from healthy goats, has been recently described [133].

Lipid-dependent species have been associated with OE in bovines [134, 135]. Different species such as *M. furfur, M. sympodialis, M. slooffiae* and *M. globosa* were phenotypically identified from cerumen or secretions of the external ear of these animals. Two examples of *M. furfur* were confirmed by rDNA sequence analysis. Two lipid-dependent strains from cows with OE from Brazil were genetically conspecific with the type strain of *M. nana*, a species described from a cat with OE [95]. However, molecular analyses showed that there were some differences between the cat strain and the cow strains. Otitis in cattle, among other factors, is commonly attributed to infestation caused by rhabditiform nematodes, ticks and mites. Recently, *Malassezia* spp. have been associated with soil nematodes using PCR techniques, and their role as possible 'vectors' for species of *Malassezia* has been suggested [136].

A lipid-dependent species has been also associated with OE in an okapi (*Okapia johnstoni*) [86]. More recently, the fungal flora of the ear canal of captive okapi with otitis has been studied but *Malassezia* spp. were not isolated [137]. However, the cultural conditions (Sabouraud's dextrose agar plates incubated at 26°C for 48 h) were not appropriate for recovering lipid-dependent yeasts, and *M. pachydermatis* colonies might not have developed well during the 2 day-period of incubation.

Malassezia sympodialis was isolated from the skin of a horse with dermatitis; the strain was identified by means of phenotypic characterization and electrophoretic karyotyping [138]. This species has been also reported in horses without skin disease together with other lipid-dependent species and non-identified isolates using phenotypic techniques [130]. The presence of a novel *Malassezia* species was reported from normal equine skin [139], which was tentatively named "*Malassezia equi*", but a valid description was not included. It was identified by 26S rDNA D1/D2 sequence analysis as a member of the genus *Malassezia*, and was found to be most closely related to *M. sympodialis*. Several isolates from horses phylogenetically related to *M. sympodialis* grouped together by D1/D2 and ITS sequence analysis and were considered conspecific strains [140]. These yeasts have been finally included in the newly described species, *M. equina* [133].

Malassezia sympodialis was also isolated from clinically-altered combs of adult chickens, but it was also frequently isolated from healthy birds, and the authors concluded that it was a member of the resident flora [141]. In this study, yeasts were identified by phenotype and no other *Malassezia* species was isolated.

10.4 Zoonotic Transmission of *M. pachydermatis*

There are many bacterial and viral diseases that are transmissible from animals to humans whereas few mycoses are usually considered as zoonoses. Dermatophytosis associated with zoophilic dermatophyte species (e.g. *Microsporum canis* and *Trichophyton verrucosum*) and cat-transmitted cases of sporotrichosis are well known. However, the role that animals play in the epidemiology of the main human mycoses is still not well known and definitive environmental niches for their fungal agents have not been completely determined [142].

Although *M. pachydermatis* is mainly adapted to animals, it has also been implicated as a zoonotic agent. The first reports of systemic infection in humans by *M. pachydermatis* were published in the 1980s [143, 144]. Except for a single eye infection report by this species, the remaining cases involved infants in neonatal intensive care units. Until that time, reports of catheter-related *Malassezia* spp. sepsis in infants were caused by the anthropophilic species *M. furfur*, a relatively uncommon cause in these cases compared with other yeasts such as *Candida* spp. [145]. In a retrospective study of 32 *Malassezia* spp. isolates from human clinical specimens received since 1984 at the Centre for Disease Control in Atlanta (USA) [146], 15 were identified as *M. pachydermatis*. These last isolates were mainly from skin, tissue fluids (e.g. eye, ear, vagina) and four were isolated from blood. These authors reported that they had received *M. pachydermatis* from neonates who were receiving fat emulsions intra-venously and from compromised adults with various syndromes. In these cases, the clinical presumption was that the patients had infections caused by *M. furfur*. At this time, the epidemiology of *M. pachydermatis* with systemic infection in humans was mysterious because man is a rare carrier of this zoophilic species [147]. Also, it was unclear whether this species was harboured by a significant number of humans or whether some individuals were simple reservoirs for commensal or infectious associations between house pets and predisposed humans [146]. Subsequently, several additional studies have reported similar neonatal infections by *M. pachydermatis* [148–150].

The zoonotic association of pets with these neonatal infections was firstly mentioned by Chang et al. [148]. These authors hypothesized that *M. pachydermatis* was introduced into the intensive care nursery on health care workers' hands after being colonised from pet dogs at home. This species persisted in the nursery through patient-to-patient transmission. The mechanism of introduction and reservoir of *M. pachydermatis* were traced by molecular typing using RFLP. A common strain caused colonisation in the dogs, colonisation and infection in the infants, and colonisation in the health care workers. Careful hand-washing practises were essential to stop this outbreak. In the neonatal intensive care unit, strains of *M. furfur* and *M. pachydermatis* may persist on incubator surfaces for prolonged periods of time [151]; in this last study, all clinical *M. pachydermatis* strains appeared to be identical by PCR fingerprinting.

Skin colonisation by *Malassezia* species in full-term healthy newborns has been also investigated [152]. *M. pachydermatis* was not isolated from the skin of human neonates, but *M. sympodialis* and *M. globosa* colonisation begins at birth and increases in the first weeks of life. This study corroborates the animal origin of *M. pachydermatis* in human infections. Carriage of *M. pachydermatis* in dog owners was evaluated by microbiological culture and nested PCR [153], corroborating the possible route of zoonotic transmission in these neonatal infections. PCR identified this species on most (approximately 93%) human participants, whereas fungal culture was less reliable at detecting detect carriage with only 3 out of the 50 owners (6%) of healthy dogs positive by this method. In contrast, 29 out of the 75 owners (38.7%) of dogs with *Malassezia* dermatitis or otitis were positive for *M. pachydermatis*. A severe systemic infection by *M. pachydermatis* in an adult with acute myeloid leukaemia that underwent an allogeneic bone marrow transplantation has been cited [154], similar to the infections previously noted in neonates.

Currently, in human medicine, there are no known reports of intra-vascular catheter-acquired infections in humans by lipid-dependent species from animals, and no reports of fungemia caused by lipid-dependent species other than *M. furfur*. In fact, nearly all the fungemia reports produced by lipid-dependent species have been described as produced by *M. furfur sensu lato*, because they were performed before the taxonomic revision of the genus *Malassezia* [111], or the differentiation of the new *Malassezia* species was not used. However, the only lipid dependent *Malassezia* species isolated from human blood included in the last revision of the genus *Malassezia* was grouped within *M. furfur sensu stricto* [111]. Molecular methods were used to identify and type the new *Malassezia* species from different sources [155] to demonstrate that all the lipid-dependent yeasts from systemic diseases were *M. furfur* with a unique AFLP genotype. However, an atypical non lipid-dependent *M. sympodialis* was recently reported in an elderly patient with central venous catheter-related infection [156].

A heightened awareness of the potential for the transfer of *Malassezia* spp. to human patients and the application of molecular typing methods might lead to the recognition of more cases in the future. On the other hand, the renewed emphasis on hand hygiene in hospitals following the emergence of nosocomial infections with multidrug-resistant bacterial pathogens should help prevent the development of zoonotic *Malassezia* spp. infections.

References

1. Weidman FD (1925) Exfoliative dermatitis in the Indian rhinoceros (*Rhinoceros unicornis*), with description of a new species: *Pityrosporum pachydermatis.* In: Fox H (ed) Rep Lab Museum Comp Pathol Zoo Soc Philadelphia, Philadelphia, pp 36–43
2. Gustafson BA (1955) Otitis externa in the dog. A bacteriological and experimental study. Thesis. Royal Veterinary College, Stockholm
3. Slooff WC (1970) *Pityrosporum* Sabouraud. In: Lodder J (ed) The yeasts, a taxonomic study. North Holland, Amsterdam, pp 1167–1186
4. Fraser G (1961) *Pityrosporum pachydermatis* Weidman of canine origin. Trans Br Mycol Soc 44:441–448
5. Baxter M (1976) The association of *Pityrosporum pachydermatis* with the normal external ear canal of dogs and cats. J Small Anim Pract 17:231–234
6. Sharma VD, Rhoades HE (1975) The occurrence and microbiology of otitis externa in the dog. J Small Anim Pract 16:241–247
7. Sinha BK, Mohapatra LN, Kumar R (1976) Studies on otitis externa in dogs. 1. Survey of aetiological agents: fungi. Mykosen 19:63–69
8. Gedek B, Brutzel K, Gerlach R et al (1979) The role of *Pityrosporum pachydermatis* in otitis externa of dogs: evaluation of a treatment with miconazole. Vet Rec 104:138–140

9. Uchida Y, Mizutani M, Kubo T et al (1992) Otitis externa induced with *Malassezia pachydermatis* in dogs and the efficacy of pimaricin. J Vet Med Sci 54:611–614
10. Uchida Y, Nakade T, Otomo K et al (1994) Efficacy of a pimaricin suspension for treating otitis externa associated with *Malassezia pachydermatis*. J Small Anim Pract 35:521–523
11. Cannon PF (1986) International Commision on the Taxonomy of Fungi (ICTF): name changes in fungi of microbiological, industrial and medical importance. Part 2. Microbiol Sci 3:285–287
12. Yarrow D, Ahearn DG (1984) Genus 7. *Malassezia* Baillon. In: Kreger Van Rij NJW (ed) The Yeasts, a taxonomic study. Elsevier, Amsterdam, pp 882–885
13. Guillot J, Guého E (1995) The diversity of *Malassezia* yeasts confirmed by rRNA sequence and nuclear DNA comparisons. Antonie van Leeuwenhoek 67:297–314
14. Guillot J, Guého E, Chévrier G et al (1997) Epidemiological analysis of *Malassezia pachydermatis* isolates by partial sequencing of the large subunit ribosomal RNA. Res Vet Sci 62:22–25
15. Dufait R (1983) *Pityrosporon canis* as the cause of canine chronic dermatitis. Vet Med Small Animal Clin 78:1055–1057
16. Larsson CE, Gandra CRP, Larsson M et al (1988) Dermatitis in dogs caused by *Malassezia (Pityrosporum) pachydermatis*. Ars Veterinaria 4:63–68
17. Scott DW, Miller WH Jr (1989) Epidermal dysplasia and *Malassezia pachydermatis* infection in West Highland white terriers. Vet Dermatol 1:25–36
18. Mason KV, Evans AG (1991) Dermatitis associated with *Malassezia pachydermatis* in 11 dogs. J Am Anim Hosp Assoc 27:13–20
19. Plant JD, Rosenkrantz WS, Griffin CE (1992) Factors associated with a prevalence of high *Malassezia pachydermatis* numbers on dog skin. J Am Vet Med Assoc 201:879–885
20. Bond R, Ferguson EA, Curtis CF et al (1996) Factors associated with elevated cutaneous *Malassezia pachydermatis* populations in dogs with pruritic skin disease. J Small Anim Pract 37:103–107
21. Guaguère E, Prélaud P (1996) A rétrospective study of 54 dogs with *Malassezia pachydermatis* dermatitis: epidemiological, clinical, cytological, and histopathological results. Pract Méd Chir Animal Comp 31:309–323
22. Mason KV (1992) Seborrhoeic dermatitis – the aetiology rediscovered. Vet Dermatol Newslett 14:6–12
23. Olivry T, DeBoer DJ, Griffin CE et al (2001) The ACVD task force on canine atopic dermatitis: forewards and lexicon. Vet Immunol Immunopathol 81:143–146
24. Bond R, Collin NS, Lloyd DH (1994) Use of contact plates for the quantitative culture of *Malassezia pachydermatis* from canine skin. J Small Anim Pract 35:68–72
25. White SD, Bourdeau P, Blumstein P et al (1998) Comparison via cytology and culture of carriage of *Malassezia pachydermatis* in atopic and healthy dogs. In: Kwochka KW, Willemse T, von Tscharner C (eds) Advances in veterinary dermatology, vol 3. Butterworth Heinemann, Oxford, pp 291–298
26. Nardoni S, Dini M, Taccini F et al (2007) Occurrence, distribution and population size of *Malassezia pachydermatis* on skin and mucosae of atopic dogs. Vet Microbiol 122:172–177
27. DeBoer DJ, Marsella R (2001) The ACVD task force on canine atopic dermatitis (XII): the relationship of cutaneous infections to the pathogenesis and clinical course of canine atopic dermatitis. Vet Immunol Immunopathol 81:239–249
28. Bond R, Saijonmaa-Koulumies LE, Lloyd DH (1995) Population sizes and frequency of *Malassezia pachydermatis* at skin and mucosal sites on healthy dogs. J Small Anim Pract 36:147–150
29. Hajsig M, Tadic V, Lukman P (1985) *Malassezia pachydermatis* in dogs: significance of its location. Vet Arhiv 55:259–266
30. Kennis RA, Rosser EJ, Olivier NB et al (1996) Quantity and distribution of *Malassezia* organisms on the skin of clinically normal dogs. J Am Vet Med Assoc 208:1048–1051

31. Bond R (1996) *Malassezia pachydermatis* colonisation and infection of canine skin. Thesis. Royal Veterinary College, University of London, London
32. Bond R, Lloyd DH (1997) Skin and mucosal populations of *Malassezia pachydermatis* in healthy and seborrhoeic basset hounds. Vet Dermatol 8:101–106
33. Guillot J, Bensignor E, Jankowski F et al (2003) Comparative efficacies of oral ketoconazole and terbinafine for reducing *Malassezia* population sizes on the skin of basset hounds. Vet Dermatol 14:153–157
34. Bond R, Rose JF, Ellis JW et al (1995) Comparison of two shampoos for treatment of *Malassezia pachydermatis*-associated seborrhoeic dermatitis in basset hounds. J Small Anim Pract 36: 99–104
35. Morris DO, Olivier NB, Rosser EJ (1998) Type-1 hypersensitivity reactions to *Malassezia pachydermatis* extracts in atopic dogs. Am J Vet Res 59:836–841
36. Kieffer M, Bergbrant IM, Faergemann J et al (1990) Immune reactions to *Pityrosporum ovale* in adult patients with atopic and seborrheic dermatitis. J Am Acad Dermatol 22:739–742
37. Lowery CA, Dickerson TJ, Janda KD (2008) Interspecies and interkingdom communication mediated by bacterial quorum sensing. Chem Soc Rev 37:1337–1346
38. Kennedy MJ (1988) Adhesion and association mechanisms of *Candida albicans*. In: McGinnis MR (ed) Current topics in medical mycology, vol 2. Springer, New York, pp 73–169
39. Bergbrant IM, Faergemann J (1994) Adherence of *Malassezia furfur* to human stratum corneum cells *in vitro*: a study of healthy individuals and patients with seborrhoeic dermatitis. Mycoses 37:5–6
40. Xu J, Saunders CW, Hu P et al (2007) Dandruff-associated *Malassezia* genomes reveal convergent and divergent virulence traits shared with plant and human fungal pathogens. Proc Natl Acad Sci USA 104:18730–18735
41. Bond R, Curtis CF, Hendricks A et al (2002) Intradermal test reactivity to *Malassezia pachydermatis* in atopic dogs. Vet Rec 150:448–449
42. Nuttal T, Halliwell R (2001) Serum antibodies to *Malassezia pachydermatis* extracts in atopic dermatitis. Vet Dermatol 12:327–332
43. Chen TA, Halliwell RE, Pemberton AD et al (2002) Identification of major allergens of *Malassezia pachydermatis* in dogs with atopic dermatitis and *Malassezia* overgrowth. Vet Dermatol 13:141–150
44. Morris DO, DeBoer DJ (2003) Evaluation of serum obtained from atopic dogs with dermatitis attributable to *Malassezia pachydermatis* for passive transfer of immediate hypersensitivity to that organism. Am J Vet Res 64:262–266
45. Gaitanis G, Menounos P, Katsambas A et al (2003) Detection and mutation screening of *Malassezia sympodialis* sequences coding for the Mal s 1 allergen implicated in atopic dermatitis. J Invest Dermatol 121:1559–1560
46. Nett CS, Reichler I, Grest P et al (2001) Epidermal dysplasia and *Malassezia* infection in two West Highland white terrier siblings: an inherited skin disorder or reaction to severe *Malassezia* infection? Vet Dermatol 12:285–290
47. Bond R, Patterson-Kane JC, Lloyd DH (2002) Intradermal test reactivity to *Malassezia pachydermatis* in healthy basset hounds and bassets with *Malassezia* dermatitis. Vet Rec 151:105–109
48. Bond R, Perrins N, Patterson-Kane JC et al (2006) Patch test responses to *Malassezia pachydermatis* in healthy basset hounds and in basset hounds with *Malassezia* dermatitis. Med Mycol 44:419–427
49. Morris DO (1999) *Malassezia* dermatitis and otitis. Vet Clin North Am Small Animal Pract 29:1303–1310
50. Keddie F, Orr A, Liebes D (1961) Direct staining on vinyl plastic tape, demonstration of the cutaneous flora of the epidermis by the strip method. Sabouraudia 1:108–111

51. Bond R, Lloyd DH, Plummer JM (1995) Evaluation of a detergent scrub technique for the quantitative culture of *Malassezia pachydermatis* from canine skin. Res Vet Sci 58:133–137
52. Gross TL, Ihrke PJ, Walder EJ et al (2005) *Malassezia* dermatitis. In: Skin diseases of the dog and cat. Clinical and histopathological diagnosis, 2nd edn. Blackwell Science, Oxford, pp 142–146
53. Maudlin EA, Scott DW, Miller WH et al (1997) *Malassezia* dermatitis in the dog: a retrospective histopathological and immunopathological study of 86 cases (1990–95). Vet Dermatol 8:191–202
54. Gupta AK, Kohli Y, Li A et al (2000) *In vitro* susceptibility of the seven *Malassezia* species to ketoconazole, voriconazole, itraconazole and terbinafine. Br J Dermatol 142:758–765
55. Schmidt A (1997) *In vitro* activity of climbazole, clotrimazole and silver sulphadiazine against isolates of *Malassezia pachydermatis*. Zentral Veterinarmed B 44:193–197
56. Schmidt A, Ruhl Horster B (1996) *In vitro* susceptibility of *Malassezia furfur* against azole compounds. Mycoses 39:309–312
57. Bourdeau P, Marchand A-M, Etore F (2004) *In vitro* activity of posaconazole and other antifungals against *Malassezia pachydermatis* isolated from dogs. Vet Dermatol 15:46
58. Velegraki A, Alexopoulos EC, Kritikou S et al (2004) Use of fatty acid RPMI 1640 media for testing susceptibilities of eight *Malassezia* species to the new triazole posaconazole and to six established antifungal agents by a modified NCCLS M27–A2 microdilution method and Etest. J Clin Microbiol 42:3589–3593
59. Lorenzini R, Mercantini R, de Bernardis F (1985) *In vitro* sensitivity of *Malassezia* spp. to various antimycotics. Drugs Exp Clin Res 11:393–395
60. Nakano Y, Wada M, Tani H et al (2005) Effects of beta-thujaplicin on anti-*Malassezia pachydermatis* remedy for canine otitis externa. J Vet Med Sci 67:1243–1247
61. Hammer KA, Carson CF, Riley TV (2000) *In vitro* activities of ketoconazole, econazole, miconazole, and *Melaleuca alternifolia* (tea tree) oil against *Malassezia* species. Antimicrob Agents Chemother 44:467–469
62. Nenoff P, Haustein UF, Brandt W (1996) Antifungal activity of the essential oil of *Melaleuca alternifolia* (tea tree oil) against pathogenic fungi *in vitro*. Skin Pharmacol 9:388–394
63. Weseler A, Geiss HK, Saller R et al (2002) Antifungal effect of Australian tea tree oil on *Malassezia pachydermatis* isolated from canines suffering from cutaneous skin disease. Schweiz Arch Tierheilkd 144:215–221
64. Nakano Y, Matsuo S, Tani H et al (2006) Therapeutic effects of beta-thujaplicin eardrops on canine *Malassezia*-related otitis externa. J Vet Med Sci 68:373–374
65. Dufait R, Dierckxsens I (1971) Otitis bij de hond. Vlaams Diergeneeskundig Tijdschrift 40: 147–152
66. Vanden Bossche H, Engelen M, Rochette F (2003) Antifungal agents of use in animal health–chemical, biochemical and pharmacological aspects. J Vet Pharmacol Ther 26:5–29
67. Studdert VP, Hughes KL (1991) A clinical trial of a tropical preparation of miconazole, polymyxin and prednisolone in the treatment of otitis externa in dogs. Aust Vet J 68:193–195
68. Rougier S, Borell D, Pheulpin S et al (2005) A comparative study of two antimicrobial/anti-inflammatory formulations in the treatment of canine otitis externa. Vet Dermatol 16:299–307
69. Bensignor E, Grandemange E (2006) Comparison of an antifungal agent with a mixture of antifungal, antibiotic and corticosteroid agents for the treatment of *Malassezia* species otitis in dogs. Vet Rec 158:193–195
70. Gupta AK, Kohli Y (2003) *In vitro* susceptibility testing of ciclopirox, terbinafine, ketoconazole and itraconazole against dermatophytes and nondermatophytes, and in vitro evaluation of combination antifungal activity. Br J Dermatol 149:296–305
71. Ratnavel RC, Squire RA, Boorman GC (2007) Clinical efficacies of shampoos containing ciclopirox olamine (1.5%) and ketoconazole (2.0%) in the treatment of seborrhoeic dermatitis. J Dermatol Treat 18:88–96

72. Gupta AK, Fleckman P, Baran R (2000) Ciclopirox nail lacquer topical solution 8% in the treatment of toenail onychomycosis. J Am Acad Dermatol 43:S70–S80
73. Lyskova P, Vydrzalova M, Mazurova J (2007) Identification and antimicrobial susceptibility of bacteria and yeasts isolated from healthy dogs and dogs with otitis externa. J Vet Med A Physiol Pathol Clin Med 54:559–563
74. Roques C, Brousse S, Panizzutti C (2006) *In vitro* antifungal efficacy of ciclopirox olamine alone and associated with zinc pyrithione compared to ketoconazole against *Malassezia globosa* and *Malassezia restricta* reference strains. Mycopathologia 162:395–400
75. Lee RE, Liu TT, Barker KS et al (2005) Genome-wide expression profiling of the response to ciclopirox olamine in *Candida albicans*. J Antimicrob Chemother 55:655–662
76. Bourdeau P, Blumstein P, Marchand A-M et al (2006) An *in vivo* procedure to evaluate antifungal agents on *Malassezia pachydermatis* in dogs: example with a piroctone olamine containing shampoo. J Mycol Méd 16:9–15
77. Lloyd DH, Bond R, Lamport I (1998) Antimicrobial activity *in vitro* and *in vivo* of a canine ear cleanser. Vet Rec 143:111–112
78. Cole LK, Kwochka KW, Kowalski JJ et al (1998) Microbial flora and antimicrobial susceptibility patterns of isolated pathogens from the horizontal ear canal and middle ear in dogs with otitis media. J Am Vet Med Assoc 212:534–538
79. Pinchbeck LR, Hillier A, Kowalski JJ et al (2002) Comparison of pulse administration versus once daily administration of itraconazole for the treatment of *Malassezia pachydermatis* dermatitis and otitis in dogs. J Am Vet Med Assoc 220:1807–1812
80. Chen TA, Hill PB (2005) The biology of *Malassezia* organisms and their ability to induce immune responses and skin disease. Vet Dermatol 16:4–26
81. Rosales MS, Marsella R, Kunkle G et al (2005) Comparison of the clinical efficacy of oral terbinafine and ketoconazole combined with cephalexin in the treatment of *Malassezia* dermatitis in dogs–a pilot study. Vet Dermatol 16:171–176
82. Vlaminck KMJA, Engelen MACM (2005) An overview of pharmacokinetic and pharmacodynamic studies in the development of itraconazole for feline *Microsporum canis* dermatophytosis. In: Hillier A, Foster AP, Kwochka KW (eds) Advances in veterinary dermatology, vol 5. Blackwell, Oxford, pp 130–136
83. Legendre AM, Rohrbach BW, Toal RL et al (1996) Treatment of blastomycosis with itraconazole in 112 dogs. J Vet Intern Med 10:365–371
84. Bond R, Anthony RM, Dodd M et al (1996) Isolation of *Malassezia sympodialis* from feline skin. J Med Vet Mycol 34:145–147
85. Bond R, Howell SA, Haywood PJ et al (1997) Isolation of *Malassezia sympodialis* and *Malassezia globosa* from healthy pet cats. Vet Rec 141:200–201
86. Guillot J, Chermette R, Guého E (1994) Prevalence of the genus *Malassezia* in the Mammalia. J Mycol Méd 4:72–79
87. Hajsig D, Hajsig M, Svoboda Vukovic D (1990) *Malassezia pachydermatis* in healthy cats. Vet Arch 60:69–73
88. Crespo MJ, Abarca ML, Cabañes FJ (2002) Occurence of *Malassezia* spp. in the external ear canals of dogs and cats with and without otitis externa. Med Mycol 40:115–121
89. Dizotti CE, Coutinho SD (2007) Isolation of *Malassezia pachydermatis* and *M. sympodialis* from the external ear canal of cats with and without otitis externa. Acta Vet Hung 55:471–477
90. Nardoni S, Mancianti F, Rum A et al (2005) Isolation of *Malassezia* species from healthy cats and cats with otitis. J Feline Med Surg 7:141–145
91. Cafarchia C, Gallo S, Capelli G et al (2005) Occurrence and population size of *Malassezia* spp. in the external ear canal of dogs and cats both healthy and with otitis. Mycopathol 160:143–149
92. Colombo S, Nardoni S, Cornegliani L et al (2007) Prevalence of *Malassezia* spp. yeasts in feline nail folds: a cytological and mycological study. Vet Dermatol 18:278–283

93. Crespo MJ, Abarca ML, Cabañes FJ (1999) Isolation of *Malassezia furfur* from a cat. J Clin Microbiol 37:1573–1574
94. Coutinho SD, Fedullo JD, Correa SH (2006) Isolation of *Malassezia* spp. from cerumen of wild felids. Med Mycol 44:383–387
95. Hirai A, Kano R, Makimura K et al (2004) *Malassezia nana* sp. nov., a novel lipid-dependent yeast species isolated from animals. Int J Syst Evol Microbiol 54:623–627
96. Perrins N, Gaudiano F, Bond R (2007) Carriage of *Malassezia* spp. yeasts in cats with diabetes mellitus, hyperthyroidism and neoplasia. Med Mycol 45:541–546
97. Åhman S, Perrins N, Bond R (2007) Carriage of *Malassezia* spp. yeasts in healthy and seborrhoeic Devon Rex cats. Med Mycol 45:449–455
98. Bond R, Stevens K, Perrins N et al (2008) Carriage of *Malassezia* spp. yeasts in Cornish Rex, Devon Rex and domestic short-haired cats: a cross-sectional survey. Vet Dermatol 19:299–304
99. Sierra P, Guillot J, Jacob H et al (2000) Fungal flora on cutaneous and mucosal surfaces of cats infected with feline immunodeficiency virus or feline leukemia virus. Am J Vet Res 61:158–161
100. Godfrey DR (1998) A case of feline paraneoplastic alopecia with secondary *Malassezia*-associated dermatitis. J Small Anim Pract 39:394–396
101. Mauldin EA, Morris DO, Goldschmidt MH (2002) Retrospective study: the presence of *Malassezia* in feline skin biopsies. A clinicopathological study. Vet Dermatol 13:7–13
102. Forster-Van Hijfte MA, Curtis CF, White RN (1997) Resolution of exfoliative dermatitis and *Malassezia pachydermatis* overgrowth in a cat after surgical thymoma resection. J Small Anim Pract 38:451–454
103. Carlotti DN, Hubert B, Delmas H et al (1993) Les mycoses superficielles chez le chat. Pract Méd Chir Animal Comp 28:241–257
104. Jazic E, Coyner KS, Loeffler DG et al (2006) An evaluation of the clinical, cytological, infectious and histopathological features of feline acne. Vet Dermatol 17:134–140
105. White SD, Bourdeau P, Blumstein P et al (1997) Feline acne and results of treatment with mupirocin in an open clinical trial: 25 cases (1994–96). Vet Dermatol 8:157–164
106. Bond R, Curtis CF, Ferguson EA et al (2000) An idiopathic facial dermatitis of Persian cats. Vet Dermatol 11:35–41
107. Scott DW (1992) Bacteria and yeast on the surface and within non-inflamed hair follicles of skin biopsies from cats with non-neoplastic dermatoses. Cornell Vet 82:371–377
108. Ordeix L, Galeotti F, Fondati A (2006) *Malassezia* spp. overgrowth in allergic cats. In: 21st Annual Congress of the ESVD-ECVD Lisbon: Eur Society of Veterinary Dermatology – European College of Veterinary Dermatology
109. Mauldin EA, Ness TA, Goldschmidt MH (2007) Proliferative and necrotizing otitis externa in four cats. Vet Dermatol 18:370–377
110. Ordeix L, Galeotti F, Scarampella F et al (2007) *Malassezia* spp. overgrowth in allergic cats. Vet Dermatol 18:316–323
111. Guého E, Midgley G, Guillot J (1996) The genus *Malassezia* with description of four new species. Antonie van Leeuwenhoek 69:337–355
112. Guillot J, Bond R (1999) *Malassezia pachydermatis*: a review. Med Mycol 37:295–306
113. Bauwens L, De Vroey Ch, De Meurichy W (1996) A case of exfoliative dermatitis in a captive Southern white rhinoceros (*Ceratotherium simum simum*). J Zoo Wildlife Med 27: 271–274
114. Kuttin ES, Müller J (1994) The fungal flora of zoo animals' ears. Mycoses 37:1–2
115. Wesche P, Bond R (2003) Isolation of *Malassezia pachydermatis* from the skin of captive rhinoceroses. Vet Rec 153:404–405
116. Guillot J, Petit F, Degorce F et al (1998) Dermatitis caused by *Malassezia pachydermatis* in a California sea lion (*Zalophus californianus*). Vet Rec 142:311–312

117. Nakagaki K, Hata K, Iwata E et al (2000) *Malassezia pachydermatis* isolated from a South American sea lion (*Otaria byronia*) with dermatitis. J Vet Med Sci 62:901–903
118. Pin D (2004) Seborrhoeic dermatitis in a goat due to *Malassezia pachydermatis*. Vet Dermatol 15:53–56
119. Dinsdale J, Rest J (1995) Yeast infection in ferrets. Vet Rec 137:647–648
120. Kuttin ES, Glas I (1985) Mycotic otitis externa in animals. Mykosen 28:61–65
121. Pinter L, Anthony R, Glumac N et al (2002) Apparent cross-infection with a single strain of *Malassezia pachydermatis* on a pig farm. Acta Vet Hung 50:151–156
122. Salkin IF, Gordon MA, Stone WB (1978) *Pityrosporum pachydermatis* in a black bear (*Ursus americanus*). Sabouraudia 16:35–38
123. Salkin IF, Stone WB, Gordon MA (1980) Association of *Malassezia* (*Pityrosporum*) *pachydermatis* with sarcoptic mange in New York State. J Wildl Dis 16:509–514
124. Radi Z (2004) Outbreak of sarcoptic mange and malasseziasis in rabbits (*Oryctolagus cuniculus*). Comp Med 54:434–437
125. Gandra RF, Gambale W, de Cássia Garcia Simão R et al (2008) *Malassezia* spp. in acoustic meatus of bats (*Molossus molossus*) of the Amazon region, Brazil. Mycophathologia 1765: 21–26
126. Breuer-Strosberg R, Hochleithner M, Kuttin E (1990) *Malassezia pachydermatis* isolation from a scarlet macaw. Mycoses 33:247–250
127. Preziosi D, Morris D, Johnston M et al (2006) Distribution of *Malassezia* organisms on the skin of unaffected psittacine birds and psittacine birds with feather-destructive behavior. J Am Vet Med Assoc 228:216–220
128. Dufait R (1985) Presence of *Malassezia pachydermatis* (syn. *Pityrosporum canis*) on the hair and feathers of domestic animals. Bull Soc Fr Mycol Méd 14:19–22
129. Midgley G, Clayton Y (1969) The yeast flora of birds and mammals in captivity. Antonie van Leeuwenhoek 35(S):E23–E24
130. Crespo MJ, Abarca ML, Cabañes FJ (2002) Occurrence of *Malassezia* spp. in horses and domestic ruminants. Mycoses 45:333–337
131. Bliss EL (1984) Tinea versicolor dermatomycosis in the goat. J Am Vet Med Assoc 184: 1512–1513
132. Uzal F, Paulson D, Eigenheer A et al (2007) *Malassezia slooffiae*-associated dermatitis in a goat. Vet Dermatol 18:348–352
133. Cabañes FJ, Theelen B, Castellá G et al (2007) Two new lipid-dependent *Malassezia* species from domestic animals. FEMS Yeast Res 7:1064–1076
134. Duarte ER, Melo MM, Hahn RC, Hamdan JS (1999) Prevalence of *Malassezia* spp. in the ears of asymptomatic cattle and cattle with otitis in Brazil. Med Mycol 37:159–162
135. Duarte ER, Batista RD, Hahn RC, Hamdan JS (2003) Factors associated with the prevalence of *Malassezia* species in the external ears of cattle from the state of Minas Gerais, Brazil. Med Mycol 41:137–142
136. Renker C, Alphei J, Buscot F (2003) Soil nematodes associated with the mammal pathogenic fungal genus *Malassezia* (Basidiomycota: Ustilaginomycetes) in Central European forests. Biol Fertil Soils 37:70–72
137. Allender M, Langan J, Citino S (2008) Investigation of aural bacterial and fungal flora following otitis in captive okapi (*Okapia johnstoni*). Vet Dermatol 19:95–100
138. Senczek D, Siesenop U, Böhm K (1999) Characterization of *Malassezia* species by means of phenotypic characteristics and detection of electrophoretic karyotypes by pulsed-field gel electrophoresis (PFGE). Mycoses 42:409–414
139. Nell A, James SA, Bond CJ et al (2002) Identification and distribution of a novel *Malassezia* species yeast on normal equine skin. Vet Rec 150:395–398
140. Cabañes FJ, Hernández JJ, Castellá G (2005) Molecular analysis of *Malassezia sympodialis*-related strains from domestic animals. J Clin Microbiol 43:277–283

141. Gründer S, Mayser P, Redmann T, Kaleta E (2005) Mycological examinations on the fungal flora of the chicken comb. Mycoses 48:114–119
142. Cabañes FJ (2008) Micosis and zoonosis: *Cryptococcus* spp. Rev Iberoam Micol 25:S1–S3
143. Larocco M, Dorenbaum A, Robinson A et al (1988) Recovery of *Malassezia pachydermatis* from eight infants in a neonatal intensive care nursery: clinical and laboratory features. Pediatr Infect Dis J 7:398–401
144. Mickelsen P, Viano-Paulson M et al (1988) Clinical and microbiological features of infection with *Malassezia pachydermatis* in high-risk infants. J Infect Dis 157:1163–1168
145. Marcon M, Powell D (1992) Human infections due to *Malassezia* spp. Clin Microbiol Rev 5:101–119
146. Guého E, Simmons R, Pruitt W et al (1987) Association of *Malassezia pachydermatis* with systemic infections of humans. J Clin Microbiol 25:1789–1790
147. Guého E, Simmons R, Ahearn D (1987) *Malassezia* spp. et infections systémiques. Bull Soc Fr Mycol Méd 16:329–332
148. Chang HJ, Miller HL, Watkins N et al (1998) An epidemic of *Malassezia pachydermatis* in an intensive care nursery associated with colonization of health care workers' pet dogs. N Engl J Med 338:706–711
149. Chryssanthou E, Broberger U, Petrini B (2001) *Malassezia pachydermatis* fungaemia in a neonatal intensive care unit. Acta Paediatr 90:323–327
150. Welbel SF, McNeil MM, Pramanik A et al (1994) Nosocomial *Malassezia pachydermatis* bloodstream infections in a neonatal intensive care unit. Pediatr Infect Dis J 13:104–108
151. van Belkum A, Boekhout T, Bosboom R (1994) Monitoring spread of *Malassezia* infections in a neonatal intensive care unit by PCR-mediated genetic typing. J Clin Microbiol 32:2528–2532
152. Bernier V, Weill F, Hirigoyen V et al (2002) Skin colonization by *Malassezia* species in neonates: a prospective study and relationship with neonatal cephalic pustulosis. Arch Dermatol 138:215–218
153. Morris DO (2005) *Malassezia pachydermatis* carriage in dog owners. Emerg Infect Dis 11:83–88
154. Lautenbach E, Nachamkin I, Schuster M (1998) *Malassezia pachydermatis* infections. New England J Med 339:270
155. Theelen B, Silvestri M, Guého E et al (2001) Identification and typing of *Malassezia* yeasts using amplified fragment length polymorphism (AFLP), random amplified polymorphic DNA (RAPD) and denaturing gradient gel electrophoresis (DGGE). FEMS Yeast Res 1:79–86
156. Kikuchi K, Fujishiro Y, Totsuka K et al (2001) A case of central venous catheter-related infection with *Malassezia sympodialis.* Nippon Ishinkin Gakkai Zasshi 42:220–222

Malassezia Database 11

Vincent A. Robert and Aristea Velegraki

Core Messages

- In the last decade, several *Malassezia*-based research fields were encouraged by the revised taxonomy of the genus, and a significant body of fundamental and applied research data was generated. Building an information system to archive, process, evaluate, and utilize these data may be regarded as a practical research tool for researchers and other professionals. This information system will allow timely access to epidemiological surveillance data and will be a tool for laboratory surveillance of *Malassezia*-associated diseases, incorporating a defined set of conventional and molecular identification and typing information, as well as chromatographic data for the characterization of *Malassezia* yeast metabolites. The notion of the *Malassezia* website interlinking to other web-based repositories, such as the multilocus sequencing typing (MLST) (http://www.mlst.net), is certainly within the scope of designing this website.

11.1 Introduction

Malassezia-related skin diseases affect a large proportion of the population globally. Suffice to mention that PV accounts for a substantial percentage of all superficial dermatomycoses, amounting up to 40% in areas with warm and humid climate [1], and dandruff/seborrheic

A. Velagraki (✉)
Mycology Laboratory, Medical School, National and Kapodistrian University of Athens, Athens 115 27, Greece
e-mail: aveleg@med.uoa.gr

T. Boekhout et al. (eds.), *Malassezia and the Skin*
DOI: 10.1007/978-3-642-03616-3_11, © Springer Verlag Berlin Heidelberg 2010

dermatitis (D/SD) affects 45–50% of the global population [2] (viz. Chaps. 3, 6 and 7). Topical antifungals constitute the mainstay of treatment, and chronic prophylaxis [3] is often required to prevent recurrences (viz. Chaps. 6, 7 and 8). As treatment regimens comprise a significant national and private health service spending, *Malassezia*-based research outcomes can provide solutions to cost-effective management strategies. For this purpose, appropriate compilation, analysis, and assessment of research-generated data would be of assistance.

Besides the involvement of *Malassezia* yeasts in the skin diseases of man and animals, their occasional implication in bloodstream infections and outbreaks in neonatal and adult intensive care units (ICU) (viz. Chaps. 3 and 8), as well as their apparent global occurrence, has made them important and interesting to study. Consequently, a wide array of investigations encouraged by the revised taxonomy of the genus (viz. Chap. 2) generated new knowledge on *Malassezia* biology, phylogeny, epidemiology, genomics, and pathophysiology (viz. corresponding Chapters). This led to the accumulation of a considerable body of data covering diverse aspects of *Malassezia* research. Building an information system to archive, process, evaluate, and utilize these data and information can therefore become a practical research tool. In that respect, the database hosted by the *Malassezia* website can accommodate biological, physiological, biochemical, and genomic information of isolates derived from cases of human and animal skin diseases and healthy carriers, as well as isolates from human invasive disease.

Making use of a database with *Malassezia* species profiles from different skin diseases, systemic manifestations, and carriage would provide researchers and professionals with timely access to epidemiological data. In addition, it could serve as a tool for laboratory surveillance of *Malassezia*-associated diseases, incorporating a defined set of conventional and molecular identification and typing data. For instance, information on a case could be included, only if the researcher/professional deposits a complete identification and typing dataset. A further prerequisite may be the deposition of the isolate(s) in a public culture collection. Indeed, in this case, sequence-based identification is superior because of its accuracy and reproducibility.

The genus *Malassezia* is relatively small. Currently, 13 species are recognized, although lately a new species that was isolated from a hamster has been genetically characterized [4]. The recognition of this imminent species *novus* was based on two markers for differentiation from other *Malassezia* species: the nucleotide sequences of the 26S rDNA and the ITS1 region. Notwithstanding that fact, important information including the phenotypic characterization of the *Malassezia* isolate, the exact genus and species of the hamster host, and the site of isolation of the yeast (lesional or healthy skin) is still pending. It is therefore obvious, that as more species are recognized more data are accumulating, requiring analysis and evaluation in order to establish their clinical significance. Retrieving information from the *Malassezia* website could immensely facilitate investigative and valuation procedures.

As previously discussed (viz. Chap. 3), the *Malassezia* ribosomal DNA (rDNA) is the most commonly studied sequencing target, namely the ITS regions ITS1 and ITS2, and the Dl-D2 regions of the large subunit (LSU) rRNA gene. A large mass of these sequencing data has been deposited in the GeneBank (http://www.ncbi.nlm.nih.gov). For many *Malassezia* researchers, sequencing data guide the species placement within the *Malassezia* genus. Nevertheless, the GeneBank being an open database accepts sequences as submitted by the depositor without further editing. As a result, some GeneBank search products may yield a number of similar species results for a single query sequence. To avoid this setback, the *Malassezia* website can provide a reference sequence database with curated sequences to be used for the accurate identification of species within the genus. A future utility of the *Malassezia* website would be to provide high resolution fingerprints of nosocomial outbreak isolates from different geographical locations, thus supporting initial recognition of possible spatio-temporal clusters of cases caused by identical genotypes. This of course would entail standardization of a consensus PCR-based typing method.

The *Malassezia* website aims to achieve a structured collection of records that could allow researchers to deposit and retrieve data related to *Malassezia* and is founded on the experience with the yeast taxonomy website (www.mycobank.org/yeast), the *Scedosporium* website (http://www.scedosporium-ecmm.com/Database/BioloMICS.aspx), and the ITS sequences database of *Candida* species (www.mycologylab.org). In this chapter, we briefly describe the major features of the website, and review the usefulness and potential problems of specialized repositories for archiving phenotypic and genotypic data. Finally, basic, step-by-step, instructions for using this website are provided, including descriptions on how to store data records and how relevant information-processing activities are automated. This aims to aid the voluntary contributions of scientists and make it a potentially useful resource for the *Malassezia* research community.

11.2 Description of Functionalities

The software used to publish the *Malassezia* data is called BioloMICS [5]. It can be used for retrieval, management, and analysis (e.g., identification, classification and statistics) of any biological material or research experiment. In 2000, the BioloMICSWeb [6] software package already allowed online Internet publication of the data prepared with BioloMICS. It also permits polyphasic identifications to be performed online against species or strain databases [7] (see www.cbs.knaw.nl/Yeast.htm; 7,800 strains, 901 species, ±350 characters), *Penicillium* [8] (see www.cbs.knaw.nl/Penicillium.htm; 58 species, ±60 characters), or *Phaeoacremonium* [9] (see www.cbs.knaw.nl/Phaeoacremonium.htm; 22 species, ±40 characters).

11.2.1 Fields and Data

A preliminary list of fields has been established mainly based on the needs of people working in the medical field. Fields/characters can be added, modified, deleted, re-ordered, re-organized and grouped at any time by the administrator in charge of the database. Therefore, the list provided in Table 11.1 is not a definitive one and can be adapted according to the needs of the *Malassezia* research community.

The complete and up to date list of fields can be found at www.mycobank.org/malassezia/Help.aspx. The way to create links between websites is also provided at the same address.

Table 11.1 Synoptic list of available fields sorted by type

List of available fields	Options or states	Type of field
Date of isolation of the strain (dd/mm/yyyy)		Date
Outcome date (dd/mm/yyyy)		Date
Latitude, longitude where isolated (radians)		Latitude, longitude
Bibliography		Link to bibliographic table
Microscopic pictures		Link to picture(s)
Miscellaneous pictures		Link to picture(s)
Associated pathology (human or animal)	Pityriasis versicolor, atopic dermatitis, seborrhoic dermatitis, psoriasis, folliculitis, bloodstream infection, colonization, commensal, site of lesion (ear, scalp, skin, etc…)	Multiple-choice
Underlying disease(s) factors	AIDS/HIV, autoimmune disease, cancer, bone marrow transplantation (BMT), hematopoeitic stem cell transplantation (HSCT), nonablative allogenic transplant, diabetes, solid organ transplant, prematurity, surgical operation, etc…	Multiple-choice
CD4 cells	Minimum–maximum	Range
ITSs sequence		Sequence
Large subunit rDNA 26-28S sequence		Sequence
Molecular typing map		Image
Macroscopic, microscopic, morphology ultrastructure, physiological and biochemical profiles (viz. Chap. 2)		Text*
Outcome of patient in NICU or ICU	Death, death due to other disease(s), cured	Single choice
Sex	Male, female	Single choice
Age of patient		Text*
Antifungal treatment: drugs, dosages, starting and ending dates		Text
Associated pathology remarks (humans or animals)		Text
Clinical data remarks (type of lesions, fever, neurological and pulmonary signs, etc…)		Text

Table 11.1 (continued)

List of available fields	Options or states	Type of field
Clinical presentation remarks		Text
Depositor reference number of the isolate		Text
Depositor's center name and address		Text
General remarks		Text
Identified by		Text
Locality of isolation (country, city, etc…)		Text
Metabolites production		Text
Name of the depositor		Text
Optimal medium and temperature for growth (°C)		Text
Other collections numbers		Text
Patient's city, country and hospital/ward		Text
Race	Caucasian, African, Asian, Indian, etc…	Text
Veterinary host	Dog, cat, other mammals, etc…	Text
Species name given by the depositor		Text
Species name given by the *Malassezia* group		Text
Strain isolated from/substrate (man, animal, etc…)		Text
Time required to obtain a visible colony (under optimal conditions) never in hours		Text
Underlying disease remarks or underlying diseases		Text

*The user is able to write in a text-box positioned next to each relevant field.

11.2.2 General Search

The "General search" option (www.mycobank.org/malassezia/BioloMICS.aspx) allows querying any text field available from the database (Table 11.1). Other fields could be queried using the "Advanced searching option."

Enter the text to be found in the text box and click on the "Search" button (Fig. 11.1).

The default option is a wild card search, but one can also select the exact match option in order to reduce the number of records found. It is also possible to perform basic joined

Fig. 11.1 General search page in the *Malassezia* database

queries using the "AND" or "OR" separators (only one at the same time). For example, "Sweden OR Greece" or "Sweden AND Greece." "Sweden AND (Japan OR Belgium)," however, is not possible in the general search, but is possible via the "Advanced search."

11.2.3 Advanced Search

The "Advanced search" option (www.mycobank.org/malassezia/BioloMICS.aspx?SearchOpt=4) allows querying any field available from the database. Queries can be performed in five steps:

Step 1. Select a field to search from the list of all available fields.

Step 2. Depending on the field selected in step 1, an entry box or one or two dropdown list(s) will be displayed and one has to select the state/keyword of interest.

Step 3. Add the selected state/keyword to the list of conditions or questions.

Step 4. The way the conditions are combined is provided in the text box of step 4. By default, conditions are appended with the "AND separator." This could be changed using the “OR,” “AND” and “NOT” keywords. One can also use the brackets to join two or several conditions together.

Step 5. Click on the “Search” button to perform the query on the current database.

One can change a given query by just changing the “OR,” “AND,” and “NOT” keywords and the bracket positions or deleting a given condition from the list and finally clicking on the “Search” button again.

11.2.4 Additional Notes on Complex Queries

“C00 AND C01 OR C02” is equivalent to (“C00 AND C01”) “OR C02” and means that questions/conditions 0 (i.e. C00) and 1 (i.e. C01) or question/condition 2 (i.e. C02) have to be fulfilled for the record to be selected. In a given database, this could allow the retrieval of a certain number of records. Another grouping of the questions, like C00 AND (C01 OR C02), will lead to a different selection of records. In this latter example, questions/conditions 1 or 2 have to be fulfilled at the same time than question 0. The use of brackets can therefore importantly change the results of the queries, even with the same original questions/conditions.

All the opened brackets have to be closed. One may remove a question manually. For example, in C00 AND C01 AND C02, one may remove the question/condition 1 and this would then look like C00 AND C02. The AND, OR, NOT could be switched. For example, one may manually change the query from (C00 AND C01) OR C02 to (C00 AND NOT C01) OR C02 or to C00 AND C01 AND C02. Also, one should be aware that a space is needed before and after the AND, OR and the NOT. Any character other than the opening “(‘ and closing ‘)” brackets should not be added.

Wrongly written advanced queries like (C00 AND C01) AND C02 or C00 AND C01AND C02 (for example) will lead to a message stating that no record was found or that the query was wrong.

11.2.5 Online Deposit of Data

Any registered user can deposit data via the online deposit form. Only registered users can deposit and modify previously deposited records. The rationale behind the registration is to avoid corruption of the database by mal-intentioned users. The curator of the database has the right to modify or delete any record that would not be relevant or contain wrong data.

11.2.6 Registration

Registration is free, can be done in a few minutes and in 2 steps:

Step 1. Go to www.mycobank.org/malassezia/DefaultPage.aspx and click on the “Create profile” button or go directly to www.mycobank.org/malassezia/RegisterNew.aspx. Enter the email identification code and click on the “Obtain password” button to get a temporary password by email.

Step 2. Follow the instructions provided in the received email and go to www.mycobank.org/malassezia/DefaultPage.aspx. Enter your email Id and provisional password and click on the "Edit my profile" button. You will then be redirected to the registration, where one can enter personal information details. The password can also be changed using the "Save changes." Registration is then complete and data can be deposited.

11.2.7 Deposit of New Record

To add new records to the database via the online deposit form, go to www.mycobank.org/malassezia/BioloMICSSelector.aspx, register and click on the "Add" button. A new empty form will be displayed. Data can then be entered. Note that the "Description field" should contain your own strain description ID and that some of the fields highlighted in red and bold are mandatory. Once the form is complete, click on the "Save button," on top of the form. As it is the case with similar websites, the curator(s) of the database will then receive a warning email confirming that a new record was deposited. The curator(s) can then review the deposited data and modify them if needed. As soon as the new record is saved, it can be viewed by any users, queried, and used for polyphasic identification, if relevant data are being introduced.

11.2.8 Modification or Deletion of Existing Record

To modify the existing records of the database via the online deposit form, go to www.mycobank.org/malassezia/BioloMICSSelector.aspx; register and type the unique identifier of the record (record description) and click on the "Search" button. Then select the record to be changed and click on the "Edit" button. Data associated with the record will then be displayed and could be modified and saved. Note that regular users can only modify their own records. Only the curator can modify any records. Deletion of a record can be done by clicking on the "Delete" button, instead of the "Edit" button.

11.3 Online Identification

The two online identification options are major elements and functionalities of the *Malassezia* website, including pairwise sequence alignments and polyphasic identification.

11.3.1 Pairwise Sequence Alignments

With "Pairwise sequence alignments," one can align any given sequence against a standard fungal sequence reference database compiled using sequences from Genebank (www.ncbi.

nlm.nih.gov) and the CBS (www.cbs.knaw.nl). The algorithm used is a modified version of Blastn [10] where best matching sequences are ordered by similarity instead of probability.

To perform such analysis, go to www.mycobank.org/malassezia/BioloMICS Sequences.aspx; accept the disclaimer and paste the sequence to identify in the text box; and click on the "Start alignment" button.

11.3.2 Polyphasic Identification

The polyphasic identification option allows for the comparison between user's unknown record and the database of *Malassezia* strains. Comparisons are based on single-, multiple-choice-, range-, latitude–longitude- and sequence fields. Only the fields where both the record to be identified, that is the unknown *Malassezia* isolate, and the reference record, that is the *Malassezia* species the unknown isolate is compared with, are filled in and considered. Different algorithms, weighting and tolerance factors are used for the different fields.

When comparing several records with one another, the BioloMICS software computes local similarity coefficients for the characters that have been selected/included in the analysis (characters can be individually included in or excluded from the analysis). More information can be found on the algorithms used for the comparisons of data at www.bio-aware.com//BioloMICSHelp/index.html. Navigate to section "Databases and characters," "Characters and fields," and then "Type of characters/fields, applicability and associated algorithms." Subsequently, a global similarity coefficient is computed by the following formula [11]:

$$S_{\mathrm{jk}} = \sum_{i=1}^{n} \frac{S_{\mathrm{i}} W_{\mathrm{i}}}{W_{\mathrm{i}}},$$

Where, S_{jk} is the global similarity coefficient for the comparison of records j and k; S_{i} is the local similarity coefficient for character I; *W*i is the weight of character I; *n* is the number of characters accounted in the similarity comparison.

The polyphasic identification option can be started from the following page www.mycobank.org/malassezia/BioloMICSID.aspx. Read the disclaimer, check the box if agreeing and fill the form with the available data. Then click on the "Identify" button on top of the form. After a few seconds, the first 30 most similar records are displayed and sorted by decreasing similarity. By clicking on the similarity value hyperlink, one can obtain a pairwise and detailed comparison between the record to identify and the selected reference record (Fig. 11.2).

In addition to the list of best matching records, one can also produce a phenetic tree based on one of the following agglomerative clustering method: unweighted pair-group method using arithmetic mean (UPGMA), unweighted pair-group method using centroids (UPGMC), weighted pair-group method using arithmetic averages (WPGMA), weighted pair-group method using centroids (WPGMC), Single and Complete Linkages, Ward, Lance & Williams flexible method, and Neighbor Joining.

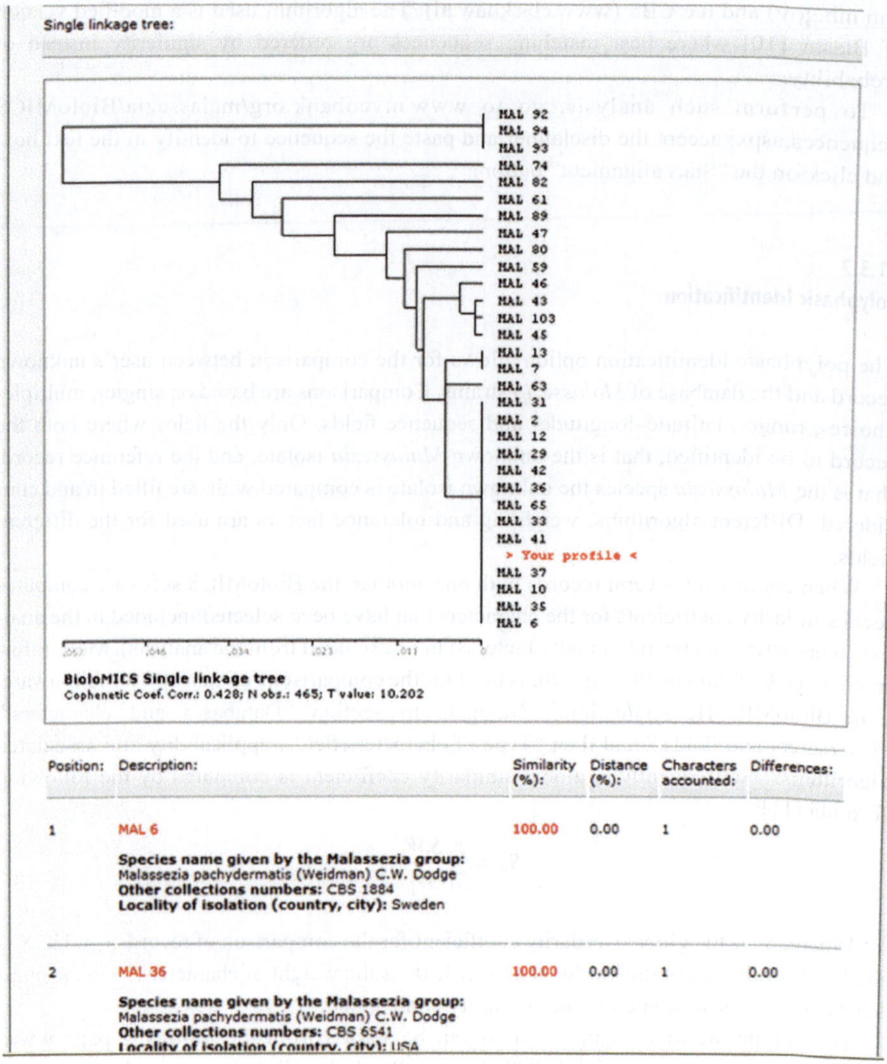

Fig. 11.2 Identification and agglomerative clustering will be possible using the *Malassezia* database

11.4
Conclusions

The currently presented version of the website is in its infancy and cannot be regarded as a reference yet. However, the basics are present to make it a dynamic and successful solution. It will be the responsibility of the user's community to contribute to it and to transform this trial into a success story. Previous experiences with similar websites proved that data depositors are enthusiastic in the early stages and tend to lose interest if the users and contributors community is not large enough. Online databases associated with collections, and managed by professional curators are more likely to be successful as well. Of course, the content and functionalities of the website are key factors for success, and feedback from the user's community should be accounted and implemented in subsequent versions. The notion of the *Malassezia* website interlinking to other web-based repositories such as the multilocus sequencing typing (MLST) (viz. Chap. 8.3) (http://www.mlst.net) is certainly within the scope of designing this website. Such link could bring together biological, physiological, chromatographical, clinical, antifungal resistance and descriptive epidemiology data with MLST maps.

References

1. Crespo Erchiga V, Gómez-Moyano E, Crespo M (2008) Pityriasis versicolor and the yeasts of genus *Malassezia*. Actas Dermosifiliogr 99:764–771
2. Xu J, Saunders CW, Hu P et al (2007) Dandruff-associated *Malassezia* genomes reveal convergent and divergent virulence traits shared with plant and human fungal pathogens. Proc Natl Acad Sci USA 104:18730–18735
3. Levin NA (2009) Beyond spaghetti and meatballs: skin diseases associated with the *Malassezia* yeasts. Dermatol Nurs 21:7–13
4. Mirhendi H, Zin A, Makimura K (2009) A new species of the genus *Malassezia* based on the sequence analysis of 26S (D1/D2) and internal transcribed spacer ribosomal DNA. Book of Abstracts. The 17th Congress of the International Society for Human and Animal Mycology, p 395
5. Robert V, Szoke S (2003) BioloMICS, Biological Manager for Identification, Classification and Statistics. Version 6.2, BioAware, Hannut, Belgium
6. Robert V (2000) BioloMICSWeb software for online publication and polyphasic identification of biological data. Version 1, BioAware, Hannut, Belgium
7. Robert V, Epping W, Boekhout T et al (2004) CBS yeasts database. Centraalbureau voor Schimmelcultures, Utrecht, The Netherlands
8. Samson RA, Frisvad JC (2004) *Penicillium* subgenus *Penicillium*: new taxonomic schemes, mycotoxins and other extrolites. Stud Mycol 49:257
9. Mostert L, Groenewald JZ, Summerbell RC et al (2005) Species of *Phaeoacremonium* associated with human infections and environmental reservoirs in infected woody plants. J Clin Microbiol 43:1752–1767
10. Altschul SF, Gish W, Miller W et al (1990) Basic local alignment search tool. J Mol Biol 215:403–410
11. Gower JC (1971) A general coefficient of similarity and some of its properties. Biometrics 27:857–871

Index

MIX
Papier aus verantwortungsvollen Quellen
Paper from responsible sources
FSC® C105338

If you have any concerns about our products,
you can contact us on
ProductSafety@springernature.com

In case Publisher is established outside the EU,
the EU authorized representative is:
Springer Nature Customer Service Center GmbH
Europaplatz 3, 69115 Heidelberg, Germany

Printed by Libri Plureos GmbH
in Hamburg, Germany